Rare Metal and Related Deposits: Geology, Geochemistry and Mineralization

Rare Metal and Related Deposits: Geology, Geochemistry and Mineralization

Guest Editors

Zhenjiang Liu
Fangfang Zhang

MDPI

Basel • Beijing • Wuhan • Barcelona • Belgrade • Novi Sad • Cluj • Manchester

Guest Editors

Zhenjiang Liu
School of Earth Sciences and Resources
China University of Geosciences
Beijing
China

Fangfang Zhang
School of Earth Sciences and Resources
China University of Geosciences
Beijing
China

Editorial Office
MDPI AG
Grosspeteranlage 5
4052 Basel, Switzerland

This is a reprint of the Special Issue, published open access by the journal *Minerals* (ISSN 2075-163X), freely accessible at: https://www.mdpi.com/journal/minerals/special_issues/819CO30X1C.

For citation purposes, cite each article independently as indicated on the article page online and as indicated below:

Lastname, A.A.; Lastname, B.B. Article Title. *Journal Name* **Year**, *Volume Number*, Page Range.

ISBN 978-3-7258-4525-5 (Hbk)
ISBN 978-3-7258-4526-2 (PDF)
https://doi.org/10.3390/books978-3-7258-4526-2

Contents

About the Editors

Zhenjiang Liu

Zhenjiang Liu serves as a Senior Lecturer at China University of Geosciences, Beijing. He earned his Ph.D. in Geology from the same institution in 2011. From 2017 to 2018, he held the position of Visiting Scholar at Indiana University Bloomington, and subsequently served as a Visitor at Durham University from 2020 to 2021. His research primarily concerns the post-formation alteration and preservation of metallic mineral deposits. He has carried out systematic investigations into the exploration prospects, post-ore denudation, and preservation of Mesozoic gold deposits in the Northwest Jiaodong region, with a particular focus on the application of thermochronology. In recent years, his research endeavors have concentrated on magmatic and magmatic–hydrothermal deposits within the Central Asian Orogenic Belt, exploring their genetic mechanisms and superimposed alterations. This includes investigations into VMS and IOA-like deposits, as well as the genetic mechanisms related to associated critical metals.

Fangfang Zhang

Fangfang Zhang is an Associate Professor at China University of Geosciences (Beijing). Her research primarily focuses on the genesis of magmatic–hydrothermal ore deposits, with an emphasis on the enrichment and mineralization of copper, molybdenum, and critical metals (e.g., Se, Te, Co). She has led or participated in numerous national projects, including projects supported by grants from the National Natural Science Foundation and the National Key R&D Program. As either the first or corresponding author, she has published multiple articles in esteemed journals such as *Economic Geology*, *Ore Geology Reviews*, *Lithos*, and the *Journal of Asian Earth Sciences*.

Preface

This Reprint compiles nine advanced studies that illuminate the geology, geochemistry, and tectonic control of rare-metal and related ore deposits across Asia, Europe, and Africa. The articles explore lithium-, tantalum-, and niobium-bearing pegmatites, sediment-hosted copper, silver-rich carbonates, and gold systems enriched in critical trace elements such as bismuth, tellurium, and antimony. They elucidate how magmatic–hydrothermal processes associated with highly fractionated granites, intracontinental subduction, crust–mantle interactions, metamorphic fluid flow, and basin inversion events dictate the concentration of metals in pegmatites, greisens, veins, skarns, and carbonate sequences.

Several contributions dissect the internal evolution of classic LCT and NYF pegmatites, examining magmatic differentiation, fluid exsolution, and metasomatic footprints that guide exploration. Others concentrate on granitic magmatism itself, utilizing zircon U-Pb-Hf isotopes and whole-rock chemistry to unravel crustal melting and mantle input. Two complementary papers on the Bailugou Zn-Pb-Ag deposit illustrate how fluid inclusion studies and multi-isotope geochemistry can differentiate metamorphic–orogenic signatures from purely magmatic ones. A robust geochronological study of Shuigou and Erdaokan links ore formation to distinct tectonic pulses, and documents mantle-derived components in gold and silver mineralization. Finally, a structural analysis of the West Congo Basin presents new codes for modeling tectonic inversion, enhancing our understanding of metal transport pathways.

The aim of this volume is twofold. First, it provides an integrated snapshot of current progress in deciphering the ore genesis, mineralizing fluids, and metallogenic chronology of deposits that supply essential metals for use in green technologies and strategic industries. Second, it provides a comparative framework that may guide future exploration strategies in various tectonic settings. By presenting a multidisciplinary collection of methods—from isotope geochemistry and fluid inclusions to structural modeling—we hope to spark dialogue among researchers, exploration geologists, and policy stakeholders. Graduate students will discover foundational case studies, seasoned investigators will discover new research perspectives, and industry professionals will discover fresh ideas for target evaluation. We trust that the insights presented here will inspire further collaborative work and promote innovative approaches to sustainable resource development.

Zhenjiang Liu and Fangfang Zhang
Guest Editors

minerals

Article

Isotope Geochemistry and Metallogenic Model of the Bailugou Vein-Type Zn-Pb-Ag Deposit, Eastern Qinling Orogen, China

Yan Yang [1,2,*], Hui Chen [1,2], Nana Guo [3], Donghao Wu [4], Zhenshan Pang [1,2] and Yanjing Chen [5,*]

1 Development and Research Center, China Geological Survey, Beijing 100037, China
2 Technical Guidance Center for Mineral Exploration, Ministry of Natural Resources of the People's Republic of China, Beijing 100037, China
3 Bureau of Natural Resources of Luanchuan County, Henan Province, Luoyang 471500, China
4 School of Earth Sciences and Resources, China University of Geosciences, Beijing 100083, China
5 Key Laboratory of Orogen and Crust Evolution, Peking University, Beijing 100871, China
* Correspondence: yyan@mail.cgs.gov.cn or yangyan520123@163.com (Y.Y.); gigyjchen@126.com (Y.C.)

Abstract: The large-scale vein-type Zn-Pb-Ag deposit in the Eastern Qinling Orogen (EQO) has sparked a long-standing debate over whether magmatism or metamorphism was the primary control or factor in its formation. Among the region's vein-type deposits, the large-sized Bailugou deposit offers a unique opportunity to study this style of mineralization. Similar to other deposits in the area, the vein-type orebodies of the Bailugou deposit are hosted in dolomitic marbles (carbonate–shale–chert association, CSC) of the Mesoproterozoic Guandaokou Group. Faults control the distribution of the Bailugou deposit but do not show apparent spatial links to the regional Yanshanian granitic porphyry. This study conducted comprehensive H–O–C–S–Pb isotopic analyses to constrain the sources of the ore-forming metals and metal endowments of the Bailugou deposit. The $\delta^{34}S_{CDT}$ values of sulfides range from 1.1‰ to 9.1‰ with an average of 4.0‰, indicating that the sulfur generated from homogenization during the high-temperature source acted on host sediments. The Pb isotopic compositions obtained from 31 sulfide samples reveal that the lead originated from the host sediments rather than from the Mesozoic granitic intrusions. The results indicate that the metals for the Bailugou deposit were jointly sourced from host sediments of the Mid-Late Proterozoic Meiyaogou Fm. and the Nannihu Fm. of the Luanchuan Group and Guandaokou Group, as well as lower crust and mantle materials. The isotopic composition of carbon, hydrogen, and oxygen collectively indicate that the metallogenic constituents of the Bailugou deposit were contributed by ore-bearing surrounding rocks, lower crust, and mantle materials. In summary, the study presents a composite geologic-metallogenic model suggesting that the Bailugou mineral system, along with other lead-zinc-silver deposits, porphyry-skarn molybdenum-tungsten deposits, and the small granitic intrusions in the Luanchuan area, are all products of contemporaneous hydrothermal diagenetic mineralization. This mineralization event transpired during a continental collision regime between the Yangtze and the North China Block (including syn- to post-collisional settings), particularly during the transition from collisional compression to extension around 140 Ma. The Bailugou lead-zinc-silver mineralization resembles an orogenic-type deposit formed by metamorphic fluid during the Yanshanian Orogeny.

Keywords: Bailgou Pb-Zn-Ag deposit; H–O–C–S–Pb isotopes; metallogenic model; collision orogeny; Eastern Qinling

1. Introduction

Vein-type Zn-Pb-Ag deposits are widely distributed globally, with some further classified as Cordilleran-type or skarn vein deposits. They are particularly common in magmatic arcs and their related porphyry systems where they may be hosted by faults or the adjacent country rocks [1]. Examples include the Magma deposit in Arizona, USA [2]; the Casapalca deposit in Peru [3]; the Tintic deposit in Utah, USA [4]; and the Taebaeksan deposit in

South Korea [5]. In contrast, other vein-type Zn-Pb-Ag deposits, known as orogenic or metamorphic hydrothermal Ag-Pb-Zn deposits are associated with crustal-scale faults and lack a direct genetic relationship to igneous intrusions; they include Keno Hill in Yukon, Canada [6]; Coeur d'Alene in Idaho, USA [7,8]; and Cierco in Spain [9]. Additionally, deposits in Eastern Qinling Orogen including Tieluping [10], Lengshuibeigou [11], Yindonggou [12], Weishancheng [13,14], Shagou [15,16], Xiaguan [17], Longmendian [18], and the Bairendaba [19] in China fall under this category. The sources and composition of ore fluids and the geological significance of these Zn-Pb-Ag deposits that have no direct association with magmatism have long been debated and remain unclear. Further exploration and research are necessary to understand the detailed ore genetic process.

The Bailugou vein-type deposit is located in the Lushi–Luanchuan WNW-trending fold-thrust belt, which is part of the Huaxiong terrane in the Eastern Qinling Orogen. It is situated on the southern margin of the North China Craton (see Figure 1). The concurrence of orogenic gold deposits and vein-type Zn-Pb (-Ag) deposits in the East Qinling area sets it apart from "gold-only" metallogenic orogenic provinces [20–22]. Moreover, some orogenic deposits in the East Qinling area are not exclusively gold but contain zinc, lead, silver, or molybdenum. For instance, the Dongchuang gold deposit is associated with lead, the Dahu gold deposit was found with molybdenum orebodies in the deep, and the Xiong'ershan gold–silver complex ore body contains an independent molybdenum vein, etc. These geological findings prompt us to consider whether the vein-type zinc-lead (-silver) deposits in the East Qinling area share the exact genetic origin of the gold deposits in this region. We are interested in their specific geological and geochemical characteristics and genetic processes. During the ore-forming process, do they have similarities and differences in metal sources, hydrothermal systems, and other metal deposits in the Luanchuan area?

Several studies have been published on isotope geochemistry in the Bailugou deposit [23–29]. However, most of the previous works were relatively superficial and lacked detailed stable isotope (S, Pb, C, O, and H) analysis of metallic mineralization and alteration, which limits the understanding of the genesis of the Bailugou ore deposit.

Here, we present the results of an S, Pb, H, O, and C isotope geochemical study on ore minerals and wall rocks from the Bailugou Zn-Pb (-Ag) ore deposit. Along with detailed field and petrographic observations, alteration characteristics, and microthermometric data of fluid inclusions [30], this paper proposes a metallogenic model for the Bailugou deposit. This study aims to enhance understanding of the nature and genesis of Bailugou and other similar vein-type deposits in the Luanchuan area, which may have direct implications for deposit exploration in the Luanchuan ore field.

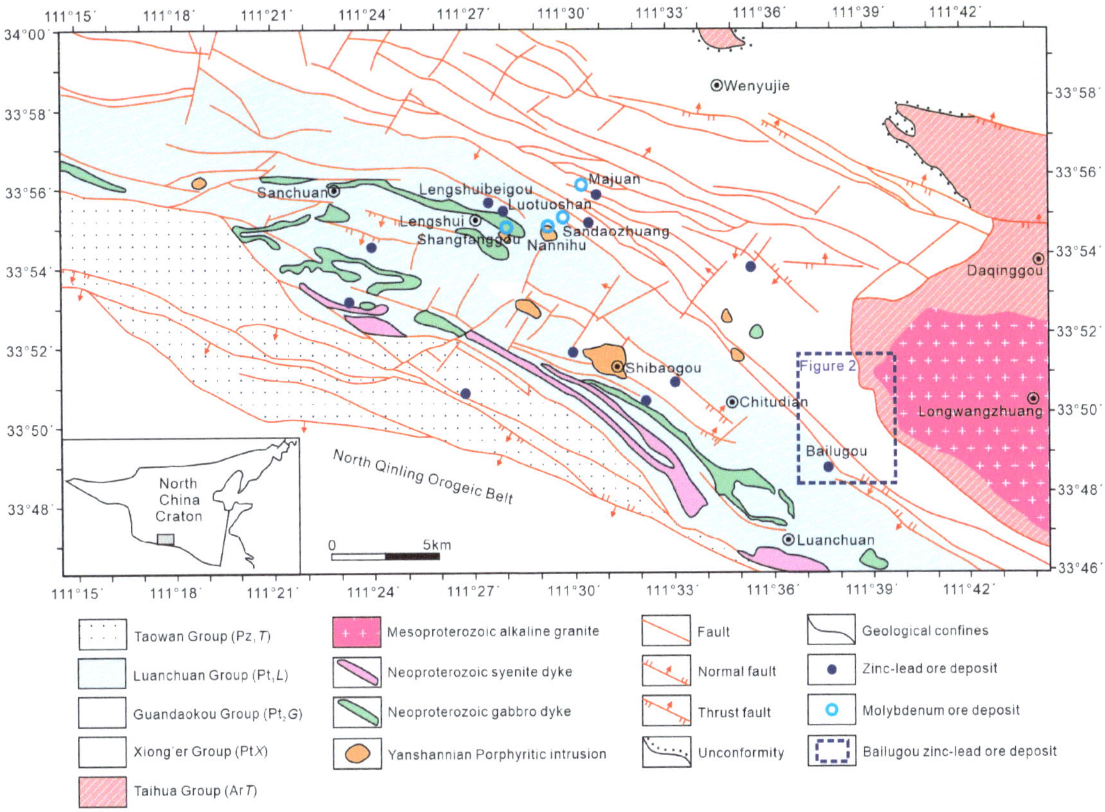

Figure 1. Geological map of the Bailugou deposit (modified from Duan et al. [31]).

2. Geological Setting and Ore Geology

2.1. Ore Geological Setting

The Bailugou Pb-Zn-Ag deposit, with proven reserves of 191.12 $\times$ 10^4 t (Pb + Zn) and 2446.79 t Ag, with average grades of Pb@11.29%, Zn@5.58%, and Ag@183.32 $\times$ 10^{-6}, respectively, is situated in the Mesoproterozoic Guandaokou Group and with no direct spatial and genetic links with the Yanshanian granitic porphyry, the endowment of the Nannihu Mo–W, and the Shangfanggou Mo–Fe super-large deposits [32,33] in the area (Figure 1). The Bailugou deposit is hosted in dolomitic marbles of the Mesoproterozoic Longjiayuan, Xunjiansi, Duguan, and Fengjiawan formations of the Guandaokou Group and the Neoproterozoic Baizhugou, Sanchuan and Nannihu formations of the Luanchuan Group (Figure 2). The main ore bodies exhibit a stratiform distribution within the host sediments and are situated at the NWW-trending faults and the NE-NNE-trending faults [34]. Their formation is governed by the interlayer fracture zones within the strata and is located to the west of the Niuxinduo domed anticline that associated with the middle-west segment of the Luanchuan fault system (Figure 1).

There are two main phases of magmatic intrusion in the region (Figure 1). The earliest phase named as the Jingningian period, comprises gabbro intruded and metamophosed during the Neoproterozoicwhere, whereas the second phase is represented by the Mesoproterozoic Longwangzhuang alkaline granite (Figures 1 and 2). Rare exposures of the Yanshanian intrusion exist in this area.

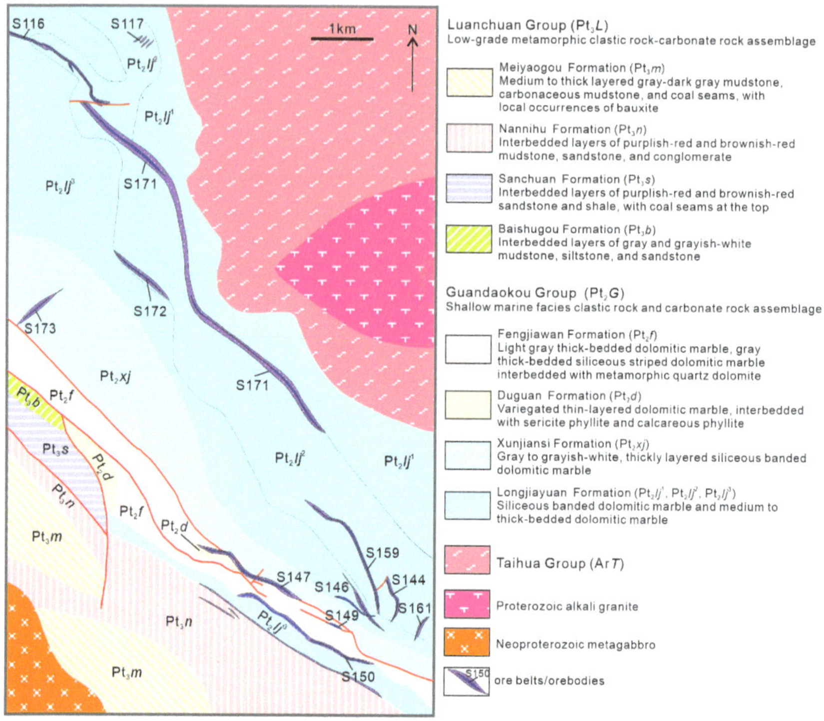

Figure 2. Geologic scheme of the Bailugou Zn–Pb deposit (modified from Yan [35]; Zhang [36]).

2.2. Ore Geology

The Bailugou ore deposit comprises four ore bodies, S150, S172, S171, and S116, along with several small, low-grade mineralized zones from south to north (Figure 2). The occurrences of mineralized zones, which are sediment-hosted, are primarily controlled by the NE-trending faults.

The mineralized zone in the south of the ore district is found within the fracture zone between the Longjiayuan Formation and Fengjiawan Formation of the Guandaokou Group, characterized by silicified dolomite marble, partially cataclastic rock, and metagabbro veins. The mineralized zone to the northwest of the ore district is situated in dolomitic marbles of the Mesoproterozoic Longjiayuan Formation, Guandaokou Group, which exhibit cataclastic features. Locally, extrusion foliation and mylonitization are present, with metagabbro veins occasionally seen intruded along the fabric (Figure 2). Four paralleled Zn-Pb-Ag ore bodies have been identified along a northwest trend, along with two ore bodies situated to the south. The occurrences of stratiform ore bodies are consistent with the bedding of the host sediment stratum. The ore grade increase with depth downwards [26].

The ore mineral assemblages primarily consist of galena, sphalerite, and pyrite, with chalcopyrite and minor amounts of argentite following. The gangue minerals include calcite, quartz, dolomite, and kaolinite, along with barite, sericite, talc, and locally carbonaceous. The ore structure predominantly exhibits various forms, including disseminated, fine-vein-disseminated, banded, massive, brecciated, vein-like, stockwork, and crystal cluster structures. The mineral texture is characterized by granular texture, metasomatic dissolution, replacement, metasomatic etching texture, cataclastic structure, bird's-eye structure, residual skeleton texture (pyrite), and growth girdle zone texture (sphalerite and dolomite).

Hydrothermal alteration in the Bailugou deposit encompasses silicification, dolomitization, carbonation, etc. The hydrothermal alteration and mineralization can be roughly divided into four stages, ranging from early to late [30]: stage 1, quartz (Q)–pyrite (Py) ± sphalerite (Sp); stage 2, Q–polymetallic sulfide (Ms); stage 3, dolomite (Dol)–Ms; and stage 4, calcite (Cc).

3. Sampling and Analytical Methods for Stable Isotope Geochemistry

Samples used in this study were collected from the main orebodies of S150, S116, S171, and S147 in the Bailugou deposit (N 33°47′36″–N 33°48′36″, E 111°38′32″–E 111°45′08″). This paper systematically and comprehensively presents the isotope geochemical characteristics of the Bailugou deposit, integrating previously published data.

Minerals for H isotope analysis were crushed and sieved using 40–60 meshes and then purified by hand-picking under a binocular microscope, whereas minerals for C, O, S, and Pb isotopes were pulverized in an agate mill under 200 meshes. All mineral separates were estimated to be more than 99% pure.

All analyses were conducted at the State Key Laboratory of Lithospheric Evolution, Institute of Geology and Geophysics, Chinese Academy of Sciences.

The oxygen isotope analysis was obtained on samples of quartz (7), dolomite (4), and whole strata rock powders (3). Samples were reacted with BrF_5 under 500 °C–700 °C for 15 h. The released O_2 from the reaction was purified using combined cold traps and converted to CO_2 gas under 700 °C for measurement of oxygen isotope ratios with a Finnigan MAT 252 mass spectrometer. The analytical precision for oxygen is ±0.2‰.

Hydrogen isotope analyses of H_2O from fluid inclusions (FIs) in 7 quartz samples and 4 dolomite powders, which were sufficiently exploded to exclude CO_2, CO, SO_2, SO_3, H_2S, O_2, H_2, N_2, CH_4, and inert gases. Then, we used dry ice to freeze the H_2O from FIs under vacuum into the chromium reactor at 800 °C to generate hydrogen gas for measurement of hydrogen isotope ratios with a Finnigan MAT 253 mass spectrometer. The analytical precision for hydrogen isotopes is ±2‰.

For the carbon isotope of FIs in the 7 quartz samples, CO_2 gas was extracted through successive purification steps and removal under a vacuum of 10^{-3} Pa using liquid nitrogen-pentane after the FIs were sufficiently exploded at the required decrepitation temperature. The 4 dolomites and 3 bulk marbles were analyzed for CO_2 gas extracted from calcite via a reaction with pure phosphoric acid under vacuum. The carbon isotopic compositions of the extracted CO_2 were determined using a Finnigan MAT 252 mass spectrometer. The precision of the carbon isotope analysis is ±0.2‰.

For sulfur analysis of 23 sulfide powders, SO_2 was extracted from samples by reacting with cuprous oxide at a ratio of 1:8 under a vacuum of 10^{-4} Pa, raising the temperature to 1000 °C to oxidize sulfide to SO_2, then collecting the SO_2 by liquid nitrogen into a sample cell (−80 °C) and purifying SO_2 to measurement with a Finnigan Delta-S mass spectrum analysis. The analytical uncertainty for $\delta^{34}S$ is ±0.2‰.

For lead isotope analysis of sulfide samples, approximately 10 to 50 mg of minerals from each of the 19 samples were first leached in acetone to remove surface contamination, then washed with distilled water and dried at 60 °C. The washed sulfides were dissolved in a dilute mix of nitric and hydrofluoric acid. Following ion-exchange chemistry, the lead in the solution was loaded onto rhenium filaments using a phosphoric acid-silica gel emitter. The Pb isotopic compositions were measured on a MAT261 thermal ionization mass spectrometer. All ratios were normalized to the values of NBS 981. The estimated precision levels for the $^{206}Pb/^{204}Pb$, $^{207}Pb/^{204}Pb$, and $^{208}Pb/^{204}Pb$ ratios were approximately 0.1%, 0.09%, and 0.30% at the 2σ level, respectively.

Hydrogen and oxygen isotope values are reported in standard notation relative to the SMOW standard. Carbon isotope values are reported in standard notation relative to the PDB standard. Sulfur isotope values are reported in standard notation relative to the V-CDT standard.

$\delta^{18}O_{SMOW}$ values of ore-forming fluid were calculated by equilibrium fractionation equations of $1000\ln\alpha_{quartz\text{-}water} = 3.38 \times 10^6\ T^{-2} - 3.40$ [37], $1000\ln\alpha_{calcite\text{-}water} = 2.78 \times 10^6\ T^{-2} - 2.89$ [38], $1000\ln\alpha_{magnetite\text{-}water} = 2.88 \times 10^6\ T^{-2} - 11.36 \times 10^3\ T^{-2} + 2.89$ [39],

$1000\ln\alpha_{\text{flogopite-water}} = 3.86 \times 10^6\ T^{-2} - 8.68 \times 10^3\ T^{-2} + 2.45$ [39], and $1000\ln\alpha_{\text{dolomite-water}} = 4.12 \times 10^6\ T^{-2} - 4.62 \times 10^3\ T^{-2} + 1.71$ [39]. $\delta^{13}C_{CO2}$ values of carbonate were calculated by equilibrium fractionation equations of $1000\ln\alpha_{\text{CO2-calcite}} = -2.4612 + (7.6663 \times 10^3\ T^{-1}) - (2.9880 \times 10^6\ T^{-2})$ [40] and $1000\ln\alpha_{\text{dolomite-calcite}} = 0.18 \times 10^6\ T^{-2} + 0.17$ [41]. The temperature using in the equations was the homogeneous peak temperature of each stage release from Yang et al. [30].

4. Results

4.1. Sulfur Isotopic Compositions

This contribution analyzed the $\delta^{34}S$ values of galenas, pyrites, and sphalerites from 23 ore samples across different alteration and mineralization stages in the Bailugou deposit. The $\delta^{34}S_{CDT}$ values of sulfides in the Bailugou deposit ranged from 1.1‰ to 9.1‰, with an average of 4.0‰. The $\delta^{34}S_{CDT}$ values from 72 ore samples, ore-bearing host sediments, and regional porphyry were collected and compared (Table 1).

Table 1. Sulfur isotope composition of ores, host sediments, and rock mass in the Bailugou ore district (%).

Geologic Body	Sample No.	Sampling Site	Sample Description	Test Mineral	$\delta^{34}S$ V-CDT	Data Sources
Bailuogu deposit	S150-4	S150 mineralized zone	Ore	Gn	−1.2	[26]
Bailuogu deposit	07-52	S 116	Ore	Gn	0.6	[27]
Bailuogu deposit	07-55	S 116	Ore	Gn	0.6	[27]
Bailuogu deposit	BQC-30-Gn	S170, Quancaigou	Ore	Gn	1.1	This paper
Bailuogu deposit	S040-2	S040	Ore	Gn	1.5	[26]
Bailuogu deposit	S040-3	S040	Ore	Gn	1.5	[26]
Bailuogu deposit	BP-6-Gn	S116, Jiudinggou	Ore	Gn	2.1	This paper
Bailuogu deposit	BLY-10-Gn	S170, Luyuangou	Ore	Gn	2.5	This paper
Bailuogu deposit	BLY-19-Gn	S170, Luyuangou	Ore	Gn	2.5	This paper
Bailuogu deposit	B1472	S147	Ore	Gn	3.8	This paper
Bailuogu deposit	BGL-6-Gn	S150, Gunligou	Ore	Gn	7.2	This paper
Bailuogu deposit	B11	S150	Ore	Gn	8.4	[25]
Bailuogu deposit	BCL-11-Gn	S171, Luchanggou	Ore	Gn	8.6	This paper
		Average of Gn		$n = 13$	3.0	
Bailuogu deposit	Sjd-23		Ore	Py	1.3	[26]
Bailuogu deposit	S116-22	S116, Jiudinggou	Ore	Py	1.3	[26]
Bailuogu deposit	BGL-13-Py	S150, Gunligou	Ore	Py	1.4	This paper
Bailuogu deposit	BP-6-Py	S116, Jiudinggou	Ore	Py	1.7	This paper
Bailuogu deposit	07-53	S 116	Ore	Py	1.7	[27]
Bailuogu deposit	07-54	S 116	Ore	Py	2.7	[27]
Bailuogu deposit	BGL-6-Py	S150, Gunligou	Ore	Py	3.1	This paper
Bailuogu deposit	BLY-19-Py	S170, Luyuangou	Ore	Py	3.9	This paper
Bailuogu deposit	BCL-11-Py	S171, Luchanggou	Ore	Py	4.0	This paper
Bailuogu deposit	BGL-14-Py	S150, Gunligou	Ore	Py	4.4	This paper
Bailuogu deposit	BGL-10	S150, Gunligou	Ore	Py	6.7	This paper
Bailuogu deposit	07-43	S150	Ore	Py	8.4	[27]
Bailuogu deposit	CM15001-K4	S150	Ore	Py	9.2	[26]
Bailuogu deposit	BGL-2	S150, Gunligou	Ore	Py	9.8	This paper
Bailuogu deposit	S150-1	S150	Ore	Py	10.9	[26]
		Average of Py		$n = 15$	4.7	
Bailuogu deposit	BGL-14-Sp	S150, Gunligou	Ore	Sp	1.4	This paper
Bailuogu deposit	07-54	S 116	Ore	Sp	1.8	[27]
Bailuogu deposit	07-56	S 116	Ore	Sp	2.0	[27]
Bailuogu deposit	BGL-13-Sp	S150, Gunligou	Ore	Sp	3.3	This paper
Bailuogu deposit	BP-6-Sp	S116, Jiudinggou	Ore	Sp	3.4	This paper

Table 1. *Cont.*

Geologic Body	Sample No.	Sampling Site	Sample Description	Test Mineral	$\delta^{34}S$ V-CDT	Data Sources
Bailuogu deposit	SXW-22	S040	Ore	Sp	3.7	[26]
Bailuogu deposit	BLY-10-Sp	S170, Luyuangou	Ore	Sp	4.3	This paper
Bailuogu deposit	BLC-11-Sp	S171, Luchanggou	Ore	Sp	5.0	This paper
Bailuogu deposit	07-43	S150	Ore	Sp	5.1	[27]
Bailuogu deposit	BLC-26	S171, Luchanggou	Ore	Sp	5.3	This paper
Bailuogu deposit	BQC-30-Sp	S170, Quancaigou	Ore	Sp	7.0	This paper
Bailuogu deposit	BGL-6-Sp	S150, Gunligou	Ore	Sp	9.1	This paper
Bailuogu deposit	S150-3	S150	Ore	Sp	9.6	[26]
Bailuogu deposit	B12	S150	Ore	Sp	10.7	[25]
Bailuogu deposit	B13	S150	Ore	Sp	12.0	[25]
	Average of Sp			$n = 15$	5.6	
	Average of ores from Bailugou deposit			$n = 43$	4.5	
Luanchuan Gp.		Baizhugou Fm.	Carbonaceous phyllite	Py	12.4	[23]
Luanchuan Gp.		Baizhugou Fm.	Carbonaceous phyllite	Py	18.6	[23]
	Average of Baizhugou Fm.			$n = 2$	15.5	
Luanchuan Gp.	SS7-4	Meiyaogou Fm.	Stone coal	Py	−12.4	[24]
Luanchuan Gp.	SS7-1-2	Meiyaogou Fm.	Stone coal	Py	−10.9	[24]
Luanchuan Gp.	SS7-1-1	Meiyaogou Fm.	Stone coal	Py	−9.7	[24]
Luanchuan Gp.	SS7-1-3	Meiyaogou Fm.	Stone coal	Py	−8.1	[24]
Luanchuan Gp.	SS7-3	Meiyaogou Fm.	Stone coal	Py	6.6	[24]
Luanchuan Gp.	SS7-2	Meiyaogou Fm.	Stone coal	Py	7.2	[24]
Luanchuan Gp.	SS7-7	Meiyaogou Fm.	Stone coal	Py	10.5	[24]
	Average of Stone coal from Meiyaogou Fm.			$n = 7$	−2.4	
Luanchuan Gp.	S-22	Meiyaogou Fm.	Marble from Shangfanggou	Py	1.7	[24]
Luanchuan Gp.	S-32	Meiyaogou Fm.	Marble from Shangfanggou	Py	2.5	[24]
Luanchuan Gp.	S-18	Meiyaogou Fm.	Marble from Shangfanggou	Py	2.6	[24]
Luanchuan Gp.	S-21	Meiyaogou Fm.	Marble from Shangfanggou	Py	2.6	[24]
Luanchuan Gp.	S-25	Meiyaogou Fm.	Marble from Shangfanggou	Py	2.7	[24]
	Average of Marble from Shangfanggou			$n = 5$	2.4	
Luanchuan Gp.	S-6	Meiyaogou Fm.	Skarn from Shangfanggou	Py	−4.9	[24]
Luanchuan Gp.	S-8	Meiyaogou Fm.	Skarn from Shangfanggou	Py	2.5	[24]
Luanchuan Gp.	S-7	Meiyaogou Fm.	Skarn from Shangfanggou	Py	2.7	[24]
	Average of Skarn from Shangfanggou			$n = 3$	0.1	
Luanchuan Gp.	N-6	Sanchuan, Meiyaogou Fm.	Mineralized wall rock from Nannihu	Py	−4.5	[24]
Luanchuan Gp.	N-9	Sanchuan, Meiyaogou Fm.	Mineralized wall rock from Nannihu	Py	−2.0	[24]
Luanchuan Gp.	N-1	Sanchuan, Meiyaogou Fm.	Mineralized wall rock from Nannihu	Py	−0.8	[24]
Luanchuan Gp.	N-3	Sanchuan, Meiyaogou Fm.	Mineralized wall rock from Nannihu	Py	0.6	[24]
Luanchuan Gp.	N-11	Sanchuan, Meiyaogou Fm.	Mineralized wall rock from Nannihu	Py	1.8	[24]
Luanchuan Gp.	N-2	Sanchuan, Meiyaogou Fm.	Mineralized wall rock from Nannihu	Py	2.9	[24]

Table 1. *Cont.*

Geologic Body	Sample No.	Sampling Site	Sample Description	Test Mineral	$\delta^{34}S$ V-CDT	Data Sources
Luanchuan Gp.	N-14	Sanchuan, Meiyaogou Fm.	Mineralized wall rock from Nannihu	Py	2.9	[24]
Luanchuan Gp.	N-8	Sanchuan, Meiyaogou Fm.	Mineralized wall rock from Nannihu	Py	3.5	[24]
Average of Sanchuan, Meiyaogou Fm.				$n = 8$	0.6	
Luanchuan Gp.	XS5-10	Sanchuan Fm.	Skarn from Sandaozhuang	Py	−2.9	[24]
Luanchuan Gp.	XS5-6	Sanchuan Fm.	Skarn from Sandaozhuang	Py	1.0	[24]
Luanchuan Gp.	XS3	Sanchuan Fm.	Skarn from Sandaozhuang	Py	1.8	[24]
Luanchuan Gp.	XS5-5	Sanchuan Fm.	Skarn from Sandaozhuang	Py	2.5	[24]
Luanchuan Gp.	XS5-8	Sanchuan Fm.	Skarn from Sandaozhuang	Py	2.8	[24]
Average of Sanchuan Fm.				$n = 5$	1.0	
Average of Luanchuan Gp.				$n = 30$	1.2	
Nannihu rock mass	N-15	Nannihu rock mass	Mineralized porphyry	Py	1.9	[24]
Nannihu rock mass	N-18	Nannihu rock mass	Mineralized porphyry	Py	2.5	[24]
Nannihu rock mass	N-41	Nannihu rock mass	Mineralized porphyry	Py	2.5	[24]
Nannihu rock mass	N-47	Nannihu rock mass	Mineralized porphyry	Py	2.6	[24]
Nannihu rock mass	N-56	Nannihu rock mass	Mineralized porphyry	Py	2.6	[24]
Nannihu rock mass	NS-2	Nannihu rock mass	Mineralized porphyry	Py	2.7	[24]
Nannihu rock mass	NS-9	Nannihu rock mass	Mineralized porphyry	Py	2.7	[24]
Nannihu rock mass	N-53	Nannihu rock mass	Mineralized porphyry	Py	2.8	[24]
Nannihu rock mass	N-22	Nannihu rock mass	Mineralized porphyry	Py	2.9	[24]
Nannihu rock mass	N-33	Nannihu rock mass	Mineralized porphyry	Py	2.9	[24]
Nannihu rock mass	N-37	Nannihu rock mass	Mineralized porphyry	Py	2.9	[24]
Nannihu rock mass	NS-11	Nannihu rock mass	Mineralized porphyry	Py	2.9	[24]
Nannihu rock mass	N-16	Nannihu rock mass	Mineralized porphyry	Py	3.0	[24]
Nannihu rock mass	N-29	Nannihu rock mass	Mineralized porphyry	Py	3.1	[24]
Nannihu rock mass	N-43	Nannihu rock mass	Mineralized porphyry	Py	3.1	[24]
Nannihu rock mass	NS-4	Nannihu rock mass	Mineralized porphyry	Py	3.1	[24]
Nannihu rock mass	N-27	Nannihu rock mass	Mineralized porphyry	Py	3.2	[24]
Average of Nannihu rock mass				$n = 17$	2.8	
Shangfanggou rock mass	S-2	Shangfanggou rock mass	Mineralized porphyry	Py	2.0	[24]
Shangfanggou rock mass	S-15	Shangfanggou rock mass	Mineralized porphyry	Py	2.9	[24]
Shangfanggou rock mass	S-13	Shangfanggou rock mass	Mineralized porphyry	Py	3.0	[24]
Shangfanggou rock mass	S-29	Shangfanggou rock mass	Mineralized porphyry	Py	3.1	[24]
Shangfanggou rock mass	S-5	Shangfanggou rock mass	Mineralized porphyry	Py	3.3	[24]
Average of Shangfanggou rock mass				$n = 5$	2.9	

Abbreviations: Gn = galena; Py = pyrite; Sp = sphalerite; Gp. = group; Fm. = formation.

4.2. Lead Isotopic Compositions

In this paper, nineteen Pb isotopes from galena, sphalerite, and pyrite samples collected from the ores of the Bailugou deposit were analyzed, yielding Pb isotopic compositions of $^{206}Pb/^{204}Pb$ = 17.676 to 18.456, with an avg. of 18.182; $^{207}Pb/^{204}Pb$ = 15.506 to 15.646, avg. of 15.572; and $^{208}Pb/^{204}Pb$ = 38.672 to 39.858, avg. of 39.473, respectively (Table 2). To systematically and comprehensively study the Pb isotope geochemical characteristics and geological significance of the Bailugou deposit, 47 sets of previously reported Pb isotopic analyses from the ores of the Bailugou deposit, regional strata, and porphyries were included (Table 2).

Table 2. Lead isotope composition of ores, strata, and rock mass from the Bailugou district.

Geologic Body	Sample No.	Sample Description	Test Mineral	$^{206}Pb/^{204}Pb$	$^{207}Pb/^{204}Pb$	$^{208}Pb/^{204}Pb$	Data Sources
Bailugou deposit	BP-6-Gn	Ore	Gn	17.803	15.633	39.344	This paper
Bailugou deposit	BQC-30-Gn	Ore	Gn	18.291	15.569	39.592	This paper
Bailugou deposit	S150-4	Ore	Gn	18.426	15.549	39.637	[35]
Bailugou deposit	07-43	Ore	Gn	18.259	15.579	39.558	[28]
Bailugou deposit	07-53	Ore	Gn	17.718	15.532	38.968	[28]
Bailugou deposit	07-56	Ore	Gn	17.725	15.531	38.990	[28]
Bailugou deposit	B11	Ore	Gn	18.218	15.526	39.264	[25]
Average of Gn			$n = 7$	18.063	15.560	39.336	
Bailugou deposit	BP-6	Ore	Py	17.701	15.506	38.915	This paper
Bailugou deposit	BLC-11-Py	Ore	Py	18.239	15.546	39.502	This paper
Bailugou deposit	BGL-13-Py	Ore	Py	18.422	15.596	39.690	This paper
Bailugou deposit	BLY-19-Py	Ore	Py	18.234	15.552	39.551	This paper
Bailugou deposit	BGL-14-Py	Ore	Py	18.456	15.600	39.727	This paper
Bailugou deposit	BLC-6-Py	Ore	Py	18.313	15.595	39.613	This paper
Bailugou deposit	BGL-2	Ore	Py	18.336	15.566	39.569	This paper
Bailugou deposit	BGL-10	Ore	Py	18.320	15.565	39.515	This paper
Bailugou deposit	S150-1	Ore	Py	18.286	15.575	39.529	[35]
Bailugou deposit	CM15001-K4	Ore	Py	18.273	15.500	39.344	[42]
Average of Py			$n = 10$	18.258	15.560	39.496	
Bailugou deposit	BLC-26	Ore	Sp	18.224	15.559	39.433	This paper
Bailugou deposit	BP-6	Ore	Sp	17.709	15.514	38.947	This paper
Bailugou deposit	BLC-11-Sp	Ore	Sp	18.284	15.611	39.714	This paper
Bailugou deposit	BLC-6-Sp	Ore	Sp	18.238	15.520	39.408	This paper
Bailugou deposit	B1472	Ore	Sp	18.351	15.604	39.715	This paper
Bailugou deposit	BQC-30-Sp	Ore	Sp	17.676	15.557	38.672	This paper
Bailugou deposit	BGL-13-Sp	Ore	Sp	18.353	15.646	39.858	This paper
Bailugou deposit	BLY-10-Sp	Ore	Sp	18.306	15.581	39.607	This paper
Bailugou deposit	BGL-14-Sp	Ore	Sp	18.211	15.551	39.624	This paper
Bailugou deposit	S150-3	Ore	Sp	17.632	15.461	38.358	[35]
Bailugou deposit	S040-2	Ore	Sp	17.606	15.520	38.496	[35]
Bailugou deposit	S040-3	Ore	Sp	17.552	15.451	38.264	[35]
Bailugou deposit	B12	Ore	Sp	17.609	15.466	38.325	[25]
Bailugou deposit	B13	Ore	Sp	17.608	15.465	38.323	[25]
Average of Sp			$n = 14$	17.954	15.536	39.053	
Average of Ores from Bailugou deposit			$n = 31$	18.070	15.551	39.257	
Luanchuan Gp.	200416		Whole rock	18.215	15.648	38.350	[42]
Luanchuan Gp.	200418		Whole rock	19.249	15.684	41.104	[42]
Luanchuan Gp.	200421		Whole rock	17.998	15.560	38.979	[42]
Luanchuan Gp.	200438		Whole rock	18.217	15.635	38.598	[42]
Luanchuan Gp.	200439		Whole rock	18.868	15.685	38.303	[42]
Luanchuan Gp.	S139-3		Whole rock	17.694	15.525	38.692	[42]
Average of whole rock from Luanchuan Gp.			$n = 6$	18.374	15.623	39.004	
Luanchuan Gp. Meiyaogou Fm.	Y02	Biotite schist	Whole rock	17.735	15.468	38.631	[25]
Luanchuan Gp. Meiyaogou Fm.	Y03	Phyllite	Whole rock	18.071	15.482	38.905	[25]
Luanchuan Gp. Meiyaogou Fm.	Y21	Biotite marble	Whole rock	17.736	15.478	38.463	[25]

Table 2. *Cont.*

Geologic Body	Sample No.	Sample Description	Test Mineral	$^{206}Pb/^{204}Pb$	$^{207}Pb/^{204}Pb$	$^{208}Pb/^{204}Pb$	Data Sources
Luanchuan Gp. Meiyaogou Fm.	Y22	Altered granite	Whole rock	17.742	15.466	38.493	[25]
Average of Meiyaogou Fm.			$n = 4$	17.821	15.474	38.623	
Luanchuan Gp. Nannihu Fm.	L8	Biotite quartz schist	Whole rock	17.700	15.486	38.366	[25]
Luanchuan Gp. Nannihu Fm.	L30	Dimicaceous quartz schist	Whole rock	18.257	15.518	38.819	[25]
Luanchuan Gp. Nannihu Fm.	L31	Graphite dimicaceous schist	Whole rock	17.862	15.492	38.640	[25]
Luanchuan Gp. Nannihu Fm.	L32	Graphite dimicaceous schist	Whole rock	18.817	15.556	39.097	[25]
Luanchuan Gp. Nannihu Fm.	L37	Dimicaceous quartz schist	Whole rock	17.612	15.485	38.432	[25]
Luanchuan Gp. Nannihu Fm.	L39	Quartz schist	Whole rock	17.590	15.455	38.333	[25]
Luanchuan Gp. Nannihu Fm.	L40	Carbon-bearing biotite Quartz schist	Whole rock	17.670	15.569	38.700	[25]
Luanchuan Gp. Nannihu Fm.	L41	Calcareous quartz schist	Whole rock	17.585	15.457	38.336	[25]
Luanchuan Gp. Nannihu Fm.	L42	Weak mineralized biotite quartz schist	Whole rock	17.594	15.462	38.359	[25]
Average of Nannihu Fm.			$n = 9$	17.854	15.498	38.565	
Average of Luanchuan Gp.			$n = 19$	18.011	15.532	38.716	
Guandaokou Gp.	20043b		Whole rock	18.103	15.634	38.316	[42]
Guandaokou Gp.	S116-1		Whole rock	18.738	15.672	39.094	[42]
Guandaokou Gp.	B3	Metagabbro	Whole rock	17.937	15.498	38.840	[25]
Guandaokou Gp.	B26	Breccialized marble	Whole rock	18.296	15.508	39.272	[25]
Guandaokou Gp.	B27	Quartz schist	Whole rock	18.086	15.501	38.667	[25]
Guandaokou Gp.	B28	Muscovite quartz schist	Whole rock	18.443	15.505	39.179	[25]
Guandaokou Gp.	B29	Muscovite quartz schist	Whole rock	18.326	15.506	39.043	[25]
Average of Guandaokou Gp.			$n = 7$	18.275	15.546	38.916	
Nannihu	DF-2	Molybdenite–galane ore	Gn	17.450	15.540	39.010	[43]
Nannihu	J-5	Biotite granodiorite	K-feldspar	17.806	15.569	38.508	[43]
Nannihu	J-4	Porphyritic granite	K-feldspar	17.894	15.482	38.093	[43]
Nannihu-Sandaozhuang	ZK705	Porphyritic biotite granodiorite	K-feldspar	17.189	15.381	37.655	[23]
Nannihu-Sandaozhuang	Nan r-1	Medium grained porphyritic granite	K-feldspar	17.499	15.427	37.843	[23]
Nannihu-Sandaozhuang	DF-2	Gn-Sp-Py vein	Gn	17.450	15.540	39.010	[23]
Nannihu-Sandaozhuang	N-013	Pyrite–fluorite–quartz vein	Py	17.605	15.421	37.710	[23]
Average of Nannihu rock mass			$n = 7$	17.556	15.480	38.261	

Table 2. *Cont.*

Geologic Body	Sample No.	Sample Description	Test Mineral	$^{206}Pb/^{204}Pb$	$^{207}Pb/^{204}Pb$	$^{208}Pb/^{204}Pb$	Data Sources
Shangfanggou	DF-1	Gn- magnetite vein in marble	Gn	16.820	14.840	37.770	[23]
Shangfanggou	DF-16	Gn-Sp in mineralized marble	Gn	17.120	15.230	37.570	[23]
Average of Shangfanggou rock mass			$n = 2$	16.970	15.035	37.670	
East Qinling molybdenum ore belt	Average of 6 smple	Minor intermediate-acid rock masses	K-feldspar	17.619	15.478	38.040	[44]

4.3. Hydrogen and Oxygen Isotopic Compositions

In this paper, the gangue minerals quartz and dolomite from different mineralization zones, various occurrences, and syngenetic with varying stages of mineralization ores of the Bailugou deposit were selected, and the mineral oxygen isotope and fluid inclusion hydrogen isotope analyses were conducted (Table 3). Meanwhile, previous results on hydrogen and oxygen isotopes of the Bailugou deposit were compiled [25,26,28,45] for comparative analysis (Table 3).

The $\delta^{18}O_{SMOW}$ values of the Bailugou deposit vary from 2.0 to 2.0‰–14.2‰ (n = 23, avg. 7.1‰), indicating that the fluid is relatively rich in ^{18}O. The δD_{SMOW} values of the Bailugou deposit range from −90‰ to −60‰ (n = 18, avg. −79‰) (Table 3).

Table 3. H-O-C isotope composition of the Bailugou deposit (‰).

Geologic Body	Sample No.	Occurrence	Test Mineral	Stage	SMOW			PDB		Tm_{tot}/°C	Data Sources
					$\delta^{18}O_{mineral}$	$\delta^{18}O_W$	δD_W	$\delta^{13}C_{mineral}$	$\delta^{13}C_{CO2}$		
Bailugou	BLC-30	S171, Luchanggou	Quartz	1	18.7	14.2	−85		−0.8	380	This paper
	BGL-5	S150, Gunligou	Quartz	1	12.7	5.8	−82		−3.4	300	This paper
	BLC-11	S171, Luchanggou	Quartz	1	12.6	7.0	−81		−0.6	340	This paper
	BLC-33	S171, Luchanggou	Quartz	2	15.0	6.0	−81		0.2	250	This paper
	BLC-32	S171, Luchanggou	Quartz	2	14.9	7.3	−78		−1.0	280	This paper
	BLC-13	S171, Luchanggou	Quartz	2	12.2	6.3	−76		−0.2	330	This paper
	BLY-12	S170, Luchanggou	Quartz	2	13.4	5.8	−74		−1.4	280	This paper
	BLC-30	S171, Luchanggou	White dolomite	3	15.4	7.4	−67	−3.1		250	This paper
	BGL-12	S150, Gunligou	Pale yellow dolomite	3	9.8	3.0	−66	−1.3		280	This paper
	BJD-2	S116, Jiudinggou	White dolomite	3	9.9	2.0	−65	−1.8		250	This paper
	BLY-21	S170, Luyuangou	Pale yellow dolomite	3	11.4	3.5	−60	−2.6		250	This paper
	S150-5	S150 vein	Quartz		14.9	6.0	−90			250 *	[45]
	Sjd-22	S171 vein	Calcite		9.4	2.1	−83	1.8	3.1	250 *	[45]
			Dolomite		12.0	5.8				295	[25]
			Dolomite		17.7	11.5				295	[25]
	B17		Dolomite		12.0	5.2		−3.0		280 *	[26]
	B19		Dolomite		17.7	10.9		−1.7		280 *	[26]
	S116-21	S116 vein	Quartz				−90				[45]
	07-45	S150 orebody	Quartz		17.9	10.1	−89			277	[28]
	07-46	S150 orebody	Quartz		17.2	8.2	−76			248	[28]
	07-48	S150 orebody	Calcite		15.6	8.7		−0.4		261	[28]
	07-49	S150 orebody	Calcite		16.1	9.0		1.3		255	[28]
	07-54	S150 orebody	Quartz		16.1	7.6	−89			260	[28]
	07-59	S150 orebody	Quartz		20.0	11.0	−82			248	[28]

Table 3. *Cont.*

Geologic Body	Sample No.	Occurrence	Test Mineral	Stage	SMOW			PDB		Tm_{tot}/°C	Data Sources
					$\delta^{18}O_{mineral}$	$\delta^{18}O_W$	δD_W	$\delta^{13}C_{mineral}$	$\delta^{13}C_{CO2}$		
Luanchuan Gp.	BGL-a	Pt_2l–f	Marble	S150	22.6			−1.2		200	This paper
	BLY-23	Pt_2l	Marble	S170	23.8			−0.4		200	This paper
	BLC-7	Pt_2l	Marble	S171	21.8			1.0		200	This paper
	ZK1613-22	Sanchuan Fm.	Marble		19.0			2			[46]
	Y21	Meiyaogou Fm.	Sp-bearing marble		15.7			0.1			[25]
	L41	Nannihu Fm.	Calcareous quartz schist		17.8			0.8			[25]
	Y02	Meiyaogou Fm.	Mica quartz schist		15.7						[47]
	Y03	Meiyaogou Fm.	Mineralized mica schist		16.7						[47]
	L8	Nannihu Fm.	Mica schist		11.8						[47]
	L30	Nannihu Fm.	Mica schist		15.1						[47]
	L31	Nannihu Fm.	Graphite mica schist		14.1						[47]
	L32	Nannihu Fm.	Graphite mica schist		13.2						[47]
	L37	Nannihu Fm.	Quartz schist		9.1						[47]
	L40	Nannihu Fm.	Graphite biotite quartz schist		10.4						[47]
	L42	Nannihu Fm.	Biotite quartz schist		10.6						[47]
Guandaokou Gp.	B27	Fengjiawan Fm.	Calcareous dimicaceous schist		14.9						[47]
	B28	Fengjiawan Fm.	Mica schist		14.6						[47]
	B29	Fengjiawan Fm.	Mica quartz schist		14.2						[47]
	B26	Fengjiawan Fm.	Breccialized marble		16.0			−1.8			[26]
	B27	Fengjiawan Fm.	Calcareous dimicaceous schist		19.3			−2.8			[26]

Notes: $\delta^{18}O_W$ data signed as * are calculated based on homogenization temperature measured from Yang et al. [30].

The $\delta^{18}O_{SMOW}$ value and δD_{SMOW} value of stage 1 are 5.8‰–14.2‰ (avg. 9.0‰) and −85‰ to −81‰ (avg. −83‰), the $\delta^{18}O_{SMOW}$ value and δD_{SMOW} value of stage 2 are 5.8‰–7.3‰ (avg. 6.4‰) and −81‰ to −74‰ (avg. −77‰), the $\delta^{18}O_{SMOW}$ value and δD_{SMOW} value of stage 3 are 2.0‰–7.4‰ (avg. 4.0‰) and −67‰ to −60‰ (avg. −65‰), the $\delta^{18}O_{SMOW}$ value and δD_W value of stage 4 are 2.1‰–6.0‰ (avg. 4.1‰) and −90‰ to −83‰ (avg. −87‰). The $\delta^{18}O_{SMOW}$ and δD_{SMOW} values exhibit a certain regularity as the stages progress: the $\delta^{18}O_{SMOW}$ values decline successively from stage 1 to stage 4, while the δD_{SMOW} values increase sequentially from stage 1 to stage 3.

4.4. Carbon Isotopic Compositions

Quartz and dolomite gangue minerals from various mineralization zones, occurrences, and syngenetic ores in the Bailugou deposit were selected to investigate the oxygen and carbon isotopic compositions. The carbon isotope of CO_2 in FIs and the oxygen isotope of quartz, as well as the carbon and oxygen isotope of dolomite, are presented in Table 3. The published data on the carbon and oxygen isotopes from the Bailugou deposit have been compiled [25,26,28,45–47] (Table 3) to provide a comprehensive and systematic study on Bailugou ore genesis and fluid evolution.

Nine carbonate samples from the Bailugou deposit yielded $\delta^{13}C_{PDB}$ values ranging from −3.1‰ to 1.8‰ (avg. −1.2‰), while seven FIs in quartz yielded $\delta^{13}C_{PDB}$ values from −3.4‰ to 0.2‰ (avg. −1.0‰). Six bulk-rock samples in the Luanchuan Fm. yielded $\delta^{13}C_{PDB}$ values from −1.2‰ to 2.0‰ (avg. 0.4‰), and two bulk-rock samples from the Guandaokou Fm. yielded $\delta^{13}C_{PDB}$ values from −2.8‰ to −1.8‰ (avg. −2.3‰).

5. Discussions

5.1. Origin and Evolution of Ore-Forming Fluids

5.1.1. Constraint from H-O Isotopes

The δD_{SMOW} value of ore-forming fluid in stage 1 of the Bailugou deposit was relatively low, indicating it to be of magmatic or hydrothermal origin. However, the $\delta^{18}O_{SMOW}$ value was relatively higher than that of magmatic water. Since the Bailugou ore bodies were found within low-permeability sedimentary rock that typically has a high ^{18}O and relatively low water content, any accompanying hydrothermal fluid can naturally exhibit a high $\delta^{18}O$ value. With hydrothermal mineralization, the $\delta^{18}O_{SMOW}$ value of the fluid system gradually declined, while the δD_{SMOW} value gradually increased, reflecting the changing process of the isotope exchange and homogenization between the original ore-forming fluid and the host sediments. Although there are currently no hydrogen and oxygen isotope data for the host strata of the Bailugou deposit, the $\delta^{18}O_{quartz}$ of Nannihu porphyritic granite (sample no. ZK-38) reported by Zhou et al. was 10.1‰ [24]. The corresponding $\delta^{18}O_{SMOW}$ value and δD_{SMOW} value were 4.1‰ (calculated at 325 °C) and −70‰, respectively. This consistency offers reasonable evidence for the aforementioned inference.

In the hydrogen and oxygen isotope plot of the Bailugou deposit (Figure 3), the samples from stage 1 were plot in the lower part of the magmatic water box and its lower right area, indicating that the ore-forming fluid was initially extracted from a high ^{18}O geological source such as sediments belonging to the Luanchuan and Guandaokou Groups, collectively known as the CSC formation (carbonaceous carbonate–shale–sandstone–chert) [10,48,49], with dolomite in the CSC exhibiting the highest $\delta^{18}O$ value [50]. However, since no intrusive bodies are exposed in the immediate vicinity of the ore deposits, an intense tectonic-thermal process is necessary to provide the high temperature required for isotopic exchange. Stage 2 is concentrated in the lower left part of magmatic water, indicating a magmatic-hydrothermal process. Stage 3 transitions from magmatic water in the upper left to the meteoric water line, indicating the precipitation of meteoric water and a gradual weakening of the magmatic-hydrothermal process. The results of the hydrogen and oxygen isotope study of the Bailugou deposit are consistent with the evolution of salinity-temperature of fluid inclusion across different stages.

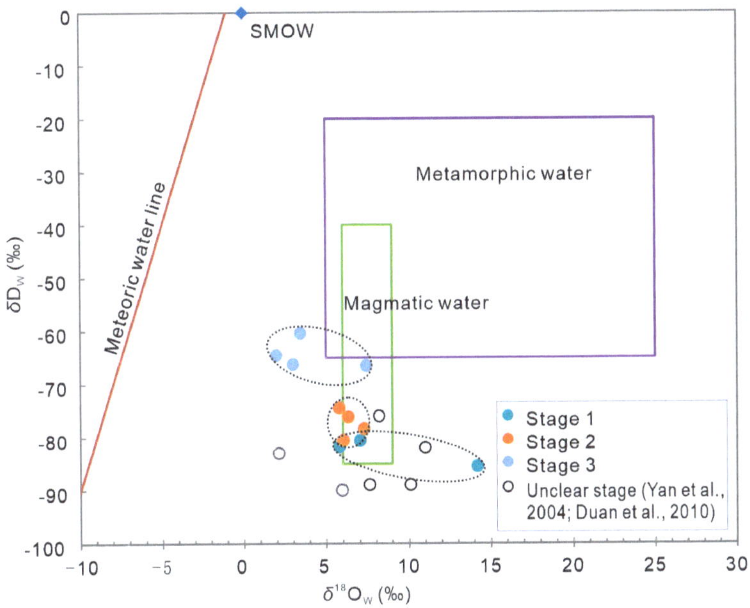

Figure 3. δD_W (‰) versus $\delta^{18}O_W$ (‰) in the ore-forming fluids of the Bailugou deposit (based on Taylor [51]). Data are from Table 3. Unclear stage from [27,45].

Based on the characteristic temperature revealed by the solid solution separation structure, homogenization temperature of FIs, and equilibrium temperature from sulfur isotope, Yang et al. [30] suggest that the temperature of mineral precipitation in the ore-forming fluid of Bailugou deposit ranges from 100 °C to 400 °C, predominantly concentrated between 350 °C and 400 °C. Both the temperature range and the peak value at each stage of the ore-forming fluid of the Bailugou deposit exhibit a trend of decreasing from early to late stages of mineralization. The fluid density changes slightly across each stage. The Bailugou deposit maintains a moderate to high density fluid during mineralization. The estimated formation depth of the Bailugou deposit is approximately 2–12 km, exhibiting typical characteristics of a fault-controlled deposit. In stage 1, the ore-forming fluid originated from a mixture of fluids, followed by fluid immiscibility between the magma-sourced fluid and the ore-bearing CSC host sediments fluid. This fluid transformed into a metamorphic hydrothermal fluid during the water–rock interactions in stage 2, gradually receding and possibly experiencing fluid boiling during the mineralization of stage 3.

Research on the evolution of salinities, homogenization temperatures [30], and hydrogen-oxygen isotopes of FIs from different mineralization stages in the Bailugou deposit indicates that the initial ore-forming fluid was triggered under intense thermal-tectonic processes. Magmatic-hydrothermal processes predominated in the early and middle stages, gradually shifting to an increasing influence of atmospheric precipitation as they approached the late stage.

5.1.2. Constraint from C-O Isotopes

The range of carbon isotopes of the Bailugou deposit closely aligns with the host surrounding sediments (Figure 4), falling within the range of marine carbonate rocks (−3‰ to 2‰ [52]). This indicates that both the Bailugou deposit and the surrounding sediments have experienced similar diagenetic alterations. This may involve processes such as carbonate dissolution, precipitation, or interactions with organic matter. A comparison of $\delta^{13}C$ values from several primary carbon reservoirs, including natural organic matter, igneous/magmatic systems, mantle, continental crust, and typical marine faces limestone with ores in Bailugou deposit, reveals that the $\delta^{13}C$ of Bailugou deposit varies between the ranges of marine limestone and magmatic rock, and the carbon was sourced from marine limestone.

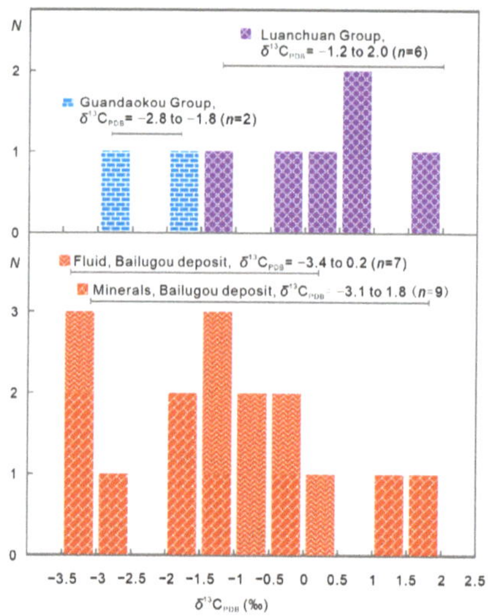

Figure 4. Compilation histograms of $\delta^{13}C_{PDB}$ (‰) values in sulfides from ores, ore-bearing host sediments, and regional porphyries of the Bailugou deposit. Data are from Table 1.

The calculated $\delta^{18}O_{SMOW}$ values for 23 carbonate samples in the Bailugou deposit range from 9.4‰ to 20.0‰ (avg. 14.5‰), 15 bulk-rock samples of Luanchuan Group range from 9.1‰ to 23.8‰ (avg. 15.8‰), and 5 bulk-rock samples of Guandaokou Group range from 14.2‰ to 19.3‰ (avg. 15.8‰). The range of oxygen isotopes in the deposit aligns closely with that of the stratum (Figure 5), and both are abundant in the ^{18}O isotope.

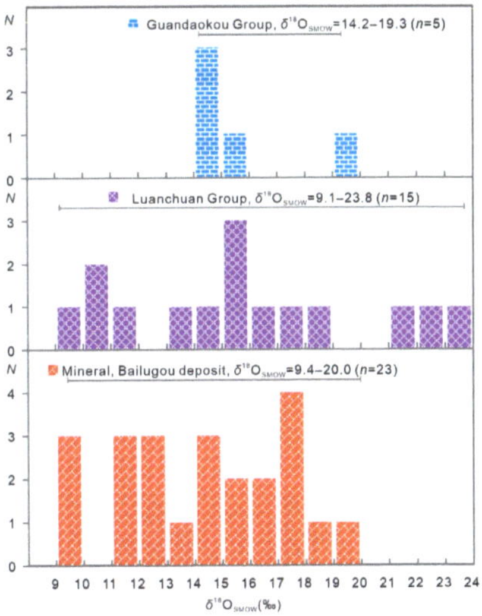

Figure 5. Compilation histograms of $\delta^{18}O_{SMOW}$ (‰) values in sulfides from ores, ore-bearing host sediments, and regional porphyries of the Bailugou deposit. Data are from Table 3.

The Luanchuan Group and Guandaokou Group are interpreted to have been deposited in a forearc sedimentary basin [53]. The primary lithology predominantly consists of medium-thick layered dolomite marble with siliceous bands, which includes conglomerate lithic sandstone, mudstone, sericite phyllite, calcareous phyllite, and metamorphic quartz dolomite, representing the CSC protolith formation that had undergone greenschist facies metamorphism. Melezhik et al. [50] revealed that dolomite in the CSC formation possesses the highest $\delta^{13}C$ and ^{18}O values. The CMF model (collisional orogeny, metallogeny, and fluid flow) [10] has deducted that the Middle-Late Proterozoic strata (Luanchuan Group, Guandaokou Group) were rich in radioactive lead and strontium, with elevated $\delta^{13}C$ and $\delta^{18}O$ values [10,48,49], which has been corroborated by studies on the carbon and oxygen isotope of the strata [47] and the relatively high $\delta^{18}O$ value of the ore-forming fluid from the Bailugou deposit [29].

5.2. Source of the Ore-Forming Constituents

5.2.1. Constraint from S Isotope

The $\delta^{34}S_{CDT}$ values of sulfides in the Bailugou deposit range from 1.1‰ to 9.1‰ with an average of 4.0‰ (as shown in Table 1). The $\delta^{34}S$ values of different sulfides showed little variation, indicating that they typically come from a similar sulfur source. The narrow range of $\delta^{34}S$ values for sulfides in the Bailugou deposit contrasts with the broader range of $\delta^{34}S$ values found in MVT deposits (−8‰ to +19‰) [54].

Previous research has demonstrated that under equilibrium conditions, the $\delta^{34}S$ values of paragenetic/coexisting sulfides (including sulfates) decrease in the order of sulfate, molybdenite, pyrite, pyrrhotite-sphalerite, chalcopyrite, to galena [39,55–57]. The $\delta^{34}S$ values of paragenetic sulfides in the Bailugou deposit exhibit the sequences $\delta^{34}S_{sphalerite}$ (avg. 5.6‰, $n = 15$) > $\delta^{34}S_{pyrite}$ (avg. 4.7‰, $n = 15$) > $\delta^{34}S_{galena}$ (avg. 3.0‰, $n = 13$), indicating that the coexisting ore mineral assemblage has not completely achieved sulfur isotope equilibrium, or that the original equilibrium has been disrupted, with only the sulfides in certain ores reaching partial equilibrium. Hydrothermal processes and sulfur cycling can significantly influence sulfur isotopic compositions. According to high-temperature equilibrium extrapolation [58], the $\delta^{34}S_{\Sigma S}$ value from two pairs of sulfur isotopes with strong correlation is approximately 2.7‰. Given that the calculated average crustal $\delta^{34}S$ is 2.2‰ [46], the elevated $\delta^{34}S_{\Sigma S}$ value of 2.7‰ in the Bailugou suggests that the sulfur present in the Bailugou deposit likely originates predominantly from the lower crust. This inference is based on the fact that the lower crust typically has higher $\delta^{34}S$ values compared to the upper crust, attributed to processes such as the differentiation and alteration of sulfide minerals over geological time. Meanwhile, the observed $\delta^{34}S$ value being higher than of the average crust indicates that while a significant portion of the sulfur originates from the lower crust, interactions with upper crustal materials (such as sedimentary rocks or altered volcanic rocks) could contribute additional sulfur with slightly different isotopic signatures, potentially altering the overall $\delta^{34}S$ composition.

Although the $\delta^{34}S$ values of individual sulfides cannot indicate their source, the $\delta^{34}S$ value of total S in the fluids may be diagnostic [59]. In hydrothermal systems, where H_2S is the dominant sulfur species in the fluids, or when the fluid redox state is below the SO_2/H_2S boundary, $\delta^{34}S$ (sulfide) = $\delta^{34}S$ (fluid) [60,61]. Therefore, the average $\delta^{34}S$ values can serve as a direct representaton of the sulfur source.

The $\delta^{34}S$ values of the ores in the Bailugou deposit are higher in ^{34}S than those of the meteorite, indicating that heavy sulfur in sulfate played a more significant role during the mineralization process. There are three proposed mechanisms for concentrating ^{34}S isotopes: (1) the addition of sulfate brine during the metallogenic process; (2) the assimilation of sulfur from sedimentary sulfate by magma, which then extracts metallogenic materials; and (3) the absorption of seawater sulfate by volcanism during volcanic-sedimentary mineralization. Additionally, some lead-zinc deposits form from the mixing of heavy sulfur in hydrogen sulfide from petroleum gas, such as the Jinding Pb-Zn deposit in northwest Yunnan, China [62].

The $\delta^{34}S_{CDT}$ values in pyrites from ore-bearing strata of the Bailugou deposit exhibit a wide range of variations (Table 1), including those enriched with isotope ^{34}S or ^{32}S. These characteristics are indicative of sulfides formed by bacterial hydrogen sulfide in environments with fluctuating sedimentary conditions [62]. The $\delta^{34}S$ values from the ores of the Bailugou deposit differ from and are more homogeneous than those of the host sediments.

The small range of variation in $\delta^{34}S$ values indicates an equilibrium sulfur source for the ore deposit, typically observed in the following scenarios: (1) sulfide deposits with genetic and paragenetic sources related to magma exhibit a small range of 5‰–6‰ (e.g., most Cu–Ni sulfide deposits), which is less than 10‰ in most cases, indicating complete homogenization of sulfur in high-temperature magma; (2) some epithermal sulfide deposits may show a narrow range of $\delta^{34}S$ values, suggesting a homogenized sulfur source; and (3) some deposits are without genetic-related magma, such as gold-quartz deposits in the Yellowknife area, Canada, which are associated with intense chemical activity modified by granitization and metamorphism [63].

The relatively uniform $\delta^{34}S$ values of the Bailugou deposit indicate sulfur homogenization of the hydrothermal fluid from the high-temperature source or during the migration process. To gain a thorough and precise understanding of the sulfur source in the Bailugou deposit, systematically collected $\delta^{34}S$ values from relevant geological formations, including marble and skarn from the Luanchuan Group, Nannihu porphyry, and Shangfanggou porphyry, have been analyzed (Table 1, Figure 6).

The $\delta^{34}S$ values of marble and skarn are more homogeneous than those of the stone coal from the Meiyaogou Formation of the Luanchuan Group. The $\delta^{34}S$ values of the Shangfanggou porphyry and Nannihu porphyry are consistent with, yet more homogeneous than, those of the Luanchuan Group. The Shangfanggou porphyry, Nannihu porphyry, and ores of the Bailugou deposit exhibit a consistent peak assemblage of $\delta^{34}S$ values at 1‰–5‰, indicating that the formation of the Bailugou deposit, Shangfanggou porphyry, and Nannihu porphyry may have undergone similar sulfur homogenization under intense high-temperature activity.

In summary, this paper suggests that the sulfur in the Bailugou deposit is generated from sulfur homogenization while the host rocks were being subjected to a high-temperature hydrothermal event.

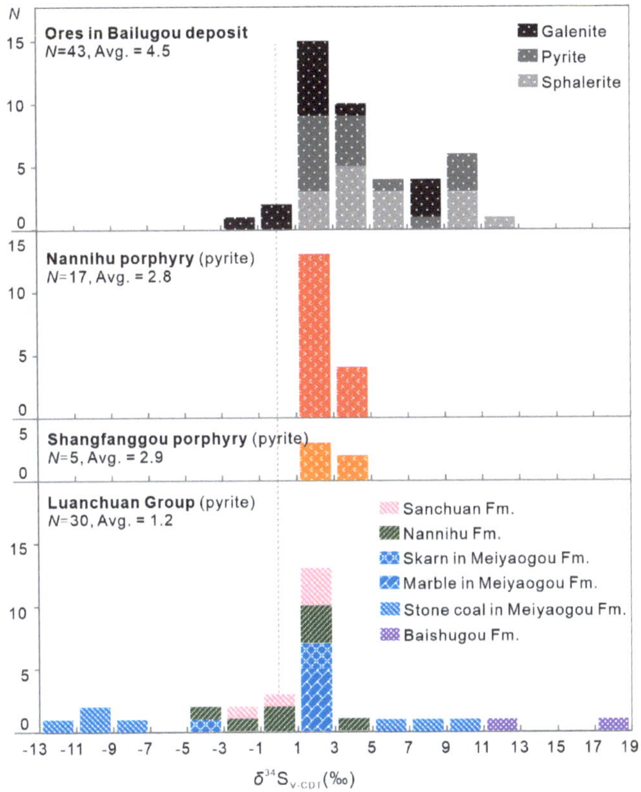

Figure 6. Compilation histograms of $\delta^{34}S$ values in sulfides from ores, ore-bearing strata, and regional porphyries of the Bailugou deposit. Data are from Table 1.

5.2.2. Constraint from Pb Isotope

Scholars have tried to study the metal source and ore genesis of the Bailugou deposit through lead isotope research (Table 2). Yan et al. [26] proposed that the lead in the Bailugou deposit originated from host sediments, given the consistency in both the distribution range and evolution trend of the Pb isotope of the Bailugou deposit and the host sediments. Meanwhile, considering the close spatial relationship between the Bailugou Pb-Zn deposit and the Yanshanian porphyry molybdenum deposit in the ore concentration area, Duan et al. [28,29] suggested that the metallogenic metal of the Bailugou deposit was more likely generated from the Yanshanian magmatic-hydrothermal processes, with the lead primarily derived from the host sediments. Based on lead isotope studies, Duan et al. [29,31] proposed that the veined lead-zinc deposit in the Luanchuan area is a magmatic-hydrothermal filling-metasomatic deposit associated with porphyry intrusion during the middle Yanshanian.

In this study, the lead isotopic compositions derived from 31 sulfide samples in the Bailugou deposit (Table 2) showed $^{206}Pb/^{204}Pb$ = 17.552–18.456 (avg. 18.077, range 4.9%), slightly exceeding 18.000; $^{207}Pb/^{204}Pb$ = 15.451–15.646 (avg. 15.550, range 12%), surpassing 15.300, indicating significant enrichment in U-radiogenic Pb; $^{208}Pb/^{204}Pb$ = 38.264–39.858 (avg. 39.260, range 4.0%), with Th-radiogenic Pb been slightly enriched, which is characteristic of the upper crust or sediment. Among these, seven galena samples yielded lead isotopic compositions of $^{206}Pb/^{204}Pb$ = 17.718–18.426 (avg. 18.063), $^{207}Pb/^{204}Pb$ = 15.526–15.633 (avg. 15.560), and $^{208}Pb/^{204}Pb$ = 38.968–39.637 (avg. 39.336); 10 pyrite samples yielded the lead isotopic compositions of $^{206}Pb/^{204}Pb$ = 17.701–18.456 (avg. 18.258), $^{207}Pb/^{204}Pb$ = 15.500–15.600 (avg. 15.560), and $^{208}Pb/^{204}Pb$ = 38.915–39.727 (avg. 39.496); 14 sphalerite samples yielded the lead isotopic compositions of $^{206}Pb/^{204}Pb$ = 17.552–18.353 (avg. 17.954), $^{207}Pb/^{204}Pb$ = 15.451–15.646 (avg. 15.536), and $^{208}Pb/^{204}Pb$ = 38.264–39.858 (avg. 39.053). The range of lead isotope compositions exhibited noticeable variation, generally suggesting a multi-stage anomalous lead evolution. Differences among the lead isotopic compositions of various sulfide samples indicate two or more metallogenic material sources.

Nineteen bulk-rock samples from the Luanchuan Group (Table 2) yielded the lead isotopic compositions of $^{206}Pb/^{204}Pb$ = 17.585–19.249 (avg. 18.011), $^{207}Pb/^{204}Pb$ = 15.455–15.685 (avg. 15.532), and $^{208}Pb/^{204}Pb$ = 38.303–41.104 (avg. 38.716). Among these, four bulk-rock samples from the Meiyaogou Fm. yielded $^{206}Pb/^{204}Pb$ = 17.735–18.071 (avg. 17.821), $^{207}Pb/^{204}Pb$ = 15.466–15.482 (avg. 15.474), and $^{208}Pb/^{204}Pb$ = 38.463–38.905 (avg. 38.623); nine bulk-rock samples from the Nannihu Fm. yielded $^{206}Pb/^{204}Pb$ = 17.585–18.817 (avg. 17.854), $^{207}Pb/^{204}Pb$ = 15.455–15.569 (avg. 15.498), and $^{208}Pb/^{204}Pb$ = 38.333–39.097 (avg. 38.565). It demonstrates the characteristic of enriching the U-lead isotope while exhibiting a deficiency in the Th-lead isotope.

Seven bulk-rock samples from the Guandaokou Group (Table 2) yielded the lead isotopic compositions of $^{206}Pb/^{204}Pb$ = 17.937–18.738 (avg. 18.275), $^{207}Pb/^{204}Pb$ = 15.498–15.672 (avg. 15.546), and $^{208}Pb/^{204}Pb$ = 38.316–39.272 (avg. 38.916). The samples also displayed a richness in U-radiogenic Pb and a low concentration of Th-radiogenic Pb.

Seven sulfides and K-feldspars from the Nannihu porphyry (Table 2) yielded lead isotopic compositions of $^{206}Pb/^{204}Pb$ = 17.189–17.894 (avg. 17.556), $^{207}Pb/^{204}Pb$ = 15.381–15.569 (avg. 15.480), and $^{208}Pb/^{204}Pb$ = 37.655–39.010 (avg. 38.261). Both are deficient in U-radiogenic Pb and Th-radiogenic Pb.

Two galena samples from the Shangfanggou porphyry (Table 2) yielded lead isotopic compositions of $^{206}Pb/^{204}Pb$ = 16.820–17.120 (avg. 16.970), $^{207}Pb/^{204}Pb$ = 14.840–15.230 (avg. 15.035), and $^{208}Pb/^{204}Pb$ = 37.570–37.770 (avg. 37.670). Both were also deficient in U-radiogenic Pb and Th-radiogenic Pb.

The samples from the Luanchuan Group and the Guandaokou Group were more abundant in radioactive lead than those from the Nannihu porphyry and the Shangfanggou porphyry. The radioactive lead of ore sulfide in the Bailugou deposit was lower than that of the Luanchuan Group and Guandaokou Group but higher than that of the Nannihu and Shangfanggou porphyries.

The single-stage Pb model ages (based on the Holmes–Houtermans method) and the characteristic values of μ, ω, v, Th/U ratio, U/Pb ratio, $\Delta\beta$, and $\Delta\gamma$ of 31 sulfides in the Bailugou deposit were calculated, respectively (Table 4). The μ values of ore lead in the Bailugou deposit ranged from 9.27 to 9.61 (avg. 9.40), significantly higher than that of common lead (8.686–9.238). The coexistence of μ values higher than 9.58 and lower than 9.58 indicates that the lead originated from crust-mantle mixed sources [64], and the metals in the ores of the Bailugou deposit were derived from a mixed source of mantle and crust. The ω values ranged from 39.12 to 44.21, higher than common lead (35.55 $\pm$ 0.59), indicating an increased maturity of lead sources. The v values ranged from 0.067 to 0.070, also higher than common lead (0.063–0.067). The Th/U ratios ranged from 4.08–4.45, which were even higher than that of common lead (3.92 $\pm$ 0.09), indicating that Th-radiogenic Pb are slightly enriched.

Table 4. Relevant parameters for lead isotope composition of ores from the Bailugou deposit.

Sample No.	Test Mineral	Model Age (Ma)	μ	ω	v	Th/U	U/Pb	$\Delta\beta$	$\Delta\gamma$	Data Sources
BP-6-Gn	Gn	633.8	9.61	44.21	0.070	4.45	0.131	19.77	50.01	This paper
BQC-30-Gn	Gn	211.4	9.41	41.34	0.068	4.25	0.127	15.59	56.63	This paper
S150-4	Gn	86.4	9.36	40.49	0.068	4.19	0.126	14.29	57.83	[35]
07-43	Gn	246.8	9.44	41.51	0.068	4.26	0.128	16.24	55.72	[28]
07-53	Gn	580.2	9.41	42.01	0.068	4.32	0.130	13.18	39.98	[28]
07-56	Gn	574.0	9.41	42.04	0.068	4.32	0.129	13.11	40.57	[28]
B11	Gn	211.0	9.34	40.00	0.068	4.15	0.127	12.78	47.88	[25]
Average of Gn, n = 7		363.4	9.43	41.66	0.068	4.28	0.128	14.99	49.80	

Table 4. *Cont.*

Sample No.	Test Mineral	Model Age (Ma)	μ	ω	v	Th/U	U/Pb	$\Delta\beta$	$\Delta\gamma$	Data Sources
BP-6	Py	563.9	9.36	41.62	0.068	4.30	0.129	11.48	38.56	This paper
BLC-11-Py	Py	220.7	9.38	41.05	0.068	4.24	0.127	14.09	54.23	This paper
BGL-13-Py	Py	149.0	9.45	41.22	0.069	4.22	0.127	17.35	59.25	This paper
BLY-19-Py	Py	231.7	9.39	41.35	0.068	4.26	0.127	14.48	55.54	This paper
BGL-14-Py	Py	129.2	9.46	41.20	0.069	4.22	0.127	17.61	60.23	This paper
BLC-6-Py	Py	227.4	9.46	41.57	0.069	4.25	0.128	17.29	57.19	This paper
BGL-2	Py	174.5	9.40	40.94	0.068	4.21	0.127	15.39	56.02	This paper
BGL-10	Py	185.1	9.40	40.81	0.068	4.20	0.127	15.33	54.58	This paper
S150-1	Py	222.4	9.43	41.18	0.068	4.23	0.128	15.98	54.95	[35]
CM15001-K4	Py	137.1	9.28	39.72	0.067	4.14	0.126	11.09	50.01	[42]
Average of Py, n = 10		224.1	9.40	41.07	0.068	4.23	0.127	15.01	54.06	
BLC-26	Sp	247.6	9.40	41.00	0.068	4.22	0.128	14.94	52.39	This paper
BP-6	Sp	565.9	9.38	41.78	0.068	4.31	0.129	12.00	39.42	This paper
BLC-11-Sp	Sp	268.0	9.50	42.34	0.069	4.31	0.128	18.33	59.89	This paper
BLC-6-Sp	Sp	188.9	9.32	40.41	0.068	4.19	0.127	12.39	51.72	This paper
B1472	Sp	211.1	9.48	41.84	0.069	4.27	0.128	17.87	59.91	This paper
BQC-30-Sp	Sp	638.0	9.47	41.24	0.069	4.21	0.131	14.81	32.08	This paper
BGL-13-Sp	Sp	260.8	9.56	42.87	0.069	4.34	0.129	20.61	63.73	This paper
BLY-10-Sp	Sp	215.3	9.44	41.44	0.068	4.25	0.128	16.37	57.03	This paper
BGL-14	Sp	247.2	9.39	41.78	0.068	4.31	0.127	14.42	57.49	This paper
S150-3	Sp	559.8	9.28	39.13	0.067	4.08	0.129	8.55	23.70	[35]
S040-2	Sp	645.9	9.41	40.52	0.068	4.17	0.130	12.39	27.38	[35]
S040-3	Sp	606.3	9.27	39.12	0.067	4.08	0.129	7.89	21.19	[35]
B12	Sp	582.4	9.30	39.18	0.067	4.08	0.129	8.86	22.83	[25]
B13	Sp	582.0	9.29	39.17	0.067	4.08	0.129	8.80	22.76	[25]
Average of Sp, n = 14		415.7	9.39	40.84	0.068	4.21	0.129	13.45	42.25	
Average of ores from Bailugou, n = 31		342.1	9.40	41.10	0.068	4.23	0.128	14.30	47.76	

Notes: All the relevant parameters were calculated based on the data in Table 4.

The $^{206}Pb/^{204}Pb$ and $^{207}Pb/^{204}Pb$ ratios of sulfides in Bailugou ores exhibited 2-3 narrow and concentrated linear distribution ranges in the $^{207}Pb/^{204}Pb$ vs. $^{206}Pb/^{204}Pb$ plot (Figure 7), respectively, indicating a two-component mixing and common lead in single-stage evolution.

The model age calculated using the Holmes–Houtermans method based on the single-stage evolution model ranged from 86.4 Ma to 645.9 Ma (Figure 8), which can be divided into two intervals: 560 Ma–646 Ma (n = 11, avg. 594 Ma) and 86 Ma–268 Ma (n = 20, avg. 204 Ma). For the interval 560 Ma–646 Ma, the corresponding μ values were 9.27–9.61 (avg. 9.38), ω values were 39.12–44.21 (avg. 40.91), Th/U ratios were 4.08–4.45 (avg. 4.22), and U/Pb ratios were 0.129–0.131 (avg. 0.130). For the interval 86 Ma–268 Ma, the corresponding μ values were 9.28–9.56 (avg. 9.41), ω values were 39.72–42.87 (avg. 41.20), Th/U ratios were 4.41–4.34 (avg. 4.24), and U/Pb ratios were 0.126–0.129 (avg. 0.127).

The age interval 560 Ma–646 Ma is significantly younger than the deposition of the Luanchuan Group (1000 Ma to 800 Ma) [66,67] and aligns with the Neoproterozoic Sinian Period (635 Ma to 542 Ma), which is older (60–150 Myr) than the peak metamorphic age (about 500 Ma) [68–71] of HP/UHP rocks in the Qinling complex, suggesting that the deposit and the strata formed under a relatively stable setting prior to continental subduction. The subdivided peak values for the interval 86 Ma–268 Ma are 200± Ma and 140± Ma, which precisely align with the statistical result of lithogenic and metallogenic ages of molybdenum deposits in the East Qinling area [72]. Considering the geological setting and previous studies on tectonic evolution of the Qinling orogen, the Bailugou rocks likely formed from the subduction of the Neo Proterozoic continental material from the South Qinling at 560 Ma–646 Ma, were dragged by northward subduction of the Shangdan ocean, and subsequently underwent two enrichments/modifications at about 200 Ma and 140 Ma, under back-arc extension and continental collision regimes, respectively.

The above indicates that Pb in the Bailugou deposit had undergone multi-stage crust–mantle mixing evolution, featuring both relatively "old" and "young" Pb. As the model age progressed, the lead isotope compositions of sulfides exhibited increasing source maturity, relatively enriching Th-radiogenic Pb while depleting U-radiogenic Pb.

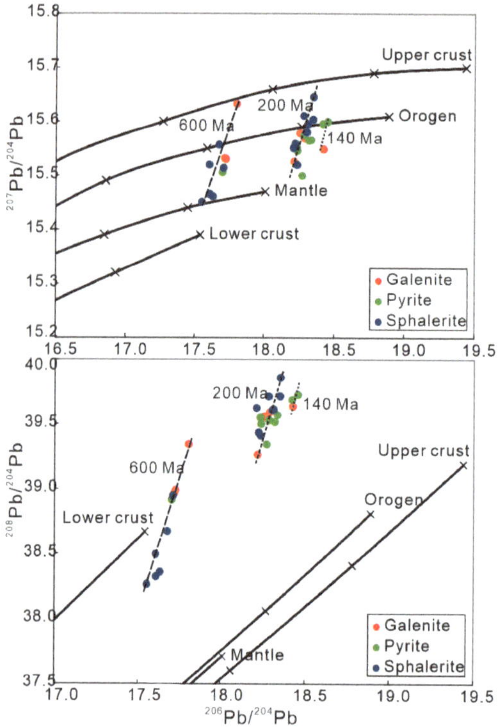

Figure 7. Values of $^{207}Pb/^{204}Pb$ vs. $^{206}Pb/^{204}Pb$ and $^{208}Pb/^{204}Pb$ vs. $^{206}Pb/^{204}Pb$ for sulfides of the Bailugou deposit plotted on diagrams proposed by Zartman and Doe [65]. Note that the different lines enclose the present Pb isotope ranges for the host rocks and granitoids in the study region. The lead isotopic data are from Table 2.

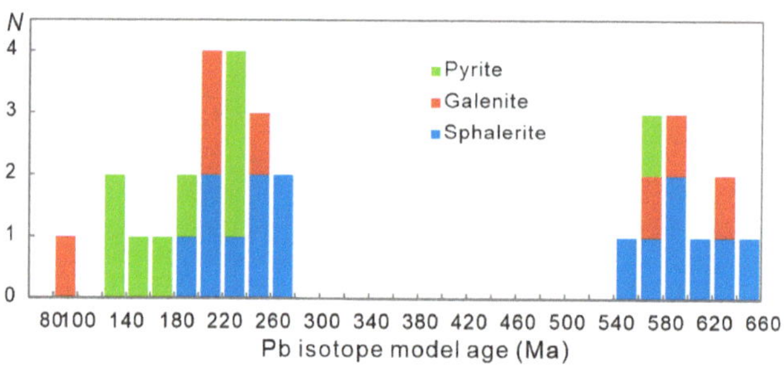

Figure 8. Histograms of the Pb isotope model age calculated by the Holmes–Houtermans method according to the single-stage evolution model of the Bailugou deposit.

Since sulfides such as pyrite and galena do not contain the parent radioactive elements of radioactive lead, such as U and Th, the lead isotope ratio remains unchanged after sulfide crystallization. Consequently, the lead isotope ratio of the ore is primarily derived from the ore-forming fluid system, which includes the metallogenic fluid, the host rock involved in water–rock interaction, and the fluid migration pathways prior to deposit formation. Thus, the lead model age is generally slightly earlier than the metallogenic age. The Bailugou deposit is speculated to have formed during the latest hydrothermal activity at 140± Ma, coinciding with the area's metallogenic age of porphyry deposits.

The lead isotopes of the Luanchuan Group, Guandaokou Group, Nannihu porphyry, and Shangfanggou porphyry were plotted synthetically to compare with the ore sulfides and to gain further insight into the genetic lead origin of the ore from the Bailugou deposit (Figure 9). In the $^{207}Pb/^{204}Pb$ vs. $^{206}Pb/^{204}Pb$ plot and $^{208}Pb/^{204}Pb$ vs. $^{206}Pb/^{204}Pb$ plot (Figure 9), the sulfides from the ores are primarily distributed within the range of the ore-hosting Luanchuan Group (Figure 9A,B) and Guandaokou Group (Figure 9C,D), confirming the established fact of strata-sourced Pb. However, some sulfides from the ores held higher $^{208}Pb/^{204}Pb$ ratios than those of the Luanchuan Group and Guandaokou Group, given that there was no accumulation of radiogenic Pb after the mineralization of ore deposit, which suggests an additional source of radiogenic Pb from the strata and granites. Furthermore, when compared with the K-feldspar and galena in Nannihu porphyry and Shangfanggou porphyry, the ore sulfides partially overlap with them but possess higher radiogenic Pb (Figure 9E,F).

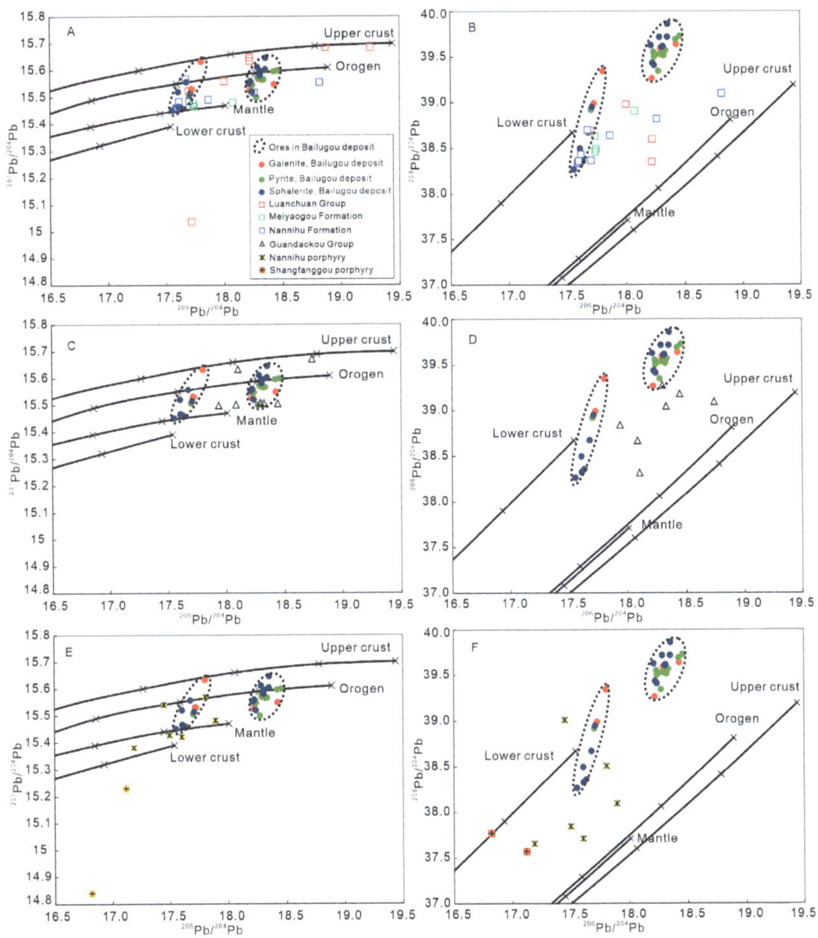

Figure 9. Lead isotope pattern for the Bailugou deposit and adjacent rocks (plotted on diagrams proposed by Zartman and Doe [65]). (**A**,**B**) the display of sulfides from the ores versus the ore-hosting Luanchuan Group; (**C**,**D**) the display of sulfides from the ores versus the ore-hosting Guandaokou Group; (**E**,**F**) the display of sulfides from the ores versus the K-feldspar and galena in Nannihu porphyry and Shangfanggou porphyry.

Previous studies on regional granite porphyries, including the lead isotope composition from Zhang et al. [73]; the Pb, Sr, and Nd isotope geochemistry from Chen and Zhang [74]; and the petrochemistry and diagenesis chronology from Bao et al. [75], collectively demonstrate that the Yanshanian granite magma in this region originated from the lower crust, with possible mixing of mantle-derived materials. Given that the $^{207}Pb/^{204}Pb$, $^{206}Pb/^{204}Pb$, and $^{208}Pb/^{204}Pb$ of strata rocks and granites should exceed those of K-feldspar and ore sulfides, the contribution from the lower crust, along with the mixing of mantle-derived materials, is essential and likely provides a high radioactive lead content. In considering the ores of the Bailugou deposit, the ore-hosting Luanchuan and Guandaokou Groups, the Nannihu porphyry, and the Shangfanggou porphyry were all products of a deep heat source acting on various geological bodies, including the strata of the Meiyaogou Fm. and the Nannihu Fm. of the Luanchuan Group, as well as the Guandaokou Group. The lower crust and mantle materials collectively supplied metallogenic materials for the Bailugou deposit. The interaction between the ore-hosting Luanchuan and Guandaokou groups, along with the Nannihu and Shangfanggou porphyries, suggests that these geological bodies were influenced by a shared deep heat source. This indicates that the thermal regime in the area was sufficiently significant to facilitate the mobilization of fluids and minerals, leading to the formation of ore deposits. The presence of deep heat sources likely resulted in extensive hydrothermal activity, which may have played a critical role in the mineralization processes observed in the Bailugou deposit.

5.2.3. Constraint from C-O Isotopes

Carbon-oxygen isotope plotting can effectively trace the source of ore-forming constituents and has been widely used [76–79]. In the plotting of $\delta^{13}C_{PDB}$ versus $\delta^{18}O_{SMOW}$ of the Bailugou deposit and adjacent rocks (Figure 10), both the host sediments (Luanchuan and Guandaokou Groups) and the Bailugou deposit vary within an area of sedimentary rock contamination/high-temperature effect, indicating that the Bailugou metallogenic hydrothermal fluid may be derived from host sediments due to the high-temperature effect. Furthermore, the plot of the mineralization stages shows a regular progression from stage 1 to stage 3. This coordinated variation in carbon and oxygen isotopes results from the mixing of the initial metallogenic hydrothermal fluid and later-injected groundwater in different proportions [46].

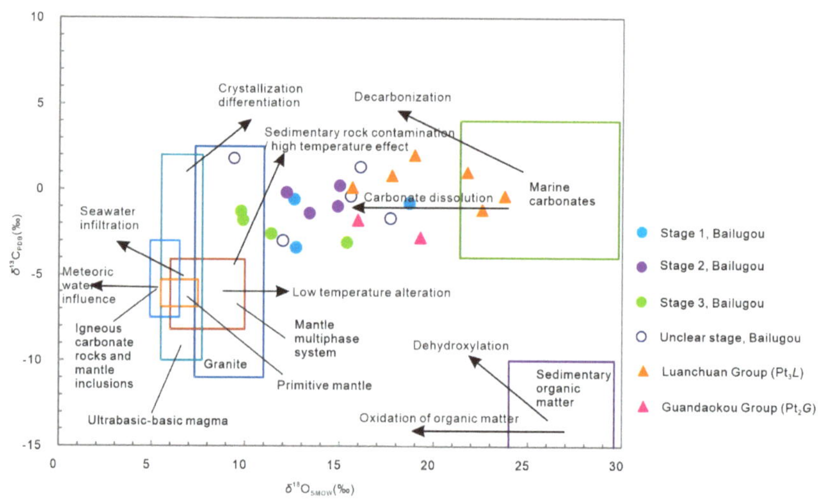

Figure 10. $\delta^{13}C_{PDB}$ (‰) versus $\delta^{18}O_{SMOW}$ (‰) of the Bailugou deposit and adjacent rocks (modified from Mao et al., [77]; Sun et al., [79]; Liu and Liu [80]; Liu et al., [81]).

The Ruyang Group in the southern margin of the North China Block shares the same paleogeographic setting as the Luanchuan and Guandaokou groups during the middle Proterozoic to early Neoproterozoic. The carbon and oxygen isotopic compositions from carbonate samples of the Luanchuan Group and Guandaokou Group in this region are similar to those of the Ruyang Group ($\delta^{18}O_{SMOW}$ = 22.5‰–24.0‰, $\delta^{13}C_{PDB}$ is about 0 [28]), which are lower than those of marine carbonate faces, indicating that it may also be synchronous result of geological events that led to molybdenum-tungsten and lead-zinc mineralization in this area.

The sulfur in the Bailugou deposit was derived from the sulfur homogenization in host sediments under a high-temperature setting. Meanwhile, the strata of the Meiyaogou Fm. and the Nannihu Fm. of the Luanchuan Group and Guandaokou Group, along with lower crust and mantle materials, collectively provide metallogenic constituents for the Bailugou deposit, as revealed by the sulfur and lead isotope. The REE distribution pattern of the Bailugou deposit [26] also shows that the ore-forming constituents may have originated from the ore-bearing strata, with no post-mineralization magmatic-hydrothermal transformation. Combined with previous studies on characteristics of fluid inclusions, as well as H–O and C–O isotopes, it is demonstrated that the metallogenic constituents of the Bailugou deposit are sourced from ore-bearing strata, lower crust, and mantle materials. This understanding aligns with the research of Gigon et al. [82], which traced metal sources for the giant McArthur River Zn-Pb deposit (Australia) using lead isotopes and suggested that leaching metals from multiple sources may be crucial for how hydrothermal systems form giant ore deposits, accounting for their substantial metal tonnage.

5.3. Process of Mineralization

Several disputes remain regarding the genesis of the Bailugou deposit. Yan [35] asserted that the Bailugou deposit is of MVT type, formed in the middle-late Proterozoic. Qi [25] contended that the Bailugou deposit is a sedimentary transformation type based on studies of ore geology, elemental geochemistry, isotopic geochemistry, and characteristics of fluid inclusions. Others have posited that the genesis of the Bailugou deposit is related to Yanshanian magmatism [27–29,31,34,83,84].

Research in this paper shows that while the Bailugou deposit shares some geological characteristics with the MVT-type deposit in the United States, notable differences also exist. The ore bodies of the Bailugou deposit are primarily controlled by interlayer fractures and exhibit significantly higher metallogenic temperatures, lower salinity of the ore-forming fluids, and smaller scale of alteration than those of the MVT type deposits. Moreover, the previous discussion regarding the source of ore-forming fluids and the origin of mineralizing materials consistently indicates that regional tectonic-thermal processes played a crucial role in mineralization of the Bailugou deposit. Additionally, a recent study by Chen et al. [85] on the texture and geochemistry of sphalerite from the Chitudian Pb-Zn-Ag deposit in this area indicates that chemical remobilization induced by tectonic deformation significantly influences trace elements distribution in sphalerite grain. Therefore, it is essential to investigate the mineralization age and tectonic background to comprehend the metallogenic mechanism of the Bailugou deposit.

Currently there are no experimental dating data available for the Bailugou deposit. The lead model age calculated in this paper is approximately 140$\pm$ Ma, although it does not constrain geochronological implications. However, the Shangfanggou, Nannihu, and Sandaozhuang superlarge porphyry-skarn molybdenum-tungsten deposits in this area were formed around 141.5 $\pm$ 7.8 Ma [86]. Meanwhile, the Lengshuibeigou large lead-zinc-silver deposit in this region was formed at 137.22 $\pm$ 2.49 Ma [35], all indicating a metallogenic age that aligns closely with about 140 Ma. The time interval from 208 Ma to 140 Ma, known as the Yanshanian (Jurassic–Early Cretaceous) period, experienced significant crustal deformation, thickening, and uplift in the Qinling orogenic belt, with the Yangtze–North China tectonic setting proposed to have shifted from compression and continental collision to extension [87–89]. Therefore, it is suggested that the Bailugou mineral system is no exception; along with other lead-zinc-silver deposits, porphyry-skarn molybdenum-tungsten deposits, and the intermediate-acid small intrusive bodies in this region, they are all products of the same tectonic-hydrothermal mineralization event under a regime of continental collision (including syn- to post-collisional settings), particularly during the transition from collisional compression to extension at about 140 Ma.

5.4. Metallogenic Model

The detailed geodynamic evolution of the Qinling Orogen, as illustrated by Chen et al. [10], improved understanding of regional and local metallogenesis. This paper proposes a metallogenic model, schematically illustrated in Figure 11, to elucidate the key tectonic processes and the genesis of the Bailugou lead-zinc deposit, as well as the other deposits in the Luanchuan polymetallic area, such as the Shangfanggou and Sandaozhuang Mo-W-Fe deposits and Lengshuibeigou Pb-Zn-Ag deposit.

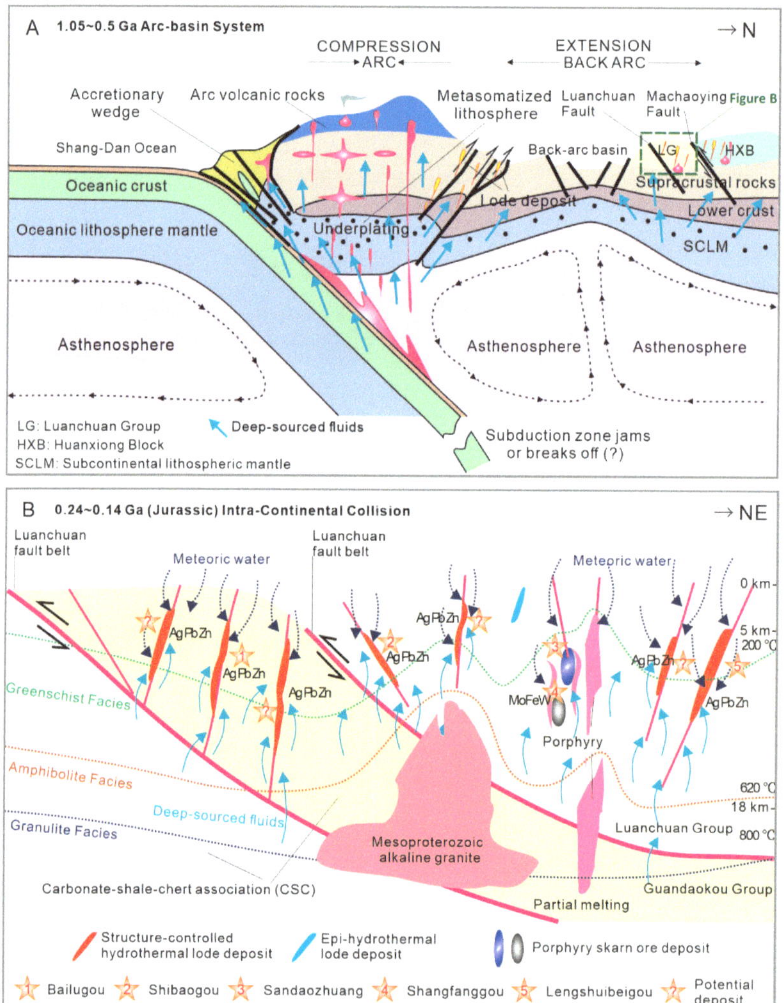

Figure 11. Tectonic evolution and genetic model for the Bailugou deposit and the Luanchuan polymetal belt. (**A**) Opening of the Erlangping Sea back arc basin and deposition of the Luanchuan

and Guandaokou Groups along the passive continental margin, while the Shang-Dan Ocean is subducted beneath the Central Qinling Terrane (figure revised from Chen et al. [10]). (**B**) Schematic representation of CMF (collisional orogeny, metallogeny, and fluid flow) illustrating the relationships between ore-hosting structures, granitoids, porphyries, and deposits in the Luanchuan area. Fluids, released from subducted slab and ocean floor sediment, or the hydrated mantle wedge, ascend along the interface between the slab and the overlying wedge or base of the lithosphere. The over-pressured ore fluids intersect deep crustal faults and then advect upwards to form orogenic deposits in second-order structures or hydraulically fractured rock bodies.

From 1.05 Ga to >0.5 Ga, an arc-back arc system and an extensional passive continental margin developed above the northward subduction of the Shan-Dan paleo-ocean (Figure 11A). The sedimentary carbonaceous CSC of the Luanchuan Group and the Guandaokou Group formed along the Luanchuan and Machaoying fault. During 240 Ma to 140 Ma, the final stages of the continental collision between the South and North China plates resulted in a stack of underthrust slabs within the Qinling Orogen (Figure 11B). The overall collision in the Qinling area commenced around 230 Ma to 220 Ma. During this collision, the lithosphere in the orogenic belt was significantly ruptured, dismantled, folded, and stacked, showing a series of intracontinental subductions or A-type subductions with varying scales, styles, and depths. The tectonic regime transitioned from Triassic–Early Jurassic compression (T_3-J, Indosinian Orogeny) through Late Jurassic–Early Cretaceous (J_2-K) transpressive compression (or transition from compression to extension) to Late Cretaceous extension (K_2, Yanshanian Orogeny), with continental collision concluding during the Late Cretaceous [30,48,49,90]. This is typical of the defined hydrothermal deposit zone in the terrane-scale CMF model (tectonic model for collisional orogeny, metallogeny, and fluid flow) [10]. During this period, large-scale hydrothermal circulation, reaction, granitic magmatism, and mineralization occurred in the Luanchuan area and the Qinling Orogenic Belt. Metallogenic ore systems that formed during this time include the Shangfanggou deposit, Sandaozhuang Mo-W-Fe deposits, the Lengshuibeigou Pb-Zn deposit [11], the Tieluping Pb–Zn deposit [10], and the Dabie–Tongbai–Erlangping Ag-Au-Pb belt [12–14].

Given the geochronological and tectonic background of the study area, the metallogenic process of the Bailugou deposit can be summarized as follows:

The increase in temperature and pressure gradients caused the melting of deep rocks (host sediments and lower crust or upper mantle), resulting in the formation of magma under elevated temperatures and metamorphic hydrothermal solutions at relatively lower temperatures. When the tectonic regime transitioned from collision to extension, CO_2-rich ore-forming metamorphic hydrothermal solutions originating from deep sources mobilized and extracted ore-forming constituents from the siliceous banded dolomite marble of the Luanchuan and Guandaokou groups, which were enriched in ore-forming elements [25,91,92]. Loaded with abundant ore-forming elements, the fluid migrated upwards through fault systems (Figure 11B) until it reached the brittle-ductile transition level (BDL). At this point, fluid immiscibility occurred due to fracturing and pressure drops, causing the ore-forming components to rapidly precipitate as rough and euhedral crystals [30]. Subsequently, meteoric water sourced from shallow depth flowed downward along fractures and participated in fluid circulation by mixing with the deep-sourced hydrothermal fluid. Boiling may have occurred during this process, leading to the accumulation and deposition of ore-forming constituents along fractures and interlayer detachment zones, resulting in extensive mineralization [30] (Figure 11B). Thus, the Bailugou Pb-Zn-Ag deposit was formed. During this process, sulfide ores inherited sulfur and lead isotope signatures derived from mantle sources, lower crust materials, and strata sources. Stage 1 of mineralization development is primarily characterized by metamorphic hydrothermal solutions originating from dehydration processes involving deep-seated materials; however, stage 4 incorporates atmospheric precipitation due to the gradual development of an open fracture system.

The geological and geochemical characteristics, along with the metallogenic environment of the Bailugou Pb-Zn-Ag deposit, not only resemble those of orogenic gold deposits in the East Qinling areas, such as Shanggong [93,94], Kangshan [95], and other global orogenic gold deposits [20,22,96,97], but also share similarities with orogenic Pb-Zn-Ag deposits identified in recent years, including Tieluping [10], Lengshuibeigou [11], Yindonggou [12], and Weishancheng ore belt [13,14]. Furthermore, integrating geological data, parameters from FIs, and the tectonic background of the study area, Yang et al. [30] first classified the genetic types of the Bailugou deposit as an orogenic metamorphic hydrothermal Pb-Zn-Ag deposit. This paper reaffirms this classification from the perspective of isotope geochemistry and a metallogenic model study.

6. Conclusions

1. Investigations of S, Pb, H–O, and C–O isotopes reveal that the ore-forming components originated from a combination of sources. The metallogenic materials for the Bailugou deposits are sourced from the ore-bearing strata, which comprise middle Proterozoic and Upper Proterozoic dolomite marble, as well as lower crust and mantle materials.
2. Lead-zinc-silver deposits, molybdenum-tungsten deposits, and small intermediate-acid intrusive bodies in the Luanchuan ore field formed within the same tectonic-magmatic-hydrothermal regime, resulting from intense large-scale magmatic-hydrothermal interaction in the Qinling Orogen during the Yanshanian period.
3. The Bailugou deposit developed concurrently with the Qinling Orogen during a transition from collisional compression to extension, under the collision regime of the Yangtze and North China continents, in the Late Triassic/Jurassic–Early Cretaceous period. The metamorphic dehydration of the inserted plates during the collision orogenic process facilitated the establishment of a hydrothermal metallogenic system in the ore belt. The ore-forming components were extracted, migrated through intense fluid–rock interaction from the Luanchuan Group and Guandaokou Group, and were eventually deposited.

Author Contributions: Conceptualization, Y.Y. and Y.C.; methodology, Y.Y.; formal analysis and investigation, Y.Y., N.G. and D.W.; writing—original draft preparation, N.G. and D.W.; writing—review and editing, Y.Y., H.C. and Y.C.; supervision, Z.P.; project administration, H.C.; funding acquisition, Y.C. All authors have read and agreed to the published version of the manuscript.

Funding: This research was funded by the National Nature Science Foundation of China (grant numbers 41702098, 42202073, and 40702013), the National Key Research and Development Program of China (grant number 2021YFC2900300), the China Geological Survey Project (grant numbers DD20230355, DD20230356, and DD20242591), and the Key Program of the National Natural Science Foundation of China (grant numbers U2244206 and 41630313).

Data Availability Statement: The data presented in this study are available on request from the corresponding author due to the internal privacy.

Acknowledgments: The authors extend gratitude to Jinping Qi, Jing Li, Jiangwei Han, and the Henan Bureau of Geological Exploration for Non-ferrous Metals for supporting this field investigation. We also thank the academic editors and anonymous reviewers for their valuable suggestions and feedback.

Conflicts of Interest: The authors declare no conflicts of interest regarding the publication of this article.

References

1. Kissin, S.A.; Mango, H. Silver vein deposits. In *Treatise on Geochemistry*, 2nd ed.; Elsevier: Oxford, UK, 2014; pp. 425–432.
2. Hammer, D.F.; Peterson, D.W. Geology of the Magma Mine area, Arizona. In *Ore Deposits of the United States, 1933–1967*; Ridge, J.D., Ed.; The American Institute of Mining, Metallurgical, and Petroleum Engineers, Inc.: New York, NY, USA, 1968; pp. 1282–1310.

3. Rye, R.O.; Sawkins, F. Fluid inclusion and stable isotope studies on the Casapalca Ag-Pb-Zn-Cu deposit, Central Andes, Peru. *Econ. Geol.* **1974**, *87*, 225–262. [CrossRef]
4. Keith, J.D.; Whitney, J.A.; Hattori, K.; Ballantyne, G.H.; Christiansen, E.H.; Barr, D.L.; Cannan, T.M.; Hook, C.J. The role of magmatic sulfides and mafic alkaline magmas in the Bingham and Tintic Mining Districts, Utah. *J. Petrol.* **1997**, *38*, 1679–1690. [CrossRef]
5. Im, H.; Jeong, J.; Shin, D. Genetic environment of W skarn and Pb-Zn vein mineralization associated with the Imog granite in the Taebaeksan Mineralized District, South Korea. *Ore Geol. Rev.* **2020**, *126*, 103721. [CrossRef]
6. Lynch, J.V.G.; Longstaffe, F.J.; Nesbitt, B.E. Stable isotopic and fluid inclusion indications of large-scale hydrothermal paleo flow, boiling, and fluid mixing in the Keno Hill Ag-Pb-Zn district, Yukon Territory, Canada. *Geochim. Cosmochim. Acta* **1990**, *54*, 1045–1059. [CrossRef]
7. Leach, D.L.; Landis, G.P.; Hofstra, A.H. Metamorphic origin of the Coeur d'Alene base-and precious-metal veins in the Belt basin, Idaho and Montana. *Geology* **1988**, *16*, 122–125. [CrossRef]
8. Leach, D.L.; Hofstra, A.H.; Church, S.E.; Snee, L.W.; Vaughn, R.B.; Zartman, R.E. Evidence for Proterozoic and Late Cretaceous-Early Tertiary ore-forming events in the Coeur d'Alene district, Idaho and Montana. *Econ. Geol.* **1998**, *93*, 347–359. [CrossRef]
9. Johnson, C.A.; Cardellach, E.; Tritlla, J.; Hanan, B.B. Cierco Pb-Zn-Ag vein deposits: Isotopic and fluid inclusion evidence for formation during the Mesozoic extension in the Pyrenees of Spain. *Econ. Geol.* **1996**, *91*, 497–506. [CrossRef]
10. Chen, Y.J.; Pirajno, F.; Sui, Y.H. Isotope geochemistry of the Tieluping silver deposit, Henan, China: A case study of orogenic silver deposits and related tectonic setting. *Miner. Depos.* **2004**, *39*, 560–575. [CrossRef]
11. Qi, J.P.; Chen, Y.J.; Ni, P.; Lai, Y.; Ding, J.Y.; Song, Y.W.; Tang, G.J. Fluid inclusion constraints on the origin of the Lengshuibeigou Pb-Zn-Ag deposit, Henan province. *Acta Petrol. Sin.* **2007**, *23*, 2119–2130. (In Chinese with English abstract)
12. Zhang, J.; Yang, Y.; Hu, H.Z.; Wang, Z.G.; Li, G.P.; Li, Z.L. C-S-Pb isotope geochemistry of the Yindonggou orogenic type silver deposit in Henan Province. *Acta Petrol. Sin.* **2009**, *25*, 2833–2842. (In Chinese with English abstract)
13. Zhang, J.; Yang, Y.; Lu, Y.H. Lead isotope geochemistry of the Weishancheng gold-silver ore belt, Henan Province, China: Implications for ore genesis. *Acta Petrol. Sin.* **2009**, *25*, 444–454. (In Chinese with English abstract)
14. Zhang, J.; Chen, Y.J.; Yang, Y.; Deng, J. Lead isotope systematics of the Weishancheng Au–Ag belt, Tongbai Mountains, central China: Implication for ore genesis. *Int. Geol. Rev.* **2011**, *53*, 656–676. [CrossRef]
15. Li, Z.K.; Li, J.W.; Zhao, X.F.; Zhou, M.F.; Selby, D.; Bi, S.J.; Sui, J.X.; Zhao, Z.J. Crustal-extension Ag-Pb-Zn veins in the Xiong'ershan district, southern North China Craton: Constraints from the Shagou Deposit. *Econ. Geol.* **2013**, *108*, 1703–1729. [CrossRef]
16. Han, J.S.; Yao, J.M.; Chen, H.Y.; Deng, X.H.; Ding, J.Y. Fluid inclusion and stable isotope study of the Shagou Ag- Pb-Zn deposit, Luoning, Henan province, China: Implications for the genesis of an orogenic lode Ag-Pb-Zn system. *Ore Geol. Rev.* **2014**, *62*, 199–210. [CrossRef]
17. Zhang, J.; Chen, Y.J.; Su, Q.W.; Zhang, X.; Xiang, S.H.; Wang, Q.S. Geology and genesis of the Xiaguan Ag–Pb–Zn orefield in Qinling orogen, Henan province, China: Fluid inclusion and isotope constraints. *Ore Geol. Rev.* **2016**, *76*, 79–93. [CrossRef]
18. Chen, X.L.; Shao, Y.J.; Lai, C.; Wang, C. Genesis of the Longmendian Ag–Pb–Zn Deposit in Henan (Central China): Constraints from Fluid Inclusions and H–C–O–S–Pb Isotopes. *Geofluids* **2020**, *2020*, 7352821. [CrossRef]
19. Li, S.D.; Wang, K.Y.; Wang, Y.C.; Zhang, X.B.; Quan, H.Y. Genesis of the Bairendaba Ag-Zn-Pb Deposit, Southern Great Xing'an Range, NE China: A Fluid Inclusion and Stable Isotope Study. *Geofluids* **2017**, *2017*, 1206587. [CrossRef]
20. Groves, D.I.; Goldfarb, R.J.; Gebre-Mariam, M.; Hagemann, S.G.; Robert, F. Orogenic gold deposits: Proposed classification in the context of their crustal distribution and relationship to other gold deposit styles. *Ore Geol. Rev.* **1998**, *13*, 7–27. [CrossRef]
21. Groves, D.I.; Goldfarb, R.J.; Robert, F.; Hart, J.R.C. Gold deposits in metamorphic belts: Overview of current understanding, outstanding problems, future research and exploration significance. *Econ. Geol.* **2003**, *98*, 1–29.
22. Kerrich, R.; Goldfarb, R.; Groves, D.; Garwin, S. The geodynamics of world class ore deposits: Characteristics, space-time distribution, and origins. *Rev. Econ. Geol.* **2000**, *13*, 501–551.
23. Luo, M.J.; Zhang, F.M.; Dong, Q.Y.; Xu, Y.R.; Li, S.M.; Li, K.H. *Molybdenum Deposits in China*; Henan Sci. & Tech. Press: Zhengzhou, China, 1991; 452p. (In Chinese)
24. Zhou, Z.X.; Li, B.L.; Guo, K.H.; Zhao, R.; Xie, Y.H. *Genesis of the Gold (Molybdenum) Deposits in Southern Margin of the Northern China Platform*; Seismological Press: Beijing, China, 1993; pp. 1–269. (In Chinese with English abstract)
25. Qi, J.P. *Geology, Geochemistry and Genesis of Vein-Type Zn-Pb-Silver Deposits in Luanchuan, Henan*; Peking University: Beijing, China, 2006; pp. 1–133. (In Chinese with English abstract)
26. Yan, C.H.; Liu, G.Y.; Peng, Y.; Song, Y.W.; Wang, J.Z.; Zhao, R.J.; Zeng, X.Y.; Lü, W.D.; Yao, X.N.; Ma, H.W.; et al. *The Metallogenic Regularity of Lead-Zinc Ore in the Southwest Hernan Province*; Geological Publishing House: Beijing, China, 2009; 369p. (In Chinese with English abstract)
27. Duan, S.G.; Xue, C.J.; Liu, G.Y.; Yan, C.H.; Feng, Q.W.; Song, Y.W.; Tu, Q.J.; Gao, Y.B.; Gao, B.Y. Geology and sulfur isotope geochemistry of lead-zinc deposits in Luanchuan district, Henan Province, China. *Earth Sci. Front.* **2010**, *17*, 375–384. (In Chinese with English abstract)
28. Duan, S.G.; Xue, C.J.; Feng, Q.W.; Gao, Y.; Liu, G.Y.; Yan, C.H.; Song, Y.W. Pb isotope compositions of vein type Pb-Zn ore deposits in Luanchuan district, Henan Province. *J. Mineral. Petrol.* **2010**, *30*, 69–78. (In Chinese with English abstract)

29. Duan, S.G.; Xue, C.J.; Liu, G.Y.; Yan, C.H.; Feng, Q.W.; Song, Y.W.; Gao, B.Y. Geology, fluid inclusions and stable isotopic geochemistry of Bailugou Pb-Zn deposit in Luanchuan area, Henan Province. *Miner. Depos.* **2010**, *29*, 810–826. (In Chinese with English abstract)
30. Yang, Y.; Chen, H.; Guo, N.N.; Wu, D.H.; Pang, Z.S.; Chen, Y.J. Genesis of Bailugou vein-type zinc-lead-silver deposit, Eastern Qinling Orogen, China: Constraints from ore geology and fluid inclusions. *Minerals* 2024, *submitted.* [CrossRef]
31. Duan, S.G.; Xue, C.J.; Chi, G.X.; Liu, G.Y.; Yan, C.H.; Feng, Q.W.; Song, Y.W. Ore Geology, Fluid inclusion, and S- and Pb-isotopic constraints on the genesis of the Chitudian Zn–Pb deposit, southern margin of the North China Craton. *Resour. Geol.* **2011**, *61*, 224–240. [CrossRef]
32. Yang, Y.; Chen, Y.J.; Zhang, J.; Zhang, C. Ore geology, fluid inclusions and four–stage hydrothermal mineralization of the Shangfanggou giant Mo–Fe deposit in Eastern Qinling, central China. *Ore Geol. Rev.* **2013**, *55*, 146–161. [CrossRef]
33. Yang, Y.; Liu, Z.J.; Deng, X.H. Mineralization mechanisms in the Shangfanggou giant porphyry-skarn Mo-Fe deposit of the east Qinling, China: Constraints from H–O–C–S–Pb isotopes. *Ore Geol. Rev.* **2017**, *81 Pt 2*, 535–547. [CrossRef]
34. Lü, W.D.; Sun, W.Z. Metallogenic condition of lead and zinc deposit in Lushi-Lanchuan terrain. *Miner. Resour. Geol.* **2004**, *18*, 507–516. (In Chinese with English abstract)
35. Yan, C.H. *Study on Inner Structure of Lead-Zinc-Silver Mineralization System in Eastern Qinling*; Geological Publishing House: Beijing, China, 2004; 144p. (In Chinese with English abstract)
36. Zhang, H.Q. Geochemical characteristics and geological significance of the Bailugou Pb–Zn–Ag deposit in Henan Province. *Miner. Explor.* **2021**, *12*, 2069–2075. (In Chinese with English abstract)
37. Clayton, R.N.; O'Neil, J.L.; Mayeda, T.K. Oxygen isotope exchange between quartz and water. *J. Geophys. Res.* **1972**, *B77*, 3057–3067. [CrossRef]
38. O'Neil, J.R.; Clayton, R.N.; Mayeda, T.K. Oxygen isotope fractionation in divalent metal carbonates. *J. Chem. Phys.* **1969**, *51*, 5547–5558. [CrossRef]
39. Zheng, Y.F.; Chen, J.F. *Stable Isotope Geochemistry*; Science Press: Beijing, China, 2000; pp. 1–316. (In Chinese)
40. Bottinga, Y. Calculated fractionation factors for carbon and hydrogen isotope exchange in the system calcite-carbon dioxide-graphite-methane-hydrogen-water vapor. *Geochim. Cosmochim. Acta.* **1969**, *33*, 49–64. [CrossRef]
41. Sheppard, S.M.F.; Epstein, S. D/H and $^{18}O/^{16}O$ ratios of minerals of possible mantel or lower crustal origin. Earth planet. *Sci. Lett.* **1970**, *9*, 232–239.
42. Liu, G.Y. *The Pb-Zn-Ag Metallogeny and Prospecting Target in the South Margin of North China Craton*; China University of Geosciences (Beijing): Beijing, China, 2007; (In Chinese with English abstract).
43. Huang, D.H.; Nie, F.J.; Wang, Y.C.; Jiang, X.J. Lead isotope composition of molybdenum deposits in East Qinling as applied to the problem of ore sureces. *Miner. Depos.* **1984**, *3*, 20–28, (In Chinese with English abstract).
44. Ma, Z.D. Stable isotope geochemical studies of various metallogenic belts in East Qinling and neighboring areas. *Miner. Depos.* **1992**, *11*, 54–64. (In Chinese with English abstract)
45. Yan, C.H.; Song, Y.W.; Liu, G.Y.; Xing, K. Geological features of Yangshuwa-Bailugou MVT lead-zinc deposit belt in Luanchuan, Henan Province. *Geol. Surv. Res.* **2004**, *27*, 249–254. (In Chinese with English abstract)
46. Liu, X.S.; Wu, C.Y.; Huang, B. Origin and evolution of the hydrothermal system in Nannihu-Sandaozhuang molybdenum (tungsten) ore deposit, Luanchuan County, Henan Province. *Geochimica* **1987**, *3*, 199–207. (In Chinese with English abstract) [CrossRef]
47. Qi, J.P.; Zhang, J.; Tang, G.J. Carbon and oxygen isotope composition of the Meso-Neoproterozoic strata south of the Xiong'er Terrane: Evidences of the CMF model. *Acta Petrol. Sin.* **2005**, *21*, 1365–1372.
48. Chen, Y.J.; Sui, Y.H.; Pirajno, F. Exclusive evidences for CMF model and a case of orogenic silver deposits: Isotope geochemistry of the Tieluping silver deposit, east Qinling orogen. *Acta Petrol. Sin.* **2003**, *19*, 551–568. (In Chinese with English abstract)
49. Chen, Y.J.; Pirajno, F.; Sui, Y.H. Geology and D–O–C isotope systematics of the Tieluping silver deposit, Henan, China: Implications for ore genesis. *Acta Geol. Sin.* **2005**, *79*, 106–119. (In Chinese with English abstract)
50. Melezhik, V.A.; Fallick, A.E.; Makarikhin, P.V. Extreme $^{13}C_{carb}$ enrichment in ca. 2.0 Ga magnesite-stromatolite-dolomite-'red beds' association in a global context: A case for the world-wide signal enhanced by a local environment. *Earth-Sci. Rev.* **1999**, *48*, 71–120. [CrossRef]
51. Taylor, H.P. The application of oxygen and hydrogen isotope studies to problems of hydrothermal alteration and ore deposition. *Econ. Geol.* **1974**, *69*, 843–883. [CrossRef]
52. Hoefs, J. *Stable Isotope Geochemistry*, 6th ed.; Springer: Berlin/Heidelberg, Germany, 2009; pp. 48–54.
53. Hu, S.X.; Lin, Q.L.; Chen, Z.M.; Li, S.M. *Geology and Metallogeny of the Collision Belt Between North China Plates*; Nanjing University Press: Nanjing, China, 1988; 558p. (In Chinese)
54. Ault, W.U.; Kulp, I.L. Sulfur isotopes and ore deposits. *Econ. Geol.* **1960**, *55*, N1. [CrossRef]
55. Ohmoto, H.; Rye, R.O. Isotopes of sulfur and carbon. In *Geochemistry of Hydrothermal Ore Deposits*; Barnes, H.L., Ed.; Wiley-Interscience: New York, NY, USA, 1979; pp. 509–567.
56. Zhang, L.G. *Geological Applicance for the Stable Isotope*; Shaanxi Science and Technology Press: Xi'an, China, 1985; pp. 1–267. (In Chinese with English abstract)
57. Han, Y.W.; Ma, Z.D.; Zhang, H.F.; Zhang, B.R.; Li, F.L.; Gao, S.; Bao, Z.Y. *Geochemistry*; Geological Publishing House: Beijing, China, 2003; pp. 1–370.

58. Pinckney, D.M.; Rafter, T.A. Fractionation of sulfur isotopes during ore-deposition in the upper Mississippi Valley Pb–Zn district. *Econ. Geol.* **1972**, *67*, 315–328. [CrossRef]
59. Ohmoto, H. Systematics of sulfur and carbon isotopes in hydrothermal ore deposits. *Econ. Geol.* **1972**, *67*, 551–578. [CrossRef]
60. Kelly, W.C.; Rye, R.O. Geologic, fluid inclusion, and stable isotope studies of the tin-tungsten deposits of Panasqueira, Portugal. *Econ. Geol.* **1979**, *74*, 1721–1822. [CrossRef]
61. Rollinson, H. *Using Geochemical Data: Evaluation, Presentation, Interpretation*; Routledge (Taylor & Francis Group): London, UK; New York, NY, USA, 2013; pp. 1–351.
62. Gao, Y.B.; Xue, C.J.; Zeng, R. Forming mechanism of H_2S in the Jinding Pb-Zn deposit, Lanping basin, northwest Yunnan Province. *J. Earth Sci. Environ.* **2008**, *30*, 367–373. (In Chinese with English abstract)
63. Wanless, R.K.; Boyle, R.W.; London, J.A. Sulfur isotope investigation of the gold-quartz deposits of the Yellowknife district. *Econ. Geol.* **1960**, *55*, 1591–1621. [CrossRef]
64. Doe, B.R.; Stacey, J.S. The application of lead isotopes to the problem of ore genesis and ore prospect evolution: A review. *Econ. Geol.* **1974**, *69*, 757–776. [CrossRef]
65. Zartman, R.E.; Doe, B.R. Plumbo tectonics—The model. *Tectonophysics* **1981**, *75*, 135–162. [CrossRef]
66. Pang, L.Y.; Zhu, X.Y.; Hu, G.H.; Qiu, Y.F.; Su, W.B.; Wang, S.Y.; Zhao, T.P. Advances in the study of Meso-Neoproterozoic stratigraphic chronology and sedimentary evolution in the southern margin of the North China Craton. *J. Stratigr.* **2021**, *45*, 180–195. [CrossRef]
67. Li, Z.S.; Jiang, R.R.; Ma, X.T.; Zhang, Y.; Li, Q.Z.; Zhang, J.D. Late Precambrian chronostratigraphic framework and tectonic evolution of the Xiong'er basin in southern North China Craton. *Acta Geol. Sin.* **2021**, *95*, 3234–3255.
68. Wang, H.; Wu, Y.B.; Gao, S.; Liu, X.C.; Gong, H.J.; Li, Q.L.; Li, X.H.; Yuan, H.L. Eclogite origin and timings in the North Qinling terrane, and their bearing on the amalgamation of the South and North China blocks. *J. Metamorph. Geol.* **2011**, *29*, 1019–1031. [CrossRef]
69. Wang, H.; Wu, Y.B.; Gao, S.; Liu, X.C.; Liu, Q.; Qin, Z.W.; Xie, S.W.; Zhou, L.; Yang, S.H. Continental origin of eclogites in the North Qinling terrane and its tectonic implications. *Precambrian Res.* **2013**, *230*, 13–30. [CrossRef]
70. Wu, Y.B.; Zheng, Y.F. Tectonic evolution of a composite collision orogen: An overview on the Qinling Tongbai Hong'an Dabie Sulu orogenic belt in central China. *Gondwana Res.* **2013**, *23*, 1402–1428. [CrossRef]
71. Liao, X.Y.; Liu, L.; Wang, Y.W.; Cao, Y.T.; Chen, D.L.; Dong, Y.P. Multistage metamorphic evolution of retrograde eclogite with a granulite facies overprint in the Zhaigen area of the North Qinling Belt, China. *Gondwana Res.* **2016**, *30*, 79–96. [CrossRef]
72. Li, N.; Chen, Y.J.; Zhang, H.; Zhao, T.P.; Deng, X.H.; Wang, Y.; Ni, Z.Y. Molybdenum deposits in East Qinling. *Earth Sci. Front.* **2007**, *14*, 186–198.
73. Zhang, H.F.; Zhang, B.R.; Zhao, Z.D.; Luo, T.C.; Chen, Y.L. Feldspar lead isotopic compositions of granitoids from the eastern Qinling orogenic belt and their tectonic significance. *Acta Geol. Sin.* **1997**, *71*, 142–149. (In Chinese with English abstract)
74. Chen, Y.L.; Zhang, B.R. Pb, Sr and Nd Isotope geochemistry of Yanshanian and granitoids on Southern Margin of North China Craton in Western Henan Province, Earth Science. *J. China Univ. Geosci.* **1994**, *19*, 375–382.
75. Bao, Z.W.; Zeng, Q.S.; Zhao, T.P.; Yuan, Z.L. Geochemistry and petrogenesis of the ore related Nannihu and Shangfanggou granite porphyries from east Qinling belt of molybdenum mineralization. *Acta Petrol. Sin.* **2009**, *25*, 2523–2536. (In Chinese with English abstract)
76. Rollinson, H.B. *Using Geochemical Data: Evolution, Presentation, Interpretation*; Longman Scientific & Technical: New York, NY, USA, 1993; pp. 270–343.
77. Mao, J.W.; He, Y.; Ding, T.P. Mantle Fluids Involved in Metallogenesis of Jiaodong (East Shandong) Gold District: Evidence of C, O and H Isotopes. *Miner. Depos.* **2002**, *21*, 121–128. (In Chinese with English abstract)
78. Liu, J.M.; Zhang, H.F.; Sun, J.G.; Ye, J. Geochemical research on C-O and Sr-Nd isotopes of mantle derived rocks from Shandong Province, China. *Sci. China Ser. D* **2004**, *47*, 171–180. [CrossRef]
79. Sun, J.G.; Hu, S.X.; Shen, K.; Yao, F.L. Research on C, O isotopic geochemistry of intermediate-basic and intermediate-acid dykes in goldfields of Jiaodong Peninsula. *Acta Petrol. Mineral.* **2001**, *20*, 47–56. (In Chinese with English abstract)
80. Liu, J.M.; Liu, J.J. Basin fluid genetic model of sediment-hosted microdisseminated gold deposits in the gold-triangle area between Guizhou, Guangxi and Yunnan. *Acta Mineral. Sin.* **1997**, *17*, 448–456. (In Chinese with English abstract)
81. Liu, J.J.; He, M.Q.; Li, Z.M.; Liu, Y.P.; Li, C.Y.; Zhang, Q.; Yang, W.G.; Yang, A.P. Oxygen and carbon isotopic geochemistry of Baiyangping silver-copper polymetallic ore concentration area in Lanping Basin of Yunnan Province and its significance. *Miner. Depos.* **2004**, *23*, 1–10. (In Chinese with English abstract)
82. Gigon, J.; Deloule, E.; Mercadier, J.; Huston, D.L.; Richard, A.; Annesley, I.R.; Wygralak, A.S.; Skirrow, R.G.; Mernagh, T.P.; Masterman, K. Tracing metal sources for the giant McArthur River Zn-Pb deposit (Australia) using lead isotopes. *Geology* **2020**, *48*, 478–482. [CrossRef]
83. Ye, H.S.; Mao, J.W.; Li, Y.F.; Yan, C.H.; Guo, B.J.; Zhao, C.S.; He, C.F.; Zheng, R.F.; Chen, L. Characteristics and metallogenic Mechanism of Mo-W and Pb-Zn-Ag deposits in Nannihu ore field, Western Henan Province. *Geoscience* **2006**, *20*, 165–174. (In Chinese with English abstract)
84. Duan, S.G.; Xue, C.J.; Feng, Q.W.; Gao, B.Y.; Liu, G.Y.; Yan, C.H.; Song, Y.W. Carbon and oxygen isotope compositions of gangue carbonates in Pb-Zn ore deposits in Luanchuan district, Southwest Henan Province. *Geoscience* **2010**, *24*, 767–775. (in Chinese with English abstract)

85. Chen, C.; Meng, L.; Xu, J.; Zhan, Q.; Zhao, T.P. Texture and geochemistry of sphalerite from the Chitudian Pb-Zn-Ag deposit, southern margin of the North China Craton: Implications for the enrichments of Cd, Ga, and In. *Ore Geol. Rev.* **2022**, *156*, 105392. [CrossRef]
86. Li, Y.F.; Mao, J.W.; Bai, F.J.; Li, J.P. Re–Os isotopic dating of molybdenites in the Nannihu molybdenum (Tungsten) ore field in the eastern Qinling and its geological significance. *Geol. Rev.* **2003**, *49*, 652–659. (In Chinese with English abstract)
87. Chen, Y.J.; Li, C.; Zhang, J.; Li, Z.; Wang, H.H. Sr and O isotopic characteristics of porphyries in the Qinling molybdenum deposit belt and their implication to genetic mechanism and type. *Sci. China Ser. D* **2000**, *43*, 82–94. [CrossRef]
88. Chen, Y.J.; Santosh, M. Triassic tectonics and mineral systems in Qinling Orogen, China. *Geol. J.* **2014**, *49*, 338–358. [CrossRef]
89. Li, N.; Pirajno, F. Early Mesozoic Mo mineralization in the Qinling Orogen: An overview. *Ore Geol. Rev.* **2017**, *81*, 431–450. [CrossRef]
90. Chen, Y.J.; Pirajno, F.; Qi, J.P.; Li, J.; Wang, H.H. Ore geology, fluid geochemistry and genesis of the Shanggong gold deposit, eastern Qinling Orogen, China. *Resour. Geol.* **2006**, *56*, 99–116. [CrossRef]
91. Li, Y.F. The Temporal-Spital Evolution of Mesozoid Granitoids in the Xiong'ershan Area and Their Relationships to Molybdenum-Gold Mineralization. Ph.D. Dissertation, China University of Geosciences, Beijing, China, 2005; pp. 1–135. (In Chinese with English summary)
92. Ye, H.S. The Mesozoic Tectonic Evolution and Pb-Zn-Ag Metallogeny in the Southern Margin of North China Craton. Ph.D. Dissertation, Chinese Academy of Geological Sciences, Beijing, China, 2006; pp. 1–225. (In Chinese with English summary)
93. Chen, Y.J.; Fu, S.G. *Gold mineralization in West Henan, China*; Seismological Press: Beijing, China, 1992; 234p. (In Chinese with English abstract)
94. Chen, Y.J. Fluidization model for continental collision in special reference to study on oreforming fluid of gold deposits in the eastern Qinling Mountains, China. *Prog. Nat. Sci.* **1998**, *8*, 385–393.
95. Wang, H.H.; Chen, Y.J.; Gao, X.L. The isotope geochemistry of the Kangshan gold deposit, Henan and its illustration of the CPMF model. *Miner. Depos.* **2001**, *20*, 190–198. (In Chinese)
96. Goldfarb, R.J.; Groves, D.I.; Cardoll, S. Orogenic gold and geologic time: A global synthesis. *Ore Geol. Rev.* **2001**, *18*, 1–75. [CrossRef]
97. Hagemann, S.G.; Cassidy, K.F. Archean orogenic lode gold deposits. *SEG Rev.* **2000**, *13*, 9–68.

minerals

MDPI

Article

Genesis of the Bailugou Vein-Type Zinc-Lead-Silver Deposit, Eastern Qinling Orogen, China: Constraints from Ore Geology and Fluid Inclusions

Yan Yang [1,2,*], Nana Guo [3], Hui Chen [1,2], Donghao Wu [4], Zhenshan Pang [1,2] and Yanjing Chen [5,*]

1 Development and Research Center, China Geological Survey, Beijing 100037, China
2 Technical Guidance Center for Mineral Exploration, Ministry of Natural Resources of the People's Republic of China, Beijing 100037, China
3 Bureau of Natural Resources of Luanchuan County, Luoyang 471500, China
4 School of Earth Sciences and Resources, China University of Geosciences, Beijing 100083, China
5 Key Laboratory of Orogen and Crust Evolution, Peking University, Beijing 100871, China
* Correspondence: yyan@mail.cgs.gov.cn (Y.Y.); gigyjchen@126.com (Y.C.)

Abstract: The Bailugou vein-type zinc-lead-silver deposit is located in the Eastern Qinling Orogen, China. There has been a long-standing debate about whether its formation is related to magmatism or metamorphism. To determine the origin of ore-forming materials and fluids, we conducted a geological and fluid inclusion investigation of the Bailugou. Field surveys show that the vein-type orebodies are controlled by faults in the dolomitic marbles of the Mesoproterozoic Guandaokou Group, and they are distal to the regional Yanshanian intrusions. Four ore stages, i.e., quartz–pyrite ± sphalerite (Stage 1), quartz–polymetallic sulfides (Stage 2), dolomite–polymetallic sulfides (Stage 3), and calcite (Stage 4), are identified through microscopic observation. The homogenization temperatures of measured fluid inclusions vary in the range of 100 °C to 400 °C, with the dominating concentration at 350 °C to 400 °C, displaying a descending trend from early to late stages. The estimated formation depth of the Bailugou deposit varies from 2 km to 12 km, which is deeper than the metallogenic limit of the epithermal hydrothermal deposit but conforms to the typical characteristics of a fault-controlled deposit. The ore-forming fluid in Stage 1 originates from a fluid mixture and experiences a phase separation (or fluid immiscibility) between the metamorphic-sourced fluid and the fluids associated with ore-bearing carbonate-shale-chert association (CSC) strata. This process results in the transition to metamorphic hydrothermal fluid due to water–rock interactions in Stage 2, culminating in gradual weakening and potential fluid boiling during the mineralization of Stage 3. Collectively, the Bailugou lead-zinc-silver mineralization resembles an orogenic-type deposit formed by metamorphic fluids in the Qinling Yanshanian intracontinental orogeny.

Keywords: metamorphic hydrothermal; orogenic zinc-lead-silver ore deposit; Bailugou; fluid inclusions; Eastern Qinling Orogen

1. Introduction

Since the general classification scheme of silver-lead veins, according to dominant mineral assemblage, have been first proposed by Beck and Weed [1], the subdivisions and ore deposits within it have been frequently reorganized, renamed, or even changed according to new geologic observations or theories, especially for those vein-type Zn–Pb–Ag deposits genetically related to the intrusions, which are further described as the Cordilleran-type or skarn vein deposits, were well documented [2–9]. The veins, which are not genetically related to felsic intrusions but more commonly occur near a crustal-scale fault, are considered late features in the tectonic evolution of an orogen by Beaudoin and Sangster [10], based on a worldwide review of silver, lead, and zinc vein deposits from six districts: the Kokanee Range and Keno Hill (Canada), Coeur d'Alène (United States), the Harz Mountains and Freiberg (Germany), and Příbram (Czechoslovak). Since then, more similar cases have been presented as orogenic or metamorphic hydrothermal Ag–Pb–Zn deposits [11–14]. Some of these vein-type deposits specially develop in Eastern Qinling Orogen, China, such as Tieluping [15], Lengshuibeigou [16], Yindonggou [17], Weishancheng [18,19], Shagou [20,21], Xiaguan [22], Longmendian [23], etc. Known as rich in the endowment of Mo–W–Fe–Ag–Pb–Zn (Figure 1), where the Nannihu Mo–W and Shangfanggou Mo–Fe super large deposits [24,25] were endowed, the Luanchuan area also developed several large to super-large vein-type zinc–lead (–silver) deposits, such as Lengshuibeigou, Bailugou, Yindonggou, Tudimiaogou, Banchang, discovered by the geological survey department of Henan Province in the 2000s [26]. The Lengshuibeigou, Bailugou, and Yindonggou zinc–lead–silver deposits were approximated with total predicted reserves of 20 Mt (Pb + Zn) and 19,000 t Ag [27]. There are still many critical issues in the genesis that are poorly understood and need to be thoroughly explored and specifically discussed. For example, what are the specific geological characteristics of vein-type zinc–lead (–Ag) deposits? What are the sources of ore-forming fluid and constituents? What is their detailed evolutionary process and genetic model? These are some of the questions that require further exploration and clarification.

Fluids are the primary carriers of mass and heat within the Earth's crust [28–30]. The flow of fluid, which can be of magmatic, metamorphic, connate, marine, or meteoric origin [31–33], leads to leaching, mobilization, and transport, finally depositing the elements of economic interest, which presents as hydrothermal ore deposits. The circulation of surface-derived fluids within the crust is enhanced by active crustal faults [34]. Thus, constraining the source of ore-forming fluids is critical to clarifying the origin of metallic elements during mineralization.

The Bailugou vein-type deposit in the Luanchuan area, Eastern Qinling Orogen, China, is located in the Lushi-Luanchuan WNW-trending fold-thrust belt, which is part of the Huaxiong terrane in the Eastern Qinling Orogen, southern margin of the North China Craton (Figure 1). The history of the discovery of the Bailugou lead-zinc deposit in Henan Province, China, goes back to the 1950s. Geologists first identified the presence of lead and zinc deposits in the area during geological prospecting activities. The mine was officially discovered and started to be further developed from the late 1950s onwards. Duan et al. studied the fluid inclusions of the Bailugou deposit briefly without dividing metallogenetic stages [35]. Although several studies about regionally developed deposits have been published [36–41], the genesis of the Bailugou vein-type zinc-lead-silver deposit remains ambiguous; moreover, the detailed evolutionary mechanism of ore-forming fluid in the Bailugou deposit still remains unclear and needs to be solved and explored. This study conducted detailed regional, field geological and petrographic observations, alteration characteristics, and microthermometric data of fluid inclusions of ore minerals and wall rock from the Bailugou Zn–Pb (–Ag) ore deposit. The present study proposes an investigation into the origin, evolution, and precipitation mechanism of ore-forming fluids at the Bailugou deposit. The primary objective is to thoroughly comprehend the nature and genesis of the Bailugou and analogous vein-type deposits in the Luanchuan area. The conclusions of this examination are anticipated to offer valuable insights for deposit exploration in the surrounding region.

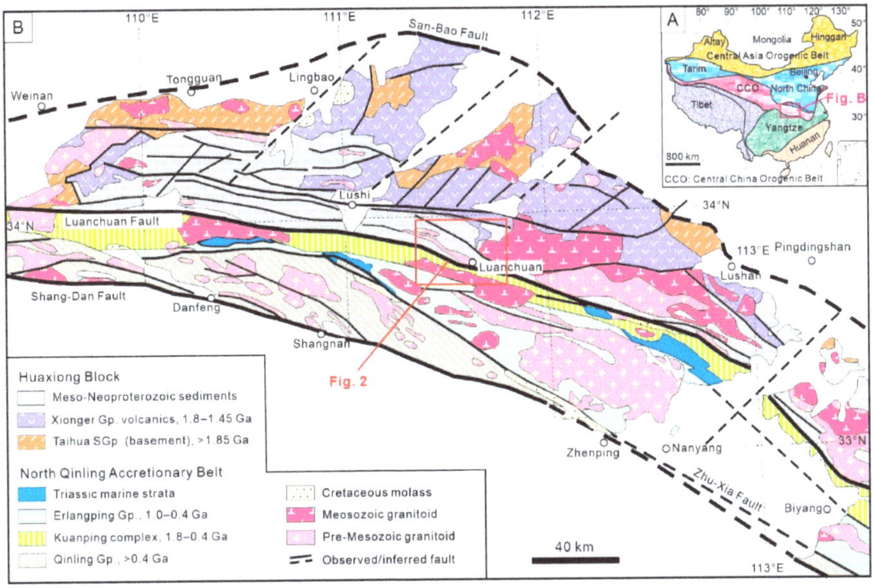

Figure 1. Geological map of the East Qinling area (modified from Li et al. [42]). (**A**) Tectonic map of China, showing the location of the Qinling Orogen; (**B**) tectonic subdivision of the Qinling Orogen, showing the location of the Bailugou Zn–Pb deposit.

2. Geological Setting

The tectonic evolution and regional metamorphism of the studied area shows that the Luanchuan Group and Guandaokou Group were sedimented under the forearc basin setting [43]. The primary lithology is mainly siliceous belt-bearing dolomite marble and medium-thick layered dolomite marble, containing conglomerate lithic sandstone, mudstone, sericite phyllite, calcareous phyllite, and metamorphic quartz dolomite, which is a CSC protolith formation that underwent greenschist facies metamorphism. The northwest fault structures in the area form the basic tectonic framework, characterized by large widths, long extensions, and dense distribution, controlling the distribution of strata in the region. A small number of tensional faults develop in the northeast direction. The intersections of these faults with the northwest-trending faults are the locations where ore bodies are hosted.

2.1. Stratigraphy

Strata in the Luanchuan area from base to top are as follows: (1) the Archean basement, Taihua Group, composed of amphibolite metamorphic complexes; (2) the Mesoproterozoic strata, the lower Xiong'er Group, consisting of andesitic-dacitic volcanic rocks, the upper Guandaokou Group, consisting of terrigenous clastic rocks and sedimentary carbonate formations; and (3) the Neoproterozoic strata of the lower Luanchuan Group, consisting of terrigenous clastic rocks, carbonate formations, and alkaline lavas, and the upper Taowan Group, composed of conglomerate and argillaceous carbonate rocks.

The Guandaokou and Luanchuan Groups extensively develop in the Bailugou Zn–Pb ore field (Figure 2) and are the primary ore-hosting strata of Mo–W–Fe and Zn–Pb–Ag ore deposits in this region. The Bailugou deposit was hosted in the dolomitic marble of the Mesoproterozoic Longjiayuan, Xunjiansi, Duguan, and Fengjiawan Formation of Guandaokou Group and the Neoproterozoic Baizhugou, Sanchuan and Nannihu Formation of Luanchuan Group (Figure 2).

The orebodies in the eastern and northern parts of the Bailugou Zn–Pb ore district are hosted in the Longjiayuan Formation (Figure 2). Meanwhile, the orebodies in the middle of the ore district are hosted in the Xunjiansi Formation, which conforms with the Longjiayuan Formation in the north and is fault-contacted with the Fengjiawan or Duguan Formation [44,45].

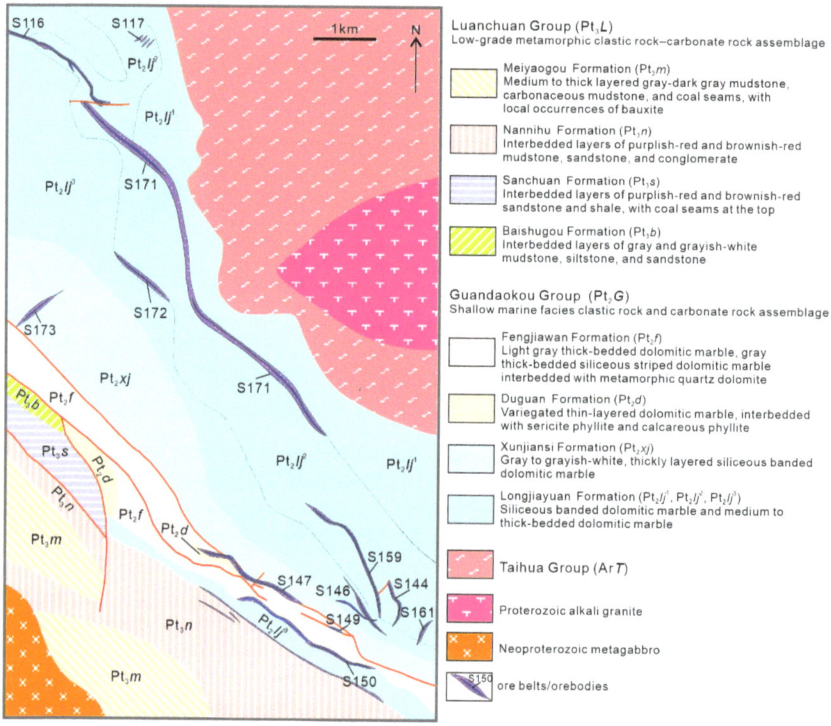

Figure 2. Geologic scheme of Bailugou Zn–Pb deposit (modified from Yan [45], Zhang [40]).

2.2. Igneous Rocks

There are two major phases of magmatic intrusion in the region. The earliest intrusion is the metagabbro in the Neoproterozoic Jingningian period, and the latter is the Mesoproterozoic Longwangzhuang alkaline granite (Figure 2). The Yanshanian intrusion was poorly outcropping.

The Longwangzhuang intrusion, which is the nearest intrusion to the Bailugou Zn–Pb ore bodies, lies in the east of the ore district, located in the core of the NWW-trending quaquaversal Niuxinduo plunging anticline. Hu et al. [43] suggested that the Longwangzhuang intrusion is A-type granite, the same age as the Luanchuan Group and formed under an extensional setting by 1235 Ma of Rb–Sr isochron age. However, according to a surface age of 1621 ± 16 Ma of zircons in Longwangzhuang potassium granite, Lu et al. [46] regarded it as the alkaline magmatism of the late period of the North China Craton break in 1800 Ma to 1600 Ma.

2.3. Structures

The ore district is located at the west limb of the Niuxinduo plunging anticline. The NWW-trending faults are the main ore-bearing structures that are well developed in the Bailugou Zn–Pb ore district (Figures 2 and 3). These faults mainly consist of SW strikes, with dips ranging from 35° to 79° and widths of 20–40 m, reaching a maximum of over

100 m [45]. A series of NE-NNE-trending faults occurred in the ore district, with SE-strike dips of 48° to 75°. The ore bodies are commonly located at the junctions of the NWW-trending faults and the NE-NNE-trending faults [44].

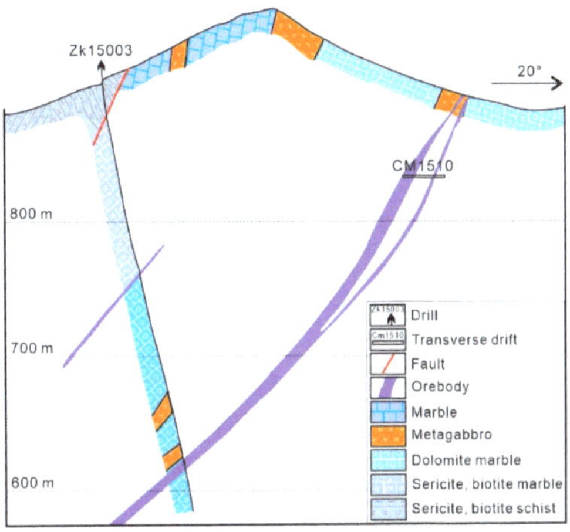

Figure 3. Geological sectional drawing of S150 mineralized zone of Bailugou deposit (modified from Yan et al. [47]).

3. Ore Geology

3.1. Ore Bodies

The Bailugou Pb-Zn-Ag deposit has a proved reserve of 198.92 × 10^4 t (Pb + Zn) and 2553.28 t Ag. The grade of the whole deposit is Pb@10.31%–16.43% with an avg. of 11.29%, Zn@2.33%–9.69% with an avg. of 5.58%, and Ag@80.18 × 10^{-6}–279.18 × 10^{-6} with an avg. of 183.32 × 10^{-6}, respectively. The Bailugou ore deposit consists of four major mineralization zones: S150 in the south, S116 in the north, S171 and S172 in the central (Figure 2), and other small, low-grade mineralized zones. Strata and NE-trending faults control the occurrences of mineralized zones.

The S150 mineralized zone is located in the Luyuangou-Gongyingou area south of the ore district. The mineralized zone occurs in the fracture zone between the Longjiayuan Formation and Fengjiawan Formation of the Guandaokou Group (Figures 2 and 3). Silicified dolomite marble was developed in the fracture zone. The fracture zone is partially cataclastic rock and metagabbro veins. The S150 mineralized zone strikes 200°–210°, dips at 50°–70°, is distributed intermittently over 3000 m, spans widths of 2–6 m, with a maximum width of 20 m. Two bedded and stratoid ore bodies have been delineated; the ore bodies' occurrences are consistent with the strata. The ore grade has an increasing tendency from surface to deep [47].

The S116 mineralized zone is located in the northwest part of the Bailugou ore district, hosted in the talc-bearing quartz dolomite marble of the middle Longjiayuan Formation, Guandaokou Group. The S116 mineralized zone widths of 5–50 m extend over 3000 m, with an occurrence as a smooth-out wave shape with branches that leap along the strike (Figures 2 and 4). The fracture zones have dipped to NE with dips of 40°–50°. The S116 mineralized zone mainly develops cataclastic rock while the compressed schistosity and mylonitization develop locally, and metagabbro veins fill occasionally. Four paralleled Zn–Pb–Ag ore bodies along the trending have been delineated. The No.1 main ore body has an occurrence as beds, stratoids, and veins, and their dips and strikes are consistent with the strata as synchronous folds. The ore body has a dip orientation of 40° and dip angles of 35°–40°, while the length and depth are 1524 m and 50–60 m, respectively, and the thickness is 0.8–10.9 m with an average of 2.5 m.

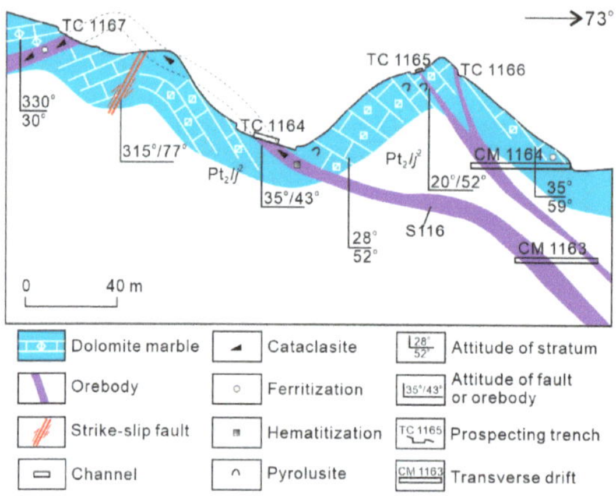

Figure 4. Geological sectional drawing of S116 mineralized zone of Bailugou deposit (modified from Yan [45]).

The S171 mineralized zone is geographically located through the Zhuyuangou, Bailugou, and Luyuangou area from west to east. The outcrop of S171 extended intermittently over 5000 m and a s2–6 m width with a maximum of 30 m. The ore bodies are located in the top part of the lower Longjiayuan Formation, distributed along NW-SE-trending as smooth-out waveforms (Figure 2). The S171 mineralized zone generally strikes NW, dips to SW with dip angles of 45° to 70°, and locally reverse dips along with the strata. The S171 mineralized zone dominantly developed taupe cataclastic silicified dolomite marble, ferritization–thematization cataclastic dolomite marble. The secondary was cataclastic with silicification, hematitization, ferritization, and iron-manganese. The wall rock was talc-bearing siliceous banding dolomite marble. There were 3 Zn–Pb–Ag ore bodies delineated at S171. The length of ore bodies was over 800 m, with stratoid and lens forms occurring—the strike of S171-1 and S171-2 ore bodies conform with the S171 mineralized zone.

The S172 mineralized zone is located on the top of the middle Longjiayuan Formation, paralleled with S171, dips to SW, with dip angles of 50°–70°. The ore bodies are longer than 300 m. The width of the ore bodies ranges from 1.0 to 2.5 m at the outcrop and 1.8–6.4 m at depth.

3.2. Mineral Assemblages

The ore minerals of S150 ore bodies include galena, sphalerite, pyrite, and minor amounts of chalcopyrite and argentite. The gangue minerals include quartz, dolomite, calcite, barite, sericite, talc, and are locally carbonaceous. The mineral grain size shows an increase with depth beneath the surface. The ore minerals of S116 ore bodies are mainly galena, followed by sphalerite, pyrite, and chalcopyrite. The gangue minerals include calcite, quartz, dolomite, and kaolinite. The S171 ore bodies developed pyritization in the deep, with ore minerals of galena, pyrite, and minor sphalerite. The gangue minerals include quartz, dolomite, calcite, talc, etc.

The ore structure is present as a disseminated, veinlet-disseminated, banded, massive, brecciated, veined, stockwork, and crystal cluster structure (Figure 5). The brecciated ores were striped or mottled early formed ores, cemented by hydrothermal ores and gangue minerals from later stages.

Figure 5. Ore structures of Bailugou deposit. (**A**) Ragged, banded distributes ores of galena (Gn), sphalerite (Sp) and dolomite (Dol); (**B**) massive ores of Sp; (**C**) Dol, quartz (Qtz) and Sp; (**D**) disseminated pyrite (Py) and Dol; (**E**) disseminated Sp, Gn and Dol; (**F**) banded Gn, Sp and Dol; (**G**) brecciated structures; (**H**) chloritization, chlorite (Chl).

The mineral texture is shown as granular texture, metasomatic dissolution, replacement, metasomatic etching texture, cataclastic structure, bird's-eye structure, residual skeleton texture (pyrite), growth girdle zone texture (sphalerite and dolomite) (Figures 5 and 6). Chalcopyrite is displayed as a beaded, emulsion droplet-like, or flake texture in the sphalerite as a solid solution separation structure only (Figure 6).

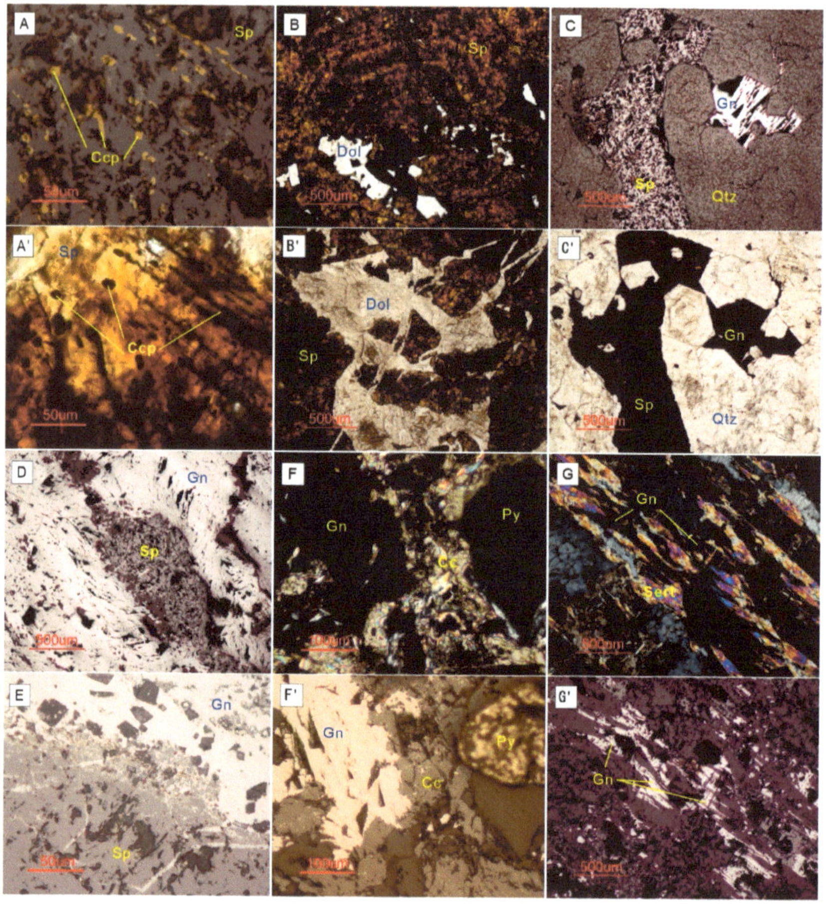

Figure 6. Ore textures of Bailugou deposit. (**A**) sphalerite (Sp)-chalcopyrite (Ccp) solid separation texture, chalcopyrite exhibit as bead-strings in sphalerite, under reflected light, (**A'**) shows under the transmission light (−); (**B**) growth girdle zone of altered sphalerite backfilled by late dolomite (Dol), under transmission light (−), (**B'**) shows fractured sphalerite backfilled by dolomite, under transmission light (−); (**C**) xenomorphic sphalerite and galena (Gn) backfilled in prior euhedral quartz (Qtz) granula, under reflected light, (**C'**) shows under the transmission light (−); (**D**) later galena etching replace sphalerite, contact as wedge-shape, under reflected light; (**E**) galena etching replaced sphalerite, contact as wedge-shape, galena stellated in sphalerite, reflect light; (**F**) pyrite (Py) and galena with late calcite, under transmission light (+), (**F'**) under reflected light; (**G**) ductile deformation of galena, Sericite (Sert) and quartz, shaping as lens, bird-eye, pressure-shadow, under transmission light (+), (**G'**) under reflected light.

3.3. Alteration and Mineralized Stages

Hydrothermal alteration in the Bailugou deposit includes silicification, dolomitization, and carbonation. The dolomitization and silicification are closely related to Pb-Zn mineralization, while the carbonation usually develops later than the main mineralization. In addition, there is a small amount of altered mineral-talc, which only developed in the wall rocks of the ore body in S116 and S171 with high hematite and limonite.

Dolomitization is very common in this area and can be roughly divided into two categories: one is regional dolomitization, which is distributed in a large range far from the ores and usually has no trace of ore mineralization; another is restricted to the mineralization range and has a close relationship with the ores.

The regional dolomitization developed in the mid-thick layer dolomite, quartz dolomite marble, or stromatolite-bearing dolomite marble; the dolomitic rocks are mainly formed in clumped, breccia-like irregular aggregate or along the sides of cracks. Dolomites of this category are idiomorphic and semi-idiomorphic crystals; the crystals are primarily arranged in a circular band with a diameter of 0.015 mm–4.45 mm and grayish-white, which can be clearly distinguished from marble by its coarser grain and lighter color. In consideration of the distribution of the regional dolomitization being related to the fault and not uniform, dolomite is developed as uneven clumps in marble, and its occurrence is inconsistent with the bedding occurrence; dolomitization can be manifested as epigenetic.

The ore-related dolomitization is distributed in a limited way with the polymetallic mineralization. Dolomites of this category are sparry white, transparent, and usually with fine irregular-shaped veins.

Based on structures and textures of ores and petrographic study, the mineralization process can be roughly ascertained from early to late as fine pyrite (Py)–quartz (Q), growth girdle zoned sphalerite (Sp), polymetallic sulfides (Ms), sphalerite-chalcopyrite solid separation + quartz + coarse euhedral pyrite + granulose dolomite (Dol)-galena, galena, magnetite vein, dolomite veinlets, and chlorination-calcite (Cc) (Figures 5 and 6).

Hydrothermal alteration and mineralization can be divided into four stages. Stage 1 is characterized by the assemblage of fine-grained pyrite + quartz ± sphalerite (Q–Py±Sp) (Figure 7). Pyrites are primarily fragmented and are usually replaced by sphalerite and galena. Stage 2 is marked by quartz-polymetallic sulfides (Q–Ms) (Figure 7). The metallic minerals in Stage 2 mainly include sphalerite with chalcopyrite solid solution, pyrite, and galena. Stage 3 is characterized by the widely distributed dolomite-polymetallic sulfides (Dol–Ms), dominated by galena, sphalerite, and minor pyrite (Figure 7). Stage 4 is mainly composed of carbonate with no sulfide (Cc), which crosscut the earlier veins and ores.

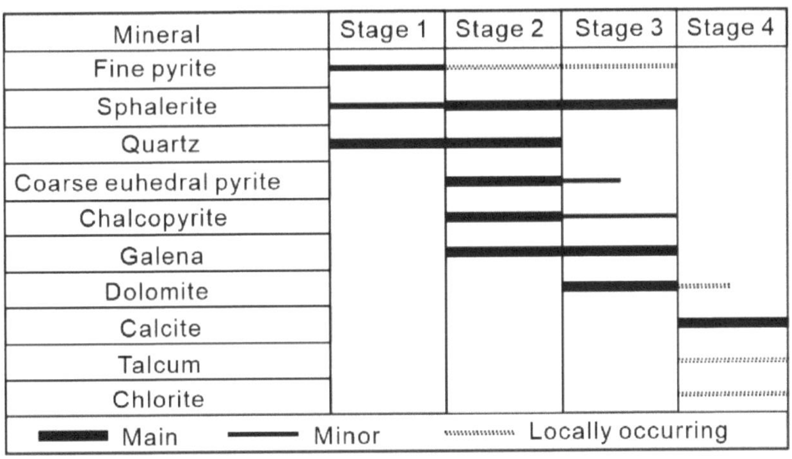

Figure 7. Paragenesis diagram of alteration assemblages showing sequence of mineralization at the Bailugou deposit.

4. Sampling and Analytical Methods for Fluid Inclusions

Samples used in this study were collected from the main orebodies of S150, S116, S171, and S147 in the Bailugou deposit (N 33°47′36″–N 33°48′36″, E 111°38′32″–E 111°45′08″). Doubly polished sample thin sections, about 0.3 mm thick, were examined. Fluid inclusions in sphalerite, dolomite, and quartz were observed, respectively (Figure 8). Fluid inclusion observations were made using a standard optical petrographic microscope to identify their genetic and composition types, vapor–liquid ratios, spatial clustering, and daughter minerals' species to document their paragenetic relationship. The volumetric fraction of phases in fluid inclusions was estimated at room temperature by reference to the volumetric chart of Roedder [48], using regular-shaped, flat inclusions whenever possible, according to the suggestion of Diamond [49]. These studies enabled successive assemblages of fluid inclusions to be defined. Representative samples of such assemblages were selected for detailed microthermometric and laser Raman spectroscopic studies.

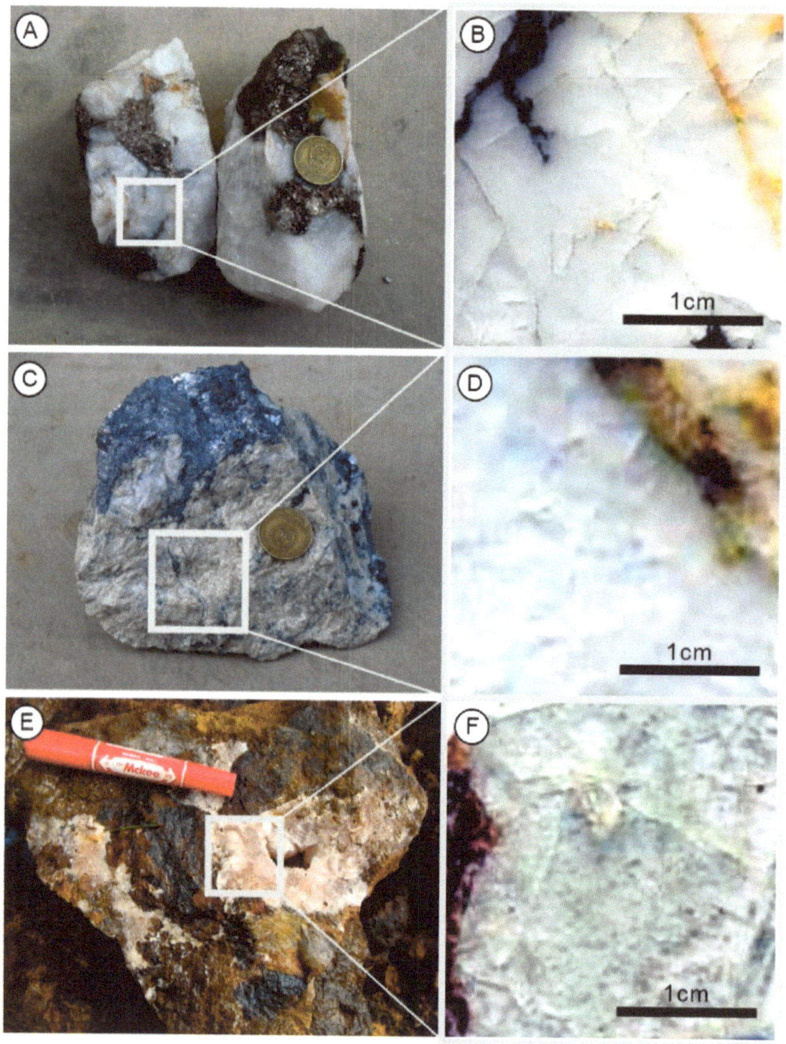

Figure 8. Petrography characteristics of FI samples at the Bailugou deposit. (**A**,**B**), quartz in stage 2 (Q–Ms); (**C**,**D**), dolomite in stage 3 (Dol–Ms); (**E**,**F**), calcite in stage 4 (Cc).

All fluid inclusion microthermometric measurements were performed on a Linkam THMSG600 heating-freezing stage mounted on a Leica microscope coupled to an image analysis system at the Fluid Inclusion Laboratory of the Key Laboratory of Mineralogy and Metallogeny, Guangzhou Institute of Geochemistry, Chinese Academy of Sciences.

The temperature range of the Linkam THMSG 600 heating-freezing stage was between 196° and 600 °C. Stage calibration was carried out at −56.6 °C, −10.7 °C, and 0 °C using synthetic FIs supplied by Fluid Inc. (Denver, CO). The uncertainty of the measurements is ±0.5 °C for temperatures between −120° and −70 °C, ±0.2 °C for the interval of –70° to +100 °C, and ±2 °C for temperatures between 100° and 500 °C. All fluid inclusions were analyzed by freezing to −180 °C and heating up at a rate of 0.2 °C/min to 5 °C/min. The heating rate was reduced from 0.2 °C/min to 0.5 °C/min near the phase change points. Melting temperatures of solid CO_2 ($T_{m,CO2}$), the freezing point of NaCl–H_2O inclusions ($T_{m,ice}$), final melting temperatures of clathrate ($T_{m,cla}$), homogenization temperatures of the CO_2 phase ($T_{h,CO2}$), dissolution temperatures of daughter minerals ($T_{m,d}$) and total homogenization temperatures of FIs (T_h) were measured.

Compositions of individual fluid inclusions that are representative of specific stages, including vapor, liquid, and daughter mineral phases, were analyzed using RM-2000 Laser Raman spectroscopy at the Fluid Inclusion Laboratory of State Key Laboratory of Lithospheric Evolution, Institute of Geology and Geophysics, Chinese Academy of Sciences. An argon laser with a wavelength of 514.5 nm was used as the laser source at a power of 1000 mW. Integration time was 10 s to 30 s, with ten accumulations for each spectral line. The spectral resolution was ±2 cm^{-1} with a beam size of 1 μm. The instrumental setting was kept constant during all analyses.

For volatile-rich fluid inclusions, CO_2 was identified by melting a solid phase in inclusions below −56.6 °C. Based on Burke [50], the molar fractions of CO_2, CH_4, and N_2 were determined.

Salinities of CO_2–H_2O, NaCl–H_2O, and halite-bearing FIs (expressed as wt%NaCl equiv.) were estimated by the clathrate melting temperatures of CO_2-clathrate [48,51], last ice melting temperatures [52–54] and halite dissolution temperatures [53,55], respectively. The density of CO_2 in C-type FIs was estimated using the Flincor program [56] and the Brown and Lamb formula [57] for the H_2O–CO_2–NaCl system. The bulk density of W-type and S-type FIs was estimated using the formula of Haas [58] and Bodnar et al. [59], the bulk density of C-type was estimated using the Flincor program [56], and the formula of Brown and Lamb [57] for the H_2O–CO_2–NaCl system.

The trapping pressure of W-type FIs was estimated using the formula of Haas [58] and the phase diagram of Bischoff [60]; the trapping pressure of C-type FIs was estimated using the Flincor program [58], and the formula of Bowers and Helgeson [61] for the H_2O–CO_2–NaCl system; trapping pressure of S-type FIs was estimated using the formula seen in Bischoff [60].

5. Results

5.1. Petrography and Types of Fluid Inclusions

Fluid inclusions (FIs) in sphalerite are tiny and small in quantity, making it challenging to observe. The overall distribution of FIs in dolomite and quartz was generally scattered and random, and there were occasional isolated distributions with no obvious directionality. Secondary FIs are clustered along trails, crosscutting grains, or grain boundaries. Most FIs were observed in calcite with about 40 μm in diameter and minor in quantity distributed sporadically. Primary FIs observed in this study mainly have negative crystal, elongated, or irregular shapes, sub-roundness, triangle, strip-type, fusiform, and square shapes. The diameters of primary inclusions of varying sizes are smaller than 2 μm, more extensive than 20 μm, and concentrated in a range of 5 μm to 12 μm.

Fluid inclusions were classified into three types based on their appearance at 25 °C and by their Raman spectra and occur in successive stages of the vein and alteration paragenesis (Figure 9).

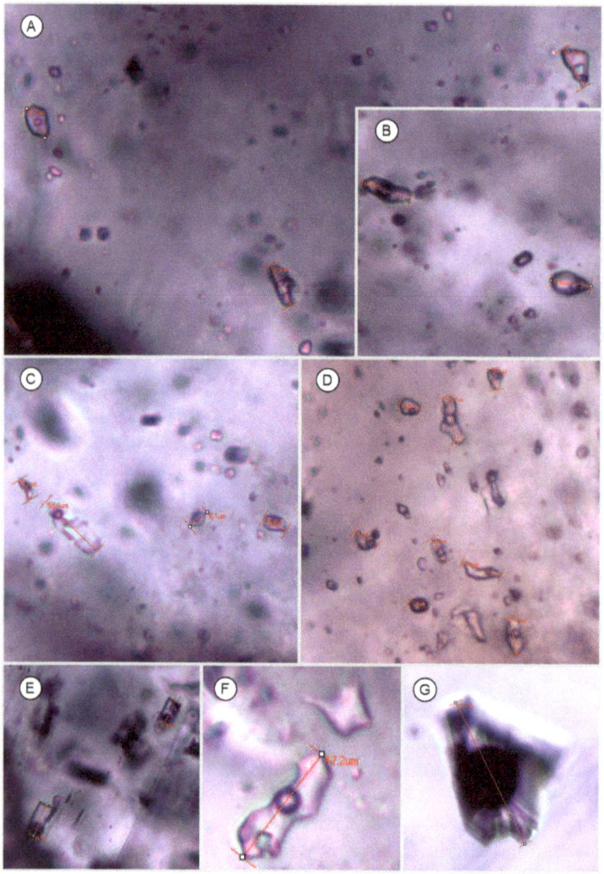

Figure 9. Microphotographs of fluid inclusions in samples from Bailugou deposit. All the pictures took under polarized light; (**A**) W-type and S-type FIs in quartz of sample BLC-30-2 (thin section) of stage 1 (Q–Py±Sp); (**B**) W-type and C-type FIs in quartz of sample BLC-30-12 (thin section) of stage 1; (**C**) S-type FIs in quartz of sample BGL-5-10 (thin section) of stage 1; (**D**) W-type FIs in quartz of sample BLC-33-2 (thin section) of stage 2 (Q–Ms); (**E**) W-type and C-S-type FIs in dolomite of sample BJD-2-11 (thin section) of stage 3 (Dol–Ms); (**F**) PL-type and S-type FI in dolomite of sample BP-8-2 (thin section) of stage 3; (**G**) W-type FI in calcite of sample B1471-3 (thin section) of stage 4 (Cc).

W-type: namely, NaCl–H_2O type FIs, which are two-phase (V_{H2O}–L_{H2O}) aqueous inclusions, can be further divided into three subtypes: (1) FIs, which are liquid-rich aqueous inclusions with 5 vol.% to 30 vol.% of a low-density vapor bubble at room temperature (Figure 9A–G); (2) Pure liquid (PL-type) FIs, which are gas-rich aqueous inclusions with gas filling degrees (ratio of V/(V + L)), most of which are smaller than 5%; and (3) Pure gas (PV-type) FIs, which are liquid-rich aqueous inclusions with gas filling degrees bigger than 80%.

S-type: namely, daughter minerals bearing FIs. Daughter minerals were detected mainly as light green cubic or roundness salt (Figure 9B–F), generally smaller than 20 vol.% S-type FIs were observed in the first three stages.

C-type: CO_2 bearing FIs, two- or three-phase aqueous-carbonic inclusions with a carbonic (vapor + CO_2 liquid) fraction of 20 vol.% to 40 vol.%. These FIs are mainly developed in the early and middle mineralized stage, about 10 vol.% to 20 vol.% of total

FIs. C-type Fis are generally of 10 μm to 15 μm in diameter and can be subdivided into two subtypes: (1) CO_2-rich three-phase (V_{CO2} + L_{CO2} + L_{H2O}) FIs at room temperature, which has CO_2 gas of 20 vol.% to 40 vol.% of the whole FI; (2) CO_2-bearing two-phase FIs, which contain a gas phase of CO_2 and H_2O ($V_{CO2+H2O}$) and a liquid phase of H_2O (L_{H2O}) (Figure 9B, E). The gas generally occupies 10 vol.% to 20 vol.% of the whole FI. Most C-type FIs in samples from the Bailugou deposit belong to this subtype.

5.2. Composition from Laser Raman Spectroscopy Detection

Laser Raman spectroscopy analyses of representative fluid inclusions have revealed that the main component of the inclusions' liquid phase is the H_2O (Figure 10A,C,D) ion occasionally. The volatile phase is of the inclusions mainly composed of CO_2, H_2O, N_2 (with Raman peaks of 2327 cm^{-1} to 2329 cm^{-1}, Figure 10A–D), with minor CH_4 (with Raman peaks of 2913 cm^{-1} to 2919 cm^{-1}, Figure 10B) by chance. Daughter minerals detected in some FIs can be identified as thenardite (with Raman peaks of 1159 cm^{-1}, Figure 10B–D).

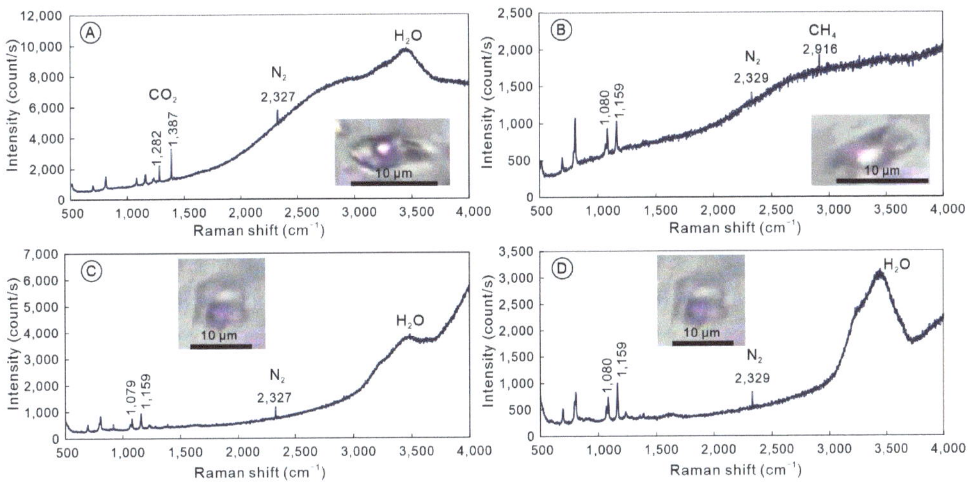

Figure 10. Representative Raman spectra of FIs from Bailugou deposit. (**A**) Raman spectrum of FI in sample BLC-30 of stage 1 (Q–Py±Sp), the compositions of volatile were mainly CO_2, N_2, and H_2O; (**B**) Raman spectrum of the same inclusion as in Figure (**A**), mainly contain N_2 and CH_4, also might contain daughter mineral of thenardite (1159 cm^{-1}); (**C**) Raman spectrum of FI in sample BLC-33 of stage 2 (Q–MS), the compositions of volatile were mainly CO_2 and N_2, and possibly with daughter mineral of thenardite nearby the detected bubble (1159 cm^{-1}); (**D**) Raman spectrum of the same inclusion as in Figure (**C**), which mainly consists of H_2O, with minor CO_2 and N_2, and possibly with daughter mineral of thenardite (1159 cm^{-1}).

5.3. Microthermometry

Microthermometric measurements were conducted on FIs in nine representative samples of different mineralization stages from the Bailugou deposit. Microthermometric data and relative parameters are summarized in Table 1 and Figure 11.

Table 1. Microthermometric data, salinity, density, pressure, and depth of the Bailugou deposit.

Stage	Sample no.	Host Minerals	Type	Size	Vapor	Tm_{CO2}	Tm_{cla}	Th_{CO2}	Tm_s	Th_{tot} [2]		Tm_{ice}	Salinity [3]		Density of CO_2	Bulk Density	Pressure	Lithostatic Depth	Hydrostatic Depth
				(μm)	(vol.%)	(°C)	(°C)	(°C) [1]	(°C)	(°C)	Counts	(°C)	(wt%NaCl eqv).	Counts	(g/cm³) [4]	(g/cm³) [5]	(MPa) [6]	(km) [7]	(km)
1	BGL-5	Qtz	W	4–10	5–30					115.6–345.3	29	−8.3 to −0.8	1.4–12.0	17		0.67–0.98	2–122	0.01–0.45	0.02–1.25
	BGL-5	Qtz	C	5–19	30–80	−61.6 to −60.2	4.5–6.7	27.2–30.6		251.5–341.5	8		6.2–9.7	8	0.56–0.67	0.74–0.93	740–1157	2.70–4.22	7.55–11.81
	BGL-5	Qtz	S	4–16	5–10, 20–30				185.2–258.3	117.5–275.4	5		31.2–35.2	4		1.07–1.28	9–32	0.03–0.12	0.09–0.32
	BLC-30	Qtz	W	5–15	5–30, 60					219.1–396.1	6	−13.8 to −4.8	7.6–17.6	3		0.76–0.92	21–183	0.08–0.67	0.22–1.86
	BLC-30	Qtz	C	7–21	5, 95	−57.1 to −33.6	−21.3 to −19.5	−8 to −5.2		236.9–268.3	3								
	BLC-30	Qtz	S	4–13	5–30				101.2–320.8	213.5–482.7	12		28.0–39.8	12		0.89–1.15	1–76	0–0.28	0.01–0.77
	BLC-13	Qtz	W	7–38	5–30					219.4–311.2	2	−9.7 to −6.7	10.1–13.6	3		0.84–1.08	0–91	0–0.33	0–0.93
	BLC-13	Qtz	S	11–17	30–60				>560	252.4–407.2	3								
2	BLC-33	Qtz	W	4–13	5–30, 60					145.5–333.5	28	−15.5 to −4.0	6.4–19.0	12		0.83–0.98	4–119	0.01–0.44	0.44–1.22
	BLC-33	Qtz	C	5–14	5–30		4.8–8.0	24.3–26.2		230.1–327.0	7		3.9–9.4	4	0.69	0.93	744	2.71	7.59
	BP-8	Qtz	W	4–18	5–30					208.7–247.5	32	−10.8 to −3.3	5.4–14.8	29		0.85–0.96	17–36	0.06–0.13	0.18–0.36
	BP-8	Qtz	S	10–13	5–10				>400	180.2–239.7	2								
3	BGL-12	Dol	W	5–11	5–30, 60					228.5–301.2	7	−7.1 to −4.6	7.3–10.6	6		0.82–0.91	25–68	0.09–0.25	0.26–0.69
	BJD-2	Dol	W	4–34	5–30, 60					106.2–281.2	18	−7.2 to −2.7	4.5–10.7	11		0.78–0.91	24–64	0.09–0.23	0.25–0.65
	BP-8	Dol	W	4–51	5–30, 60					110.1–272.4	22	−8.0 to 0.0	0–11.7	21		0.77–0.97	1–54	0.01–0.20	0.01–0.56
4	B1471	Cc	W	42–88	20–30					120.8–194.6	3								

Abbreviations: Qtz = Quartz; Cc = Calcite; Dol = Dolomite; Tm_{CO2} = melting temperature of solid CO_2; Tm_{ice} = ice-melting temperature; Tm_{cla} = clathrate dissociation temperature; Th_{CO2} = homogenization temperature of CO_2; Tm_s = melting temperature of daughter mineral; Th_{tot} = total homogenization temperature; L and V stand for liquid and vapor to which the FIs homogenize, respectively. [1] All the CO_2 phase of C-type FIs partial homogenize to liquid. [2] Most FIs totally homogenize to liquid and few to vapor occasionally. [3] Salinity of W-type estimated by final ice melting temperature [52–54], salinity of C-type estimated by clathrate melting temperature [48,51], salinity of S-type estimated by halite dissolution temperatures [53,55]. [4] Density of CO_2 in C-type FIs estimated using the Flincor program [56] and the formula of Brown and Lamb [57] for the H_2O-CO_2-NaCl system. [5] Bulk density of W-type and S-type FIs estimated using the formula of Haas [58] and Bodnar et al. [59], bulk density of C-type estimated using the Flincor program [56], and the formula of Brown and Lamb [57] for the H_2O-CO_2-NaCl system. [6] Trapping pressure of C-type FIs estimated using the Flincor program [56] and the formula of Bowers and Helgeson [61] for the H_2O-CO_2-NaCl system. [7] The density of overlying rocks is supposed to be 2.8 g/cm³.

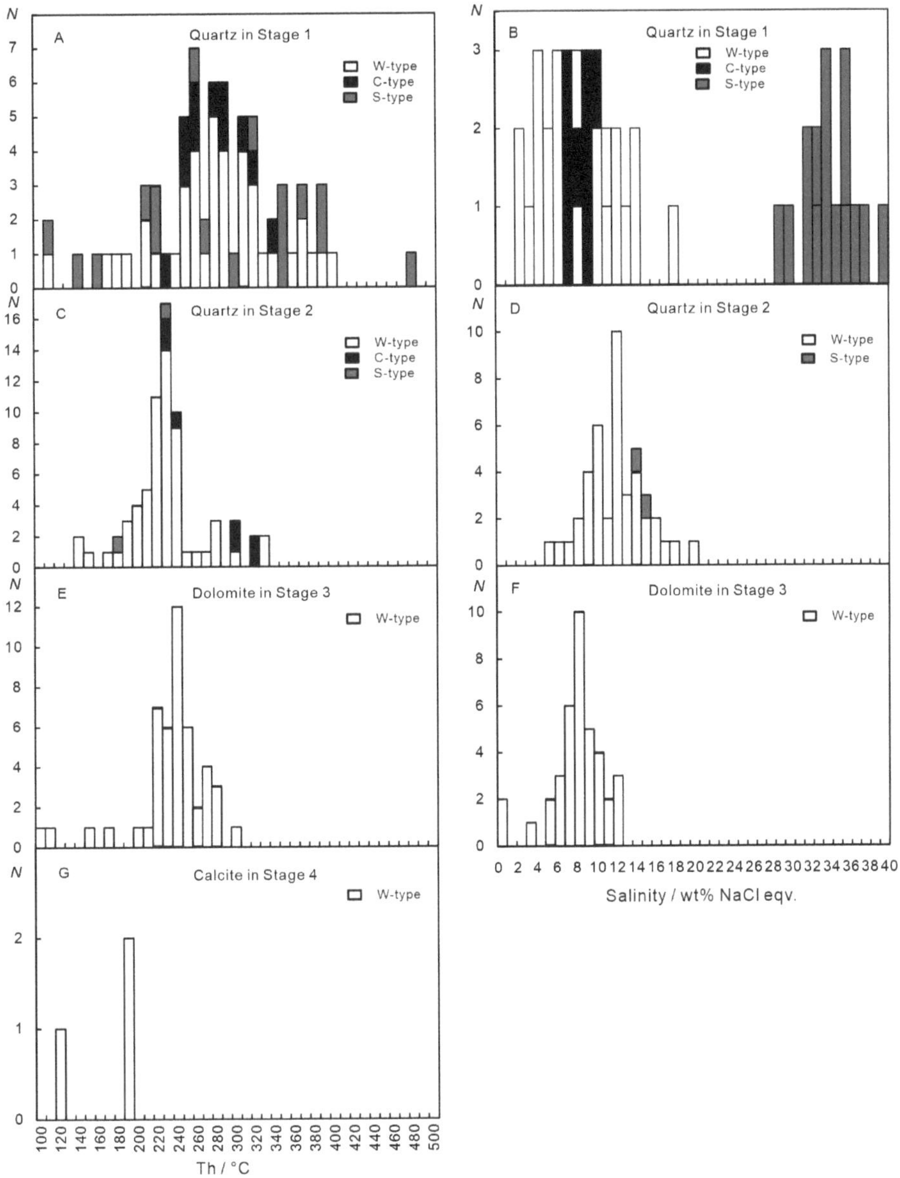

Figure 11. Histograms of homogenization temperatures and salinities of fluid inclusions of the Bailugou deposit.

In Stage 1, observation of FIs in quartz shows that mainly W-type FIs were detected with a size of 4 μm to 38 μm and a phase accounting for 5 vol.% to 30 vol.% normally. The W-type FIs' ice-melting temperatures are between −13.8 °C and −0.8 °C (Table 1), yielding salinities as 1.4 wt.% to 17.6 wt.% NaCl equiv. (Table 1, Figure 11), which are homogenized to liquid with the Th_{tot} values ranging from 115.6 °C to 396.1 °C, clustered in the intervals of 250 °C to 350 °C and peaking at 300 °C (Figure 11). Their densities range from 0.56 g/cm^3 to 0.67 g/cm^3. Beyond W-type FIs, there is a small amount of C-type FIs and S-type FIs. The clathrate dissociation temperature of C-type FIs is 4.5 °C to 6.7 °C, with corresponding salinities of 6.2 wt.% to 9.7 wt.% NaCl equiv. (Table 1, Figure 11). The carbonic phase is partially homogenized to liquid at temperatures between 27.2 °C and 30.6 °C, while homogenized to liquid at temperatures ranging from 251.5 °C to 341.5 °C with bulk densities ranging from 0.74 g/cm^3 to 0.93 g/cm^3. The melting temperatures of daughter minerals observed in S-type FIs range from 101.2 °C to 320.8 °C, while the vapor phases are homogenized to liquid at 117.5 °C to 482.7 °C, corresponding to salinities of 28.0 wt.% to 39.8 wt.% NaCl equiv. (Table 1, Figure 11).

In Stage 2, the observation of FIs in quartz shows that main W-type FIs were detected with a size of 4 μm to 18 μm and phase accounting for 5 vol.% to 30 vol.% normally. The ice-melting temperatures of W-type FIs are between −15.5 °C and −3.3 °C (Table 1), yielding salinities of 5.4 wt.% to 19.0 wt.% NaCl equiv. (Table 1, Figure 11), which are homogenized to liquid with the Th_{tot} values ranging from 145.5 °C to 333.5 °C, visibly clustered in the intervals of 200 °C to 250 °C and peaking at 250 °C (Figure 11). Their densities range from 0.83 g/cm^3 to 0.98 g/cm^3. Besides W-type FIs, there is a small amount of C-type FIs and S-type FIs. The clathrate dissociation temperature of C-type FIs is 4.8 °C to 8.0 °C, with corresponding salinities of 3.9 wt.% to 9.4 wt.% NaCl equiv. (Table 1, Figure 11). The carbonic phase is partially homogenized to liquid at temperatures between 24.3 °C and 26.2 °C while homogenized to liquid at temperatures ranging from 230.1 °C to 327.0 °C with a bulk density of 0.93 g/cm^3 (Table 1, Figure 11). Two S-type FIs in quartz in Stage 2 are observed. Still, the daughter minerals have higher melting temperatures than 550 °C, which has not been observed because it exceeded the heating limit of the Linkam THMSG600 heating-freezing units. In contrast, the vapor phase is homogenized to liquid at 180.2 °C and 239.7 °C (Table 1, Figure 11), close to the Th_{tot} of W-type and C-type. It is indicated that the trapping fluid is probably either a supersaturated solution, the daughter minerals were forecrystallized before being captured, or the inclusions were stretched in nature or during laboratory heating.

In Stage 3, W-type FIs are the main FI types detected with 4 μm–51 μm and a phase accounting for 5 vol.% to 30 vol.% normally. The W-type FIs' ice-melting temperatures are between −8.0 °C and 0 °C (Table 1), yielding salinities of 0 wt.% to 11.7 wt.% NaCl equiv. (Table 1, Figure 11), which are homogenized to liquid with Th_{tot} values ranging from 106.2 °C to 396.1 °C, clustered in the intervals of 200 °C to 300 °C and peaking at 250 °C (Figure 11). Their densities range from 0.77 g/cm^3 to 0.97 g/cm^3.

In Stage 4, only a tiny amount of W-type FIs are detected in calcite, which homogenized to liquid with the Th_{tot} values ranging from 120.8 °C to 194.6 °C (Table 1), clustered mainly in the intervals of 100 °C to 200 °C (Figure 11). W-type FIs are about 42 μm to 88 μm and phase accounting for 20 vol.% to 30 vol.% typically. The W-type FIs' ice-melting temperatures are failed observations because of the crystalline nature of host minerals. Still, the estimated salinity and density should be lower than Stage 1 to Stage 3, according to the behavior characteristics of those FIs during microthermometric measurements.

Based on the microthermometric measurements, the temperature range and its peak value at Stage 1, Stage 2, Stage 3, and Stage 4 of the Bailugou deposit show a decreasing trend from the early to late mineralization stage. The overall mineralization temperature range is about 400 °C to 100 °C (Table 1, Figure 11), belonging to a medium-temperature deposit.

6. Discussion

6.1. Fluid Salinity, Density, Pressure, and Ore-Forming Depth

6.1.1. Fluid Salinity and Density Estimated by FIs

The salinities of W-type FIs in Stage 1, Stage 2, and Stage 3 of the Bailugou deposit are estimated based on the microthermometric data, showing as 1.4 wt.% to 17.6 wt.% NaCl equiv, 5.4 wt.% to 19.0 wt.% NaCl equiv, 0 wt.% to 11.7 wt.% NaCl equiv, respectively (Table 1, Figure 11), which is relatively low and has an extensive variation range. The drift of salinity range indicates relatively low to high salinity levels during the initial mineralization stage, which remains relatively high in the later stage, indicating the fluids continued to be concentrated and may have been influenced by hydrothermal processes, then showed a decline in salinity levels, even with some inclusions showing no salinity at all, which indicates a transition to lower salinity conditions. The salinities of C-type FIs in Stage 1 and Stage 2 are estimated as 6.2 wt.% to 9.7 wt.% NaCl equiv and 3.9 wt.% to 9.4 wt.% NaCl equiv, showing that the salinity reduced gradually from the early to late stages of mineralization. The salinity of S-type FIs in Stage 1 is estimated as 28.0 wt.% to 39.8 wt.% NaCl equiv, showing a character of medium-high salinity of initial ore-forming fluid.

The fluid density in Stage 1 is 0.67 g/cm^3 to 0.98 g/cm^3 for W-type, 0.74 g/cm^3 to 0.93 g/cm^3 for C-type, and 1.07 g/cm^3 to 1.28 g/cm^3 for S-type. The fluid density in Stage 2 is 0.83 g/cm^3 to 0.98 g/cm^3 for W-type and 0.93 g/cm^3 for C-type. The fluid density in Stage 3 is 0.77 g/cm^3 to 0.97 g/cm^3 for W-type. Overall, the fluid density changes slightly through each stage. The Bailugou deposit remains a medium-high-density fluid during mineralization.

In summary, the medium-high density and salinity often suggests that the initial ore-forming fluid is derived from magmatic or metamorphic sources. Since the lack of outcroppings of intrusion in this area, the metamorphic source was more accepted. The coexistence of medium-high density and elevated temperatures meant more minerals could be dissolved and greater quantities of metals were transported, which are critical for the formation of economically viable mineral deposits. Furthermore, the density of the fluid can also provide insights into the pressure conditions during mineralization, and high density initial fluids in the Bailugou deposit are associated with great depths and pressures, suggesting that the mineralization occurred in a deep geological setting.

6.1.2. Ore-Forming Pressure and Depth

Trapping pressures and mineralization depths of Stage 1 are estimated according to Th_{tot} values of W-type FIs, Th_{CO2} values, homogenization behaviors, and the proportion of the CO_2 phase in the C-type FIs and the melting temperatures of daughter minerals in S-type FIs, respectively (Table 1). The minimum trapping pressure for W-type FIs in Stage 1 is 0 MPa to 183 MPa, and the corresponding mineralization depths under lithostatic to supra-lithostatic pressures and hydrostatic pressures are 0 to 0.67 km and 0 to 1.86 km, respectively (Table 1). The minimum trapping pressures for C-type FIs in Stage 1 is 740 MPa to 1157 MPa, and the corresponding mineralization depths under lithostatic to supra-lithostatic pressures and hydrostatic pressures are 2.70 km to 4.22 km and 7.55 km to 11.81 km, respectively (Table 1). The minimum trapping pressures for S-type FIs in Stage 1 are 1 MPa to 76 MPa, and the corresponding ore-forming depths under lithostatic to supra lithostatic pressures and hydrostatic pressures are 0 to 0.28 km and 0 to 0.93 km, respectively (Table 1).

The estimated minimum trapping pressures for W-type FIs in Stage 2 are 4 MPa to 119 MPa, and the corresponding mineralization depths under lithostatic to supra lithostatic pressures and hydrostatic pressures are 0.01 km to 0.44 km and 0.18 km to 1.22 km, respectively (Table 1). The minimum trapping pressures for C-type FIs in Stage 2 is 744 MPa, and the corresponding ore-forming depths under lithostatic to supra lithostatic pressures and hydrostatic pressures are 2.71 km and 7.59 km, respectively (Table 1).

The estimated minimum trapping pressures for W-type FIs in Stage 3 range from 1 MPa to 68 MPa, corresponding to ore-forming depths under lithostatic to supra-lithostatic pressures. Hydrostatic pressures range from 0.01 km to 0.25 km and 0.01 km to 0.69 km, respectively (Table 1).

The maximum value of the mineralization pressure revealed by the minimum trapping pressure of W-type FIs in Stage 1, Stage 2, and Stage 3 of the Bailugou deposit is 183 MPa, 119 MPa, 68 MPa, which remains in decline accordingly from Stage 1 to Stage 2 and Stage 3. The pressures revealed by the C-type FIs in Stages 1 and 2 are 1157 MPa and 744 MPa, showing that the fluid pressure reduced from Stage 1 to Stage 2. The pressure revealed by the S-type FIs in Stage 1 is 76 MPa, showing that the captured pressure is relatively low, and the S-type FIs formed under the pressure reduction process.

The estimated maximum value of the formation depth of Stage 1, Stage 2, and Stage 3 revealed by the W-type FIs of the Bailugou deposit is 0.67 km, 0.44 km, 0.25 km under lithostatic pressure and 1.86 km, 1.22 km, 0.69 km under hydrostatic pressure. The depth revealed by C-type FIs from Stage 1 to Stage 2 is 4.22 km, 2.71 km under lithostatic pressure, 11.81 km, and 7.59 km under hydrostatic pressure, respectively. Summarily, the estimated formation depth of the Bailugou deposit was characterized as deep with a significant variation, which is about 2 km to 12 km and is much greater than the metallogenic limit of the epithermal hydrothermal gold deposit. Meanwhile, the ore formation depth varies greatly, a typical characteristic of fault-controlled deposits, as fluids can penetrate along faults at depths of up to 10 to 15 km [62,63]. The ductile-brittle transition appears as a structural, rheological, and thermal boundary separating an upper crust dominated by the percolation of meteoric, marine, and basin fluids and a lower crust marked by the circulation of magmatic and metamorphic fluids [64].

6.2. *Temperature of Mineralization*

6.2.1. Characteristic Temperature Revealed by Solid Solution Separation Structure

The chalcopyrite developed in sphalerite as a solid solution separation structure, or known as chalcopyrite disease [65], in the Bailugou deposit (Figure 6), reveals a corresponding formation temperature of 350° to 400 °C [66]. The eutectoid temperature point of solid solution minerals can be regarded as a geological thermometer; this temperature can be considered the lowest temperature for the separate mineralizing of sphalerite and chalcopyrite synchronously. Considering the guest mineral chalcopyrite is generally displayed as a beaded or emulsion droplet-like texture in the sphalerite as a solid solution separation structure while a few displayed as flakiness or leaf-like textures, indicating that, as the mineralizing of host mineral sphalerite, the temperature slowly decreased to the eutectoid point, micro-fine guest mineral chalcopyrite began to form in irregular orientation. After that, the temperature dropped sharply so that granular chalcopyrite could not gather and remain in the original separated location nearby, forming a random arrangement of emulsion droplets. With the continuous and slow temperature decline, a small amount of chalcopyrite gathered as linear, flakiness, and leaf-like droplets in cleavage in sphalerite with the directional arrangement. Meanwhile, the lack of board-like, knot-like, and polysynthetic guest mineral chalcopyrite in the solid solution separation structure confirms that the formation temperature is not too high. The formation of droplet structures may be related to the flow characteristics of the mineralizing fluid. The fluid movement within the ore body may have been relatively intense, causing mineral particles to precipitate in small, dispersed forms rather than aggregating into larger, flat lamellar or skeletal forms.

6.2.2. Homogenization Temperature of Fluid Inclusions

The homogenization temperature and peak value of fluid inclusions of Stage 1 to Stage 4 in the Bailugou deposit gradually decreased while the mineralization stages evolved. The overall mineralization temperature ranges between 400 °C and 100 °C (Table 1, Figure 11).

6.2.3. Equilibrium Temperature from Sulfur Isotope

The equilibrium temperatures calculated according to $\delta^{34}S$ values [67] by sulfide pairs such as pyrite–sphalerite, pyrite–galena, and sphalerite–galena of the Bailugou deposit are mainly clustered in the range of 351 °C to 585 °C, except two lower temperatures, 71 °C and 79 °C (Table 2). The $\Delta\delta^{34}S$ values of the two fractionations for the equilibrium temperatures of 71 °C and 79 °C are 3.0‰ and 5.9‰, respectively, which are relatively high. Excluding this temperature range for it may be the result of the destruction of the original equilibrium. Thereby, the reasonable sulfur isotope equilibrium temperature of the Bailugou deposit should be 351 °C to 585 °C, with $\Delta\delta^{34}S$ values of 1.3‰ to 1.9‰. This temperature is slightly higher than that of the S150 orebody calculated by Yan [45], which is 304 °C to 309 °C, and slightly higher than the homogenization temperature of FIs (100 °C to 400 °C). Still, it is consistent with the common understanding that metal minerals' formation temperature is slightly higher than the homogenization temperature of FIs.

Table 2. Sulfur isotope geothermometer for Bailugou deposit.

Sample no.	Sampling Site	$\delta^{34}S_{CDT}$/‰			$\Delta\delta^{34}S$/‰	Temperature /°C	Formula Source
		$\delta^{34}S_{Py}$	$\delta^{34}S_{Sp}$	$\delta^{34}S_{Gn}$			
BGL-14	S150, Gunligou	4.4	1.4		3.0	71	[68]
BGL-6	S150, Gunligou		9.1	7.2	1.9	351	[69]
BLY-10	S170, Luyuangou		4.3	2.5	1.8	364	
BLY-19	S170, Luyuangou	3.9		2.5	1.4	585	
BP-6	S116, Jiudinggou		3.4	2.1	1.3	476	
BQC-30	S170, Quancaigou		7.0	1.1	5.9	79	

Based on the above summarizations from the characteristic temperature revealed by the solid solution separation structure, the homogenization temperature of fluid inclusions, and the equilibrium temperature from sulfur isotope, this paper suggests that the temperature of mineral precipitation in the ore-forming fluid of Bailugou deposit is between 400 °C and 100 °C, dominating concentrations from 400 °C to 350 °C.

6.3. *Origin, Evolution, and Precipitation Mechanism of Ore-Forming Fluids*

Studies on the evolution of salinities, homogenization temperatures, and hydrogen-oxygen isotopes of fluid inclusions of different mineralization stages in the Bailugou deposit show that the initial ore-forming fluid was triggered under intense thermal-tectonic processes, associated with significant heat flow and tectonic movements, such as subduction, continental collision, or rifting. These processes dominated during the early and middle stages, which can lead to the generation of high-temperature fluids and create pathways for fluid migration. As it trended to the late stage, atmospheric precipitation gradually increased, and the contribution of surface waters to the mineralizing fluid grows significantly.

6.3.1. Fluid Immiscibility, Mixture, and Boiling

The evolution pattern of homogenization temperatures and salinities of the Bailugou deposit varies with the mineralization stages.

The fluid salinities in Stage 1 can be divided into two ends (Figure 12A). The lower salinities were mainly contributed by W-type and C-type FIs, with relatively higher homogenization temperatures (350 °C to 150 °C) in the whole. In comparison, the higher salinities were primarily contributed by S-type FIs with lower homogenization temperatures (150 °C). This pattern shows that the W-type, C-type, and S-type FIs in Stage 1 were not formed by fluid boiling but were more likely by fluid mixture followed by immiscibility. Considering the difference in homogenization temperatures, the W-type and C-type FIs should be formed from hydrothermal fluid, and the S-type FIs should be formed from the high-salinity fluid extracted from wall rocks. Although the Yanshanian (Jurassic–Cretaceous) granites have not been recognized in the Bailugou deposit, there are many intermediate-acid small intrusive bodies developed in this area (Figure 1). Bao et al. [70] speculated that there might exist a magma chamber deep in this area. Meanwhile, it was proved that, with the decrease in pressure and temperature, de-gassing, boiling, or immiscibility accompanied by the intruding of the granite can cause CO_2 release, and consequently, fluid condensation, *p*H increase, and oxygen fugacity decrease, followed by the massive precipitation of ore minerals [71–75], which should be the sufficient causes of the formation of the W-type and C-type FIs in the initial ore-forming fluid of the Bailugou deposit—the ore-bearing strata, the Longjiayuan, Sanchuan, Nannihu,

Meiyaogou, Baishugou Fm. of the Luanchuan Group and the Guandaokou Group, regionally developed dolomite marble could be the potential extracting sources of high salinity fluid. Meanwhile, the high enrichment of Pb, Zn, Ag, Mo, and W in ore-bearing strata showed by the geochemical characters [27,76,77] also verifies that the Luanchuan and Guandaokou Groups should be the favorite source for these metals.

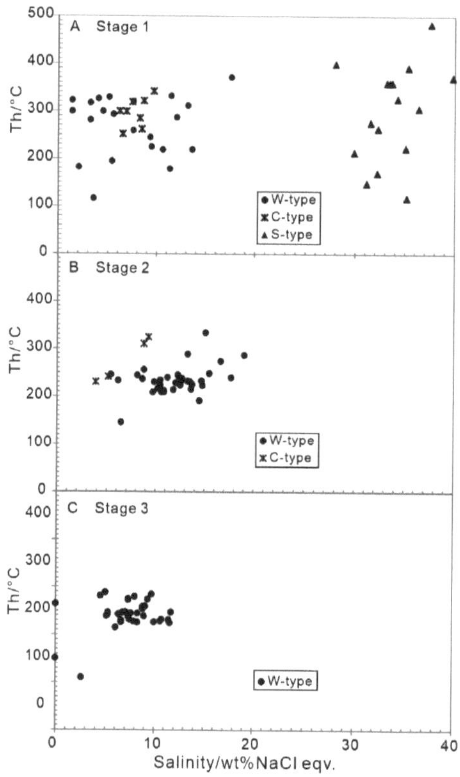

Figure 12. Homogenization temperatures versus salinities of fluid inclusions in different mineralization stages. (**A**) Stage 1; (**B**) Stage 2; (**C**) Stage 3.

The fluid salinities in Stage 2 tend to be more homogeneous, ranging from 5 wt.% to 18 wt.% NaCl equiv, which is comparable to the lower end of Stage 1's salinity (Figure 12B). The S-type FI assemblages with Th_{tot} values of daughter minerals over 550 °C observed in quartz in Stage 2 indicate the intensive fluid mixture of the hydrothermal fluid and the high-salinity fluid extracted from the ore-bearing strata.

The salinity of the fluid in Stage 3 was further decreased than in Stage 2. The observation of the C-type FI in Stage 3 indicates the fluid boiling, which was also confirmed by the observed petrography of FIs assemblages (Figure 9E,F) and the texture characteristics of ore minerals in this stage (Figure 6G). Still, the C-type FI is rare in number; therefore, no salinity of this type of FI in Stage 3 has been revealed (Figure 12C). The homogenization temperatures from Stage 2 to Stage 3 shown consist of non-significant decreases, which indicates the C-type FIs were released from fluid boiling caused by pressure reduction. The C-type FIs decrease with the evolution of the mineralization stages, which reveals gradual decreases in the pressure and shallower depth at the Bailugou deposit.

6.3.2. Genesis of the Bailugou Vein-Type Zinc-Lead-Silver Deposit

Combined with the geology, parameters from FIs, and the tectonic background of the study area, this paper first conducted the genetic types of the Bailugou deposit can be classified as orogenic metamorphic hydrothermal Pb–Zn–Ag deposit with the mineralization process as follows:

During Triassic to late Cretaceous, led by the subduction event in eastern China, the Late Mesozoic tectonic regime in this area was characterized by a transition from compression through transpressive compression (or transition from compression to extension) to extension, ending with continental collision during the late Cretaceous [15,67,78–80] (Figure 13A,B), along with intense tectonic-magmatic events [81–87]. The lithosphere in eastern China experienced a strong thinning process from the Late Jurassic to the Early Cretaceous [81,87,88], accompanied by these tectonic processed, large-scale granitic magmatic activity, such as small porphyry and large batholith, which were both developed in this intracontinental orogeny, known as the Yanshanian period [89]. The asthenospheric upwelling and underplating triggered deep penetration of the lower crust and the deeply sourced CO_2-rich ore-forming metamorphic hydrothermal solution was formed by the liquation process of deep rocks (strata and lower crust) under a high temperature, which is relatively slightly lower than the formation of magma caused by a gradient increase in temperature and pressure, triggered by the tectonic regime turned from collision to extension. The metamorphic hydrothermal solution mobilized and extracted ore-forming constituents from the siliceous banded strata, which were rich in metallogenic elements [27,76,77], dolomite marble and dolomite marble of the Luanchuan and Guandaokou Groups. Loading with the abundant metallogenic elements, the ore-forming fluid migrated upward along fault systems. It turned immiscible when it reached the brittle-ductile transition level (BDL, Figure 13C). This step leads to the ore-forming components rapidly precipitating as rough and euhedral crystals, due to fracturing and pressure, drops. Meanwhile, it then rises to a shallow surface, joins with atmospheric precipitation due to the gradually developed open fracture system, and the low flow meteoric water along with the fractures and participate mixed with the deeply sourced hydrothermal, where fluid boiling might happen, resulting in the ore-forming constituents accumulated and deposited along with the fractures and interlayer detachment zones, formed the Bailugou Pb–Zn–Ag deposit (Figure 13). The research of H–O–C–S–Pb isotope geochemistry from Yang et al. [67] also support this view.

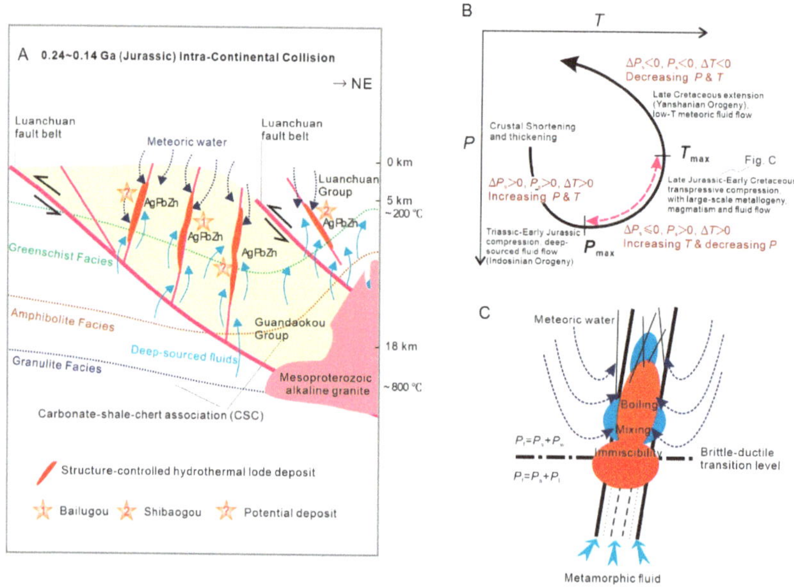

Figure 13. Tectonic evolution and genetic model for the Bailugou deposit. (**A**) Schematic representation of CMF (collisional orogeny, metallogeny and fluid flow) showing the relationships between ore-hosting structures, granitoids, porphyries and deposits in Bailugou area. Fluids, released from subducted slab and sediment or from the hydrated mantle wedge at granulite facies, travel up-dip along the interface between the slab and the overlying wedge or base of the lithosphere at amphibolite facies. The over-pressured ore fluids intersect deep crustal faults and then advent upwards to form orogenic deposits in second-order structures or hydraulically fractured rock bodies at greenschist facies. (**B**) *P*–*T*–*t* path for collisional orogeny showing a three-stage geodynamic process and the *P*–*T* conditions for fluid generation, petrogenesis and metallogenesis from Triassic to Cretaceous in Luanchuan area, East Qinling Orogeny. (**C**) Cartoons illustrating the formation process of Bailugou ore deposit. Abbreviations: P_f, fluid pressure; P_s, structural overpressure; P_l, lithostatic pressure; P_w, hydrostatic pressure; ΔT, variation in temperature; ΔP_s, variation in structural overpressure.

7. Conclusions

(1) The Bailugou lead-zinc-silver deposit is situated within the siliceous banded dolomite marble and dolomite marble of the Luanchuan and Guandaokou Groups of the Middle-Late Proterozoic. This deposit is firstly categorized as an orogenic metamorphic epi-mesothermal deposit, arising from complex, multi-stage hydrothermal processes.

(2) During an intense tectonic-thermal process, the initial ore-forming fluid is triggered. This fluid underwent a transformation from a metamorphic hydrothermal solution to a meteoric hydrothermal solution as it progresses through the mineralization stages. Throughout this process, it experiences fluid immiscibility, mixing, and boiling with metamorphic water.

(3) The genesis of the ore-forming constituents appears to be multi-sourced, as inferred from analyses of fluid inclusions. The ore-forming materials of Bailugou deposit derives mainly from a combination of the ore-bearing strata Meiyaogou Formation and Nannihu Formation within the Luanchuan and Guandaokou Groups.

Author Contributions: Conceptualization, Y.Y. and Y.C.; methodology, Y.Y.; formal analysis and investigation, Y.Y., N.G. and D.W.; writing—original draft preparation, N.G. and D.W.; writing—

review and editing, Y.Y., H.C. and Y.C.; supervision, Z.P.; project administration, H.C.; funding acquisition, Y.C. All authors have read and agreed to the published version of the manuscript.

Funding: This research was funded by the National Nature Science Foundation of China, grant number 41702098, 42202073 and 40702013, the National Key Research and Development Program of China, grant number 2021YFC2900300, the China Geological Survey Project, grant number DD20230355, DD20230356 and DD20242591 and the Key Program of the National Natural Science Foundation of China, grant number U2244206 and 41630313.

Data Availability Statement: The data presented in this study are available on request from the corresponding author. The data are not publicly available due to Privacy.

Acknowledgments: The authors extend gratitude to Jinping Qi, Jing Li, Jiangwei Han and the Henan Bureau of Geological Exploration for Non-ferrous Metals for supporting this field investigation. We also thank the academic editor Giorgio Garuti, managing editor Kimura Xia, assistant editor Cassie Zhou and anonymous reviewers for their valuable suggestions and feedback.

Conflicts of Interest: The authors declare no conflicts of interest regarding the publication of this article.

References

1. Beck, R.; Weed, W.H. *The Nature of Ore Deposits*; McGraw-Hill: New York, NY, USA, 1909; p. 685.
2. Lindgren, W. *Mineral Deposits*; McGraw-Hill: New York, NY, USA, 1933; p. 930.
3. Hammer, D.F.; Peterson, D.W. Geology of the Magma Mine Area, Arizona. In *Ore Deposits of the United States*; Ridge, J.D., Ed.; The American Institute of Mining, Metallurgical, and Petroleum Engineers, Inc.: New York, NY, USA, 1968; Volume 1933–1967, pp. 1282–1310.
4. Rye, R.O.; Sawkins, F. Fluid inclusion and stable isotope studies on the Casapalca Ag-Pb-Zn-Cu deposit, Central Andes, Peru. *Econ. Geol.* **1974**, *87*, 225–262. [CrossRef]
5. Guilbert, J.M.; Park, C.F. *The Geology of Ore Deposits*; Freeman: San Francisco, CA, USA, 1986; p. 985.
6. Graybeal, F.T.; Smith, D.M., Jr.; Vikre, P.G. The geology of silver deposits. In *Handbook of Strata-Bound and Stratiform Ore Deposits*; Wolf, K.H., Ed.; Springer: Berlin/Heidelberg, Germany, 1986; Volume 14, pp. 1–184.
7. Keith, J.D.; Whitney, J.A.; Hattori, K.; Ballantyne, G.H.; Christiansen, E.H.; Barr, D.L.; Cannan, T.M.; Hook, C.J. The role of magmatic sulfides mafic alkaline magmas in the Bingham Tintic Mining Districts Utah. *J. Petrol.* **1997**, *38*, 1679–1690. [CrossRef]
8. Kissin, S.A.; Mango, H. Silver vein deposits. In *Treatise on Geochemistry*, 2nd ed.; Elsevier: Oxford, UK, 2014; pp. 425–432.
9. Im, H.; Jeong, J.; Shin, D. Genetic environment of W skarn and Pb-Zn vein mineralization associated with the Imog granite in the Taebaeksan Mineralized District, South Korea. *Ore Geol. Rev.* **2020**, *126*, 103721. [CrossRef]
10. Beaudoin, G.; Sangster, D.F. A descriptive model for silver-lead-zinc veins in clastic metasedimentary terranes. *Econ. Geol.* **1992**, *87*, 1005–1021. [CrossRef]
11. Lynch, J.V.G.; Longstaffe, F.J.; Nesbitt, B.E. Stable isotopic and fluid inclusion indications of large-scale hydrothermal paleo flow, boiling, and fluid mixing in the Keno Hill Ag-Pb-Zn district, Yukon Territory, Canada. *Geochim. Cosmochim. Acta* **1990**, *54*, 1045–1059. [CrossRef]
12. Leach, D.L.; Landis, G.P.; Hofstra, A.H. Metamorphic origin of the Coeur d'Alene base-and precious-metal veins in the Belt basin, Idaho and Montana. *Geology* **1988**, *16*, 122–125. [CrossRef]
13. Leach, D.L.; Hofstra, A.H.; Church, S.E.; Snee, L.W.; Vaughn, R.B.; Zartman, R.E. Evidence for Proterozoic and Late Cretaceous-Early Tertiary ore-forming events in the Coeur d'Alene district, Idaho and Montana. *Econ. Geol.* **1998**, *93*, 347–359. [CrossRef]
14. Johnson, C.A.; Cardellach, E.; Tritlla, J.; Hanan, B.B. Cierco Pb-Zn-Ag vein deposits: Isotopic and fluid inclusion evidence for formation during the Mesozoic extension in the Pyrenees of Spain. *Econ. Geol.* **1996**, *91*, 497–506. [CrossRef]
15. Chen, Y.J.; Pirajno, F.; Sui, Y.H. Isotope geochemistry of the Tieluping silver deposit, Henan, China: A case study of orogenic silver deposits and related tectonic setting. *Miner. Depos.* **2004**, *39*, 560–575. [CrossRef]
16. Qi, J.P.; Chen, Y.J.; Ni, P.; Lai, Y.; Ding, J.Y.; Song, Y.W.; Tang, G.J. Fluid inclusion constraints on the origin of the Lengshuibeigou Pb-Zn-Ag deposit, Henan province. *Acta Petrol. Sin.* **2007**, *23*, 2119–2130. (In Chinese with English Abstract)
17. Zhang, J.; Yang, Y.; Hu, H.Z.; Wang, Z.G.; Li, G.P.; Li, Z.L. C-S-Pb isotope geochemistry of the Yindonggou orogenic type silver deposit in Henan Province. *Acta Petrol. Sin.* **2009**, *25*, 2833–2842. (In Chinese with English Abstract)
18. Zhang, J.; Yang, Y.; Lu, Y.H. Lead isotope geochemistry of the Weishancheng gold-silver ore belt, Henan Province, China: Implications for ore genesis. *Acta Petrol. Sin.* **2009**, *25*, 444–454. (In Chinese with English Abstract)
19. Zhang, J.; Chen, Y.J.; Yang, Y.; Deng, J. Lead isotope systematics of the Weishancheng Au–Ag belt, Tongbai Mountains, central China: Implication for ore genesis. *Int. Geol. Rev.* **2011**, *53*, 656–676. [CrossRef]
20. Li, Z.K.; Li, J.W.; Zhao, X.F.; Zhou, M.F.; Selby, D.; Bi, S.J.; Sui, J.X.; Zhao, Z.J. Crustal-extension Ag-Pb-Zn veins in the Xiong'ershan district, southern North China Craton: Constraints from the Shagou Deposit. *Econ. Geol.* **2013**, *108*, 1703–1729. [CrossRef]
21. Han, J.S.; Yao, J.M.; Chen, H.Y.; Deng, X.H.; Ding, J.Y. Fluid inclusion and stable isotope study of the Shagou Ag-Pb-Zn deposit, Luoning, Henan province, China: Implications for the genesis of an orogenic lode Ag-Pb-Zn system. *Ore Geol. Rev.* **2014**, *62*, 199–210. [CrossRef]

22. Zhang, J.; Chen, Y.J.; Su, Q.W.; Zhang, X.; Xiang, S.H.; Wang, Q.S. Geology and genesis of the Xiaguan Ag–Pb–Zn orefield in Qinling orogen, Henan province, China: Fluid. inclusion and isotope constraints. *Ore Geol. Rev.* **2016**, *76*, 79–93. [CrossRef]
23. Chen, X.L.; Shao, Y.J.; Lai, C.; Wang, C. Genesis of the Longmendian Ag–Pb–Zn Deposit in Henan (Central China): Constraints from Fluid Inclusions and H–C–O–S–Pb Isotopes. *Geofluids* **2020**. [CrossRef]
24. Yang, Y.; Chen, Y.J.; Zhang, J.; Zhang, C. Ore geology, fluid inclusions and four–stage hydrothermal mineralization of the Shangfanggou giant Mo–Fe deposit in Eastern Qinling, central China. *Ore Geol. Rev.* **2013**, *55*, 146–161. [CrossRef]
25. Yang, Y.; Liu, Z.J.; Deng, X.H. Mineralization mechanisms in the Shangfanggou giant porphyry-skarn Mo-Fe deposit of the east Qinling, China: Constraints from H–O–C–S–Pb isotopes. *Ore Geol. Rev.* **2017**, *81 Pt 2*, 535–547. [CrossRef]
26. Qi, J.P.; Song, Y.W.; Li, S.Q.; Chen, F.K. Single-grain Rb-Sr isotopic composition of the Xigou Pb-Zn-Ag deposit, Luanchuan, Henan province. *Acta Petrol. Sin.* **2009**, *25*, 2843–2854, (In Chinese with English Abstract).
27. Qi, J.P. Geology, Geochemistry and Genesis of Vein-Type Zn-Pb-Silver Deposits in Luanchuan, Henan. Ph.D. Thesis, Peking University, Beijing, China, 2006; pp. 1–133, (In Chinese with English Abstract).
28. Fyfe, W.S.; Price, N.J.; Thompson, A.B. *Fluids in the Earth's Crust*; Elsevier: Amsterdam, The Netherlands, 1978; p. 383.
29. Etheridge, M.A.; Wall, V.J. The role of the fluid phase during regional metamorphism and deformation. *J. Metamorph. Geol.* **1983**, *1*, 205–226. [CrossRef]
30. Thompson, A.B.; Connolly, J.A.D. Migration of metamorphic fluid: Some aspects of mass and heat transfer. *Earth-Sci. Rev.* **1992**, *32*, 107–121. [CrossRef]
31. Taylor, H.P., Jr. The application of oxygen and hydrogen isotope studies to problems of hydrothermal alteration and ore deposition. *Econ. Geol.* **1974**, *69*, 843–883. [CrossRef]
32. Taylor, H.P., Jr. Oxygen and hydrogen isotope relationships in hydrothermal mineral deposits. In *Geochemistry of Hydrothermal Ore Deposits*, 3rd ed.; Barnes, H.L., Ed.; Wiley: New York, NY, USA, 1997; pp. 229–302.
33. Arndt, N.; Ganino, C. Hydrothermal deposits. In *Metals and Society, An Introduction to Economic Geology*; Springer: Berlin/Heidelberg, Germany, 2012; pp. 73–112.
34. Steven, M.; Heather, A.S.; Timothy, B. Active fault and shear processes and their implications for mineral deposit formation and discovery. *J. Struct. Geol.* **2010**, *32*, 151–165. [CrossRef]
35. Duan, S.G.; Xue, C.J.; Liu, G.Y.; Yan, C.H.; Feng, Q.W.; Song, Y.W.; Gao, B.Y. Geology, fluid inclusions and stable isotopic geochemistry of Bailugou Pb–Zn deposit in Luanchuan area, Henan Province. *Miner. Depos.* **2010**, *29*, 810–825. (In Chinese with English Abstract)
36. Duan, S.G.; Xue, C.J.; Liu, G.Y.; Yan, C.H.; Feng, Q.W.; Song, Y.W.; Tu, Q.J.; Gao, Y.; Gao, B.Y. Ceology and sulfur isotope geochemistry of lead-zinc deposits in Luanchuan district,Henan Province, China. *Earth Sci. Front.* **2010**, *17*, 375–384. (In Chinese with English Abstract)
37. Duan, S.G.; Xue, C.J.; Feng, Q.W.; Gao, B.Y.; Liu, G.Y.; Yan, C.H.; Song, Y.W. Geology, fluid inclusions and S, Pb isotopic geochemistry of the Chitudian Pb- Zn deposit in Luanchuan, Henan Province. *Geol. China* **2011**, *38*, 427–441. (In Chinese with English Abstract)
38. Duan, S.G.; Xue, C.J.; Chi, G.X.; Liu, G.Y.; Yan, C.H.; Feng, Q.W. Ore geology, fluid inclusion, and S- and Pb-isotopic constraints on the genesis of the Chitudian Zn–Pb deposit, southern margin of the North China Craton. *Resour. Geol.* **2011**, *61*, 224–240. [CrossRef]
39. Cao, H.W.; Zhang, S.T.; Santosh, M.; Zheng, L.; Tang, L.; Li, D.; Zhang, X.H.; Zhang, Y.H. The Luanchuan Mo–W–Pb–Zn–Ag magmatic–hydrothermal system in the East Qinling metallogenic belt, China: Constrains on metallogenesis from C–H–O–S–Pb isotope compositions and Rb–Sr isochron ages. *J. Asian Earth Sci.* **2015**, *111*, 751–780. [CrossRef]
40. Zhang, H.Q. Geochemical characteristics and geological significance of the Bailugou Pb–Zn–Ag deposit in Henan Province. *Miner. Explor.* **2021**, *12*, 2069–2075. (In Chinese with English Abstract)
41. Xu, Y.; Wang, G.; Gao, M.; Yang, W.; Yang, S.; Yun, H.; Wu, P.; Guo, N.; Feng, Y. Genesis of the Shibaogou Mo–Pb–Zn deposit in the Luanchuan ore district, China: Constraints from geochronology, fluid inclusion, and H–O–S–Pb isotopes. *Front. Earth Sci.* **2023**, *10*, 1032183. [CrossRef]
42. Li, N.; Chen, Y.J.; Zhang, H.; Zhao, T.P.; Deng, X.H.; Wang, Y.; Ni, Z.Y. Molybdenum deposits in East Qinling. *Earth Sci. Front.* **2007**, *14*, 186–198, (In Chinese with English Abstract).
43. Hu, S.X.; Lin, Q.L.; Chen, Z.M.; Li, S.M. *Geology and Metallogeny of the Collision Belt Between North China Plates*; Nanjing University Press: Nanjing, China, 1988; p. 558. (In Chinese)
44. Lü, W.D.; Sun, W.Z. Metallogenic condition of lead and zinc deposit in Lushi-Lanchuan terrain. *Miner. Resour. Geol.* **2004**, *18*, 507–516, (In Chinese with English Abstract).
45. Yan, C.H. *Study on Inner Structure of Lead-Zinc-Silver Mineralization System in Eastern Qinling*; Geological Publishing House: Beijing, China, 2004; p. 144. (In Chinese with English Abstract)
46. Lu, S.N.; Li, H.K.; Li, H.M.; Song, B.; Wang, S.Y.; Zhou, H.Y.; Chen, Z.H. U-Pb isotopic ages and their significance of Alkaline Granite in the southern margin of the North China Craton. *Geol. Bull. China* **2003**, *22*, 762–768. (In Chinese with English Abstract)
47. Yan, C.H.; Liu, G.Y.; Peng, Y.; Song, Y.W.; Wang, J.Z.; Zhao, R.J.; Zeng, X.Y.; Lü, W.D.; Yao, X.N.; Ma, H.W.; et al. *The Metallogenic Regularity of Lead-Zinc Ore in the Southwest Hernan Province*; Geological Publishing House: Beijing, China, 2009; p. 369. (In Chinese with English Abstract)
48. Roedder, E. Fluid inclusions. Reviews in mineralogy. *Miner. Soc. Am.* **1984**, *12*, 644.

49. Diamond, L.W. Review of the systematics of CO_2–H_2O fluid inclusions. *Lithos* **2001**, *55*, 69–99. [CrossRef]
50. Burke, E.A.J. Raman microspectrometry of fluid inclusions. *Lithos* **2001**, *55*, 139–158. [CrossRef]
51. Bozzo, A.T.; Chen, H.S.; Kass, J.R.; Barduhn, A.J. The properties of hydrates of chlorine and carbon dioxide. *Desalination* **1975**, *16*, 303–320. [CrossRef]
52. Potter, R.W.; Clynne, M.A.; Brown, D.L. Freezing point depression of aqueous Sodium Chloride Solutions. *Econ. Geol.* **1978**, *73*, 284–285. [CrossRef]
53. Hall, D.L.; Sterner, S.M.; Bodner, R.J. Freezing point depression of NaCl-KCl-H_2O solutions. *Econ. Geol.* **1988**, *83*, 197–202. [CrossRef]
54. Bodnar, R.J. Reviced equation and table for determining the freezing point depression of H_2O-NaCl solutions. *Geochim. Cosmochim. Acta* **1993**, *57*, 683–684. [CrossRef]
55. Sterner, S.M.; Hall, D.L.; Bodnar, R.J. Synthetic fluid inclusions—V. Solubility relations in the system NaCl-KCl-H_2O under vapor-saturated conditions. *Geochim. Cosmochim. Acta* **1988**, *52*, 989–1005. [CrossRef]
56. Brown, P.E. FLINCOR: A microcomputer program for the reduction and investigation of fluid-inclusion data. *Am. Mineral.* **1989**, *74*, 1390–1393.
57. Brown, P.E.; Lamb, W.M. P-V-T properties of fluids in the system H_2O-CO_2-NaCl: New graphical presentations and implications for fluid inclusion studies. *Geochim. Cosmochim. Acta* **1989**, *53*, 1209–1221. [CrossRef]
58. Haas, J.L. Physical properties of the coexisting phases and thermochemical properties of the H_2O component in boiling NaCl solutions. *US Geol. Surv. Bull.* **1976**, *1421A*, 73.
59. Bodnar, R.J. A method of calculating fluid inclusion volumes based on vapor bubble diameters and *PVTX* properties of inclusion fluids. *Econ. Geol.* **1983**, *78*, 535–542. [CrossRef]
60. Bischoff, J.L. Densities of liquids and vapors in boiling NaCl-H_2O solutions: A *PVTX* summary from 300 °C to 500 °C. *Amer. J. Sci.* **1991**, *291*, 309–338. [CrossRef]
61. Bowers, T.S.; Helgeson, H.C. Calculation of the thermodynamic and geochemical consequences of nonideal mixing in the system H_2O-CO_2-NaCl on phase relations in geologic systems: Equation of state for H_2O-CO_2-NaCl fluids at high pressures and temperatures. *Geochim. Cosmochim. Acta* **1983**, *47*, 1247–1275. [CrossRef]
62. Wickham, S.M.; Peters, M.T.; Fricke, H.C.; O'Neil, J.R. Identification of magmatic and meteoric fluid sources and upward- and downward-moving infiltration fronts in a metamorphic core complex. *Geology* **1993**, *21*, 81–84. [CrossRef]
63. Famin, V.; Philippot, P.; Jolivet, L.; Agard, P. *Evolution of Hydrothermal Regime along a Crustal Shear Zone*; Tectonics: Tinos Island, Greece, 2004; Volume 23, pp. 1–23.
64. Siebenaller, L.; Boiron, M.C.; Vanderhaeghe, O.; Hibsch, C.; Jessell, M.W.; André-Mayer, A.S.; France-Lanord, C.; Photiades, A. Fluid record of rock exhumation across the brittle ductile transition during formation of a metamorphic core complex (Naxos Island, Cyclades, Greece). *J. Metamorph. Geol.* **2013**, *31*, 313–338. [CrossRef]
65. Paul, B.B.; Philip, M.B. Chalcopyrite disease in sphalerite; pathology and epidemiology. *Am. Mineral.* **1987**, *72*, 451–467.
66. Shang, J.; Lu, J.W.; Peng, X.L.; Zhang, Y. *Mineragraphy*; Geological Publishing House: Beijing, China, 2007; p. 156. (In Chinese with English Abstract)
67. Yang, Y.; Chen, H.; Guo, N.N.; Wu, D.H.; Pang, Z.S.; Chen, Y.J. Isotope Geochemistry and Metallogenic Model of the Bailugou Vein-Type Zn-Pb-Ag Deposit, Eastern Qinling Orogen, China. *Minerals*, 2024; *submitted*.
68. Sakai, H. Isotopic properties of sulfur compounds in hydrothermal processes. *Geochem. J.* **1968**, *2*, 29–49. [CrossRef]
69. Ohmoto, H.; Rye, R.O. Isotopes of Sulfur and Carbon. In *Geochemistry of Hydrothermal Ore Deposits*; Barnes, H.L., Ed.; Wiley-Interscience: New York, NY, USA, 1979; pp. 509–567.
70. Bao, Z.W.; Zeng, Q.S.; Zhao, T.P.; Yuan, Z.L. Geochemistry and petrogenesis of the ore related Nannihu and Shangfanggou granite porphyries from east Qinling belt of molybdenum mineralization. *Acta Petrol. Sin.* **2009**, *25*, 2523–2536, (In Chinese with English Abstract).
71. Fogel, R.A.; Rutherford, M.J. The solubility of carbon dioxide in rhyolitic melts: A quantitative FTIR study. *Am. Mineral.* **1990**, *75*, 1311–1326.
72. Giggenbach, W.F. The origin and evolution of fluids in magmatic-hydrothermal systems. In *Geochemistry of Hydrothermal Ore Deposits*; Barnes, H.L., Ed.; Kidston Gold Mines Company Report; Kidston Gold Mines Company: Einasleigh, QLD, Australia, 1997; p. 82.
73. Lowenstern, J. B A review of the contrasting behavior of two magmatic volatiles: Chlorine and carbon dioxide. *J. Geochem. Explor.* **2000**, *69–70*, 287–290. [CrossRef]
74. Lowestern, J.B. Carbon dioxide in magmas and implications for hydrothermal systems. *Miner. Depos.* **2001**, *36*, 490–502. [CrossRef]
75. Chen, Y.J.; Ni, P.; Fan, H.R.; Pirajno, F.; Lai, Y.; Su, W.C.; Zhang, H. Diagnostic fluid inclusions of different types hydrothermal gold deposits. *Acta Petrol. Sin.* **2007**, *23*, 2085–2108. (In Chinese with English Abstract)
76. Li, Y.F. The Temporal-Spital Evolution of Mesozoic Granitoids in the Xiong'ershan Area and Their Relationships to Molybdenum-Gold Mineralization. Ph. D. Dissertation, University of Geosciences, Beijing, China, 2005; pp. 1–135. (In Chinese with English Summary)
77. Ye, H.S. The Mesozoic tectonic evolution and Pb-Zn-Ag metallogeny in the southern margin of North China Craton. Ph.D. Dissertation, Chinese Academy of Geological Sciences, Beijing, China, 2006; pp. 1–225. (In Chinese with English Summary)

78. Chen, Y.J.; Sui, Y.H.; Pirajno, F. Exclusive evidences for CMF model and a case of orogenic silver deposits: Isotope geochemistry of the Tieluping silver deposit, east Qinling orogen. *Acta Petrol. Sin.* **2003**, *19*, 551–568. (In Chinese with English Abstract)
79. Chen, Y.J.; Pirajno, F.; Sui, Y.H. Geology and D–O–C isotope systematics of the Tieluping silver deposit, Henan, China: Implications for ore genesis. *Acta Geol. Sin.* **2005**, *79*, 106–119. (In Chinese with English Abstract)
80. Chen, Y.J.; Pirajno, F.; Qi, J.P.; Li, J.; Wang, H.H. Ore geology, fluid geochemistry and genesis of the Shanggong gold deposit, eastern Qinling Orogen, China. *Resour. Geol.* **2006**, *56*, 99–116. [CrossRef]
81. Yang, F.; Santosh, M.; Kim, S.W. Mesozoic magmatism in the eastern North China Craton: Insights on tectonic cycles associated with progressive craton destruction. *Gondwana Res.* **2018**, *60*, 153–178. [CrossRef]
82. Li, N.; Chen, Y.-J.; Santosh, M.; Pirajno, F. Late Mesozoic granitoids in the Qinling Orogen, central China, and tectonic significance. *Earth. Sci. Rev.* **2018**, *182*, 141–173. [CrossRef]
83. Xue, F.; Wang, G.W.; Santosh, M.; Yang, F.; Shen, Z.W.; Kong, L.; Guo, N.; Zhang, X.; Jia, W. Geochemistry and geochronology of ore-bearing and barren intrusions in the Luanchuan ore fields of East Qinling metallogenic belt, China: Diverse tectonic evolution and implications for mineral exploration. *J. Asian Earth Sci.* **2018**, *157*, 57–77. [CrossRef]
84. Xue, F.; Santosh, M.; Tsunogae, T.; Yang, F. Geochemical and isotopic imprints of early cretaceous mafic and felsic dyke suites track lithosphere- asthenosphere interaction and craton destruction in the North China Craton. *Lithos* **2019**, *326–327*, 174–199. [CrossRef]
85. Xue, F.; Santosh, M.; Tsunogae, T.; Yang, F.; Zhou, H. The Genesis of high Ba-Sr adakitic rocks: Insights from an Early Cretaceous volcanic suite in the central North China Craton. *Geol. J.* **2020**, *55*, 5398–5416. [CrossRef]
86. Tang, L.; Hu, X.K.; Santosh, M.; Zhang, S.T.; Spencer, C.J.; Jeon, H.; Zhao, Y.; Cao, H.-W. Multistage processes linked to tectonic transition in the Genesis of orogenic gold deposit: A case study from the Shanggong lode deposit, East Qinling, China. *Ore Geol. Rev.* **2019**, *111*, 102998. [CrossRef]
87. Yang, F.; Xue, F.; Santosh, M.; Wang, G.; Kim, S.W.; Shen, Z.; Jia, W.; Zhang, X. Late Mesozoic magmatism in the East Qinling Orogen, China and its tectonic implications. *Geosci. Front.* **2019**, *10*, 1803–1821. [CrossRef]

88. Yang, F.; Santosh, M.; Tang, L. Extensive crustal melting during craton destruction: Evidence from the Mesozoic magmatic suite of Junan, eastern North China Craton. *J. Asian Earth Sci.* **2018**, *157*, 119–140. [CrossRef]
89. Zhang, G.W.; Meng, Q.R.; Yu, Z.P.; Li, J.; Wang, Y.; Liu, H.; Chen, X.; Zhao, M.; Wu, T.; Xu, L.; et al. Orogenesis and dynamics of Qinling orogen. *Sci. China* **1996**, *26*, 193–200. (In Chinese)

minerals

MDPI

Article

Geochronology and Geochemistry of Granitic Pegmatites from Tashidaban Li Deposit in the Central Altun Tagh, Northwest China

Kai Kang [1,2,3], Yince Ma [1,2,3], Peng Zhang [4], Hang Li [5,*], Xuehai Wang [1,2,3], Zhaoxia Liao [1,2,3], Lei Niu [6], Jianzhong Chen [4], Xingzhong Liu [4] and Xingwang Xu [1,2,3,*]

1 Key Laboratory of Mineral Resources, Institute of Geology and Geophysics, Chinese Academy of Sciences, Beijing 100029, China; kangkai@mail.iggcas.ac.cn (K.K.)
2 Innovation Academy for Earth Science, Chinese Academy of Sciences, Beijing 100029, China
3 University of Chinese Academy of Sciences, Beijing 100049, China
4 No. 3 Geological Party, Xinjiang Bureau of Geology and Mineral Exploration and Development, Korla 841000, China
5 College of Architecture and Civil Engineering, Sanming University, Sanming 365004, China
6 Exploration and Development Research Institute, Zhongyuan Oilfield Company, SINOPEC, Puyang 457001, China
* Correspondence: lihang16@mails.ucas.ac.cn (H.L.); xuxw@mail.iggcas.ac.cn (X.X.)

Abstract: The Central Altun orogenic system is a result of the amalgamation of multiple micro-continental blocks and island arcs. This complex system originated from subduction–accretion–collision processes in the Proto-Tethys Ocean during the Early Paleozoic. Research has reported the discovery of several Li-Be granitic pegmatite deposits in the Central Altun Block, including the North Tugeman granitic pegmatite Li-Be deposit, Tugeman granitic pegmatite Be deposit, Tashisayi granitic pegmatite Li deposit, South Washixia granitic pegmatite Li deposit, and Tamuqie granitic pegmatite Li deposit. The Tashidaban granitic pegmatite Li deposit has been newly discovered along the northern margin of the Central Altun Block. Field and geochemical studies of the Tashidaban granitic pegmatite Li deposit indicate: (1) Spodumene pegmatites and elbaite pegmatites, as Li-bearing granitic pegmatites that form the Tashidaban granitic pegmatite Li deposit, intrude into the two-mica schist, and marble of the Muzisayi Formation of the Tashidaban Group. (2) Columbite–tantalite group minerals and zircon U-Pb dating results indicate that the mineralization age of Tashidaban Li granitic pegmatites is 450.2 ± 2.4 Ma with a superimposed magmatic event at around 418–422 Ma later. (3) Whole-rock geochemical results indicate that the Kumudaban rock sequence belongs to the S-type high-K to calc-alkaline granites and the Tashidaban Li granitic pegmatites originated from the extreme differentiation by fractional crystallization of the Kumdaban granite pluton.

Keywords: Altun Tagh; rare-metal granitic pegmatite; CGMs and zircon U-Pb dating; magma evolution

1. Introduction

Lithium is an essential metal in modern technologies with indispensable applications [1–3], which is widely used in various industries such as glass ceramics, non-ferrous metallurgy, high-energy batteries, nuclear fusion (reactors), and aerospace [1,4]. As an essential source of lithium, petrological studies of the granitic pegmatites is of paramount importance [5–7].

Several large-scale rare-metal mineralization belts have been discovered in China with distributions in the West Kunlun-Songpan-Ganzi orogenic belt, Tianshan orogenic belt, Altai orogenic belt, Qinling-Dabie orogenic belt, Himalaya orogenic belt, Sanjiang orogenic belt, Wuyi-Yunkai orogenic belt, and Jiangnan orogenic belt in recent years [8,9]. Extensive work has been carried out by previous researchers on the formation age of pegmatites [10–17], petrogenesis [18–21], mineralogical characteristics [22–25], magma evolution [26–31], ore-forming fluids [32–34], and genesis mechanisms of granitic pegmatites [35–40], which significantly enhanced the understanding of rare-metal pegmatites in China.

The Altun Orogenic Belt (AOB) has undergone a complex tectonic evolution from the Archean to Paleoproterozoic with several stages: the formation of the craton and crystalline basement, stable continental margin sedimentation (Middle Paleoproterozoic), continental expansion (Late Neoproterozoic to Early Paleozoic), multiple micro-continents subduction–collision (Paleozoic), and intra-plate evolution (Late Paleozoic to Mesozoic) [41,42]. Tectonic collisions and continental accretion have led to a thickening of the crust in the region, creating conditions that favor the generation of significant magma volumes. Consequently, this has facilitated the formation of highly differentiated pegmatitic magmas [9,37,43]. In recent years, multiple survey results pointed out the significant rare-element mineralization potential in the area. Currently, there are nine rare-metal granitic pegmatite deposits of medium to large scale that have been discovered in the region with mineralization age in 499–445 Ma. These deposits are categorized from southwest to northeast as the Washixia district, the Ayake district, the Kumusayi district, and the Tashidaban district [44–49]. The Washixia district and the Tashidaban district have been preliminarily identified as large to super-large scale. In the northern margin of the AOB, the Tashidaban granitic pegmatite Li deposit is newly discovered in the Tashidaban district with little relevant reports available, currently.

In this study, we firstly presented the geological characteristics of the Tashidaban granitic pegmatite Li deposit and constrained the mineralization age of the Li granitic pegmatites through columbite group minerals (CGMs) and zircon U-Pb dating. The whole-rock geochemical elucidates the petrogenesis of the Kumudaban granitoids and the origin of the Li granitic pegmatites in the Tashidaban granitic pegmatite Li deposit. This study provides a better understanding of the tectonic evolution of the region and facilitates exploration campaigns for Li in the study area.

2. Regional Geological Background

The AOB is a complex tectonic belt subjected to extensive magmatic activity and formed by multiple episodes of subduction–collision in the Early Paleozoic, which traverses the northern margin of the Qinghai-Tibet Plateau in the NE–SW direction. It is sandwiched between the Tarim Block, the Qaidam Block, and the Qilian-Kunlun Orogenic Belt. The tectonic units of the Altun are divided into the Dunhuang block, the North Altun subduction–accretion complex belt (NASB), the Central Altun Block (CAB), the South Altun subduction–collision complex belt, and the Apa-mangya Ophiolitic Mélange Belt from north to south (Figure 1) [50–55].

The Tashidaban granitic pegmatite Li deposit is located on the northern margin of the CAB. The exposed stratigraphy in this area includes the Mesoproterozoic Changcheng System Bashikuergan Group (Ch*B*.), Jixian System Taxidaban Group Muzisayi Formation (Jxm) and Jinyanshan Formation (Jxj), Qingbaikou System Suoerkuli Group (Qb*S*.) and Quaternary sediments. The Bashikuergan Formation consists of a suite of meta-clastic rocks, carbonate, and basic volcanic rocks. Whole-rock Sm-Nd isotopic data from plagioclase amphibolite indicate a Mesoproterozoic age [56]. The Taxidaban Group is a continental clastic–carbonate formation formed at the edge of the marine basin during the tectonic stability period after the consolidation of the transitional basement of the Changcheng System. The U-Pb age of detrital zircons constrains the sedimentary age from the Late Mesoproterozoic to Early Neoproterozoic [57]. The main lithology of the Muzisayi Formation is composed of two-mica schist, quartzite, phyllite, dolomite and garnet-bearing granulite. The main lithology of the Jinyanshan Formation is composed of dolomite, limestone, and marble. The Suoerkuli Group consists of a suite of littoral-to-shallow clastic rocks, carbonate, and volcanic sedimentary rocks. The depositional age indicated by the TIMS U-Pb age of meta-rhyolite (930 Ma) is Neoproterozoic [58].

The regional magmatic activity is characterized by multiple episodes and diverse types of processes exhibiting a northeast to east–northeast trend. The early Paleozoic magmatic rock types are biotite granite and muscovite granite, with the emplacement age between 522 and 432 Ma [42,44,46–48,59–62].

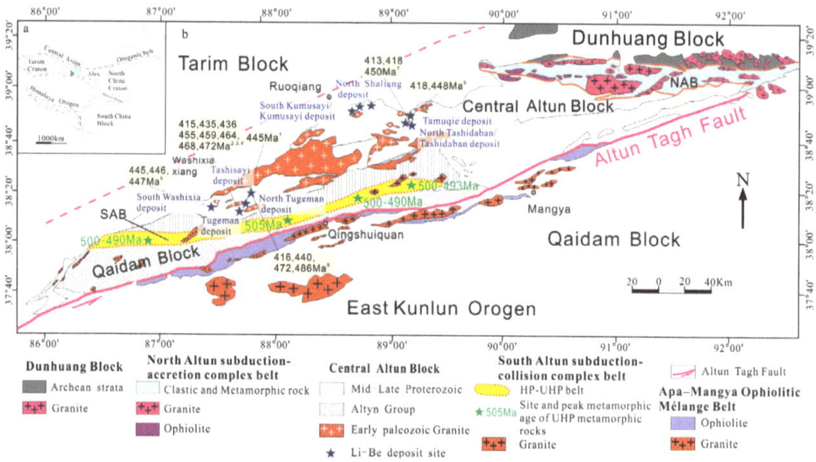

Figure 1. (**a**) Schematic map showing the position of the Altun, NW China, and (**b**) Geological map of Altun, NW China, after [45]. Igneous crystallization and mineralization ages of granitic pegmatite deposits are from: 1—[44]; 2—[46]; 3—[63]; 4—[48]; 5—[64]; 6—[65]; 7—this study.

3. Geology of the Tashidaban Granitic Pegmatite Li Deposit

The Tashidaban Li district includes the Tamuqie granitic pegmatite Li deposit, the North Tashidaban granitic pegmatite Li deposit, and the Tashidaban granitic pegmatite Li deposit. An inferred resource of over 400,000 tons of Li_2O in the district was assessed by the No. 3 Geological Party, Xinjiang Bureau of Geology and Mineral Exploration and Development. The exposed strata in the Tashidaban granitic pegmatite Li deposit include the Bashikuergan Group in the Mesoproterozoic Changcheng System, Jixian System Taxidaban Group in the Muzisayi Formation, and the Jinyanshan Formation. Early Paleozoic Kumudaban pluton is located in the Northwest of the Tamuqie granitic pegmatite Li deposit and intruded the Changcheng System Bashikuergan Group. The pluton consists of fine- to medium-grained biotite monzogranite with the crystallization age of 451~445 Ma [59,62]. The Kumudaban granite facies also includes muscovite granite, albite granite, and tourmaline granite (Figure 2).

The biotite granite (sample 23TD07-5/6) is grey-black and fine- to medium-grained. Quartz (25–30%), plagioclase (30–35%), K-feldspar (20–25%), and biotite (10%) are the primary components with accessory zircon and ilmenite. Biotite occurs as flakes between feldspar and quartz grains and exhibits weak chlorite alteration (Figure 3a). The muscovite granite (sample 21TD22-4/5) is grey-white and medium-grained. Quartz (20–25%), plagioclase (30–35%), K-feldspar (20–25%), and muscovite (10%) are the primary components with accessory garnet (3%), biotite (3%), zircon, and ilmenite. The photomicrograph of Figure 3b depicts muscovite occurring in the form of flakes interspersed between feldspar and quartz grains, exhibiting weak sericite alteration along its edges. The albite granite (sample 23TD06-6/7) is grey-white and fine- to medium-grained. Quartz (30–35%), plagioclase (35–40%), K-feldspar (5–10%), and muscovite (15%) are the primary components with accessory garnet and zircon (Figure 3c). The tourmaline granite (sample 23TD03-1/2) is grey-white and fine- to medium-grained. Quartz (30–35%), plagioclase (30–35%), K-feldspar (5–10%), muscovite (15%), and tourmaline (10%) are the primary components with accessory garnet and zircon. The tourmaline exhibits weak zoning (Figure 3d).

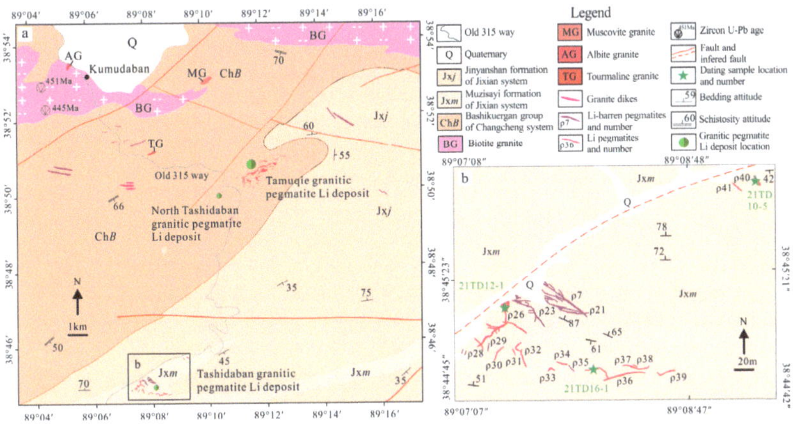

Figure 2. Geological map of (**a**) Kumudaban-Tashidaban area and (**b**) the Tashidaban granitic pegmatite Li deposit.

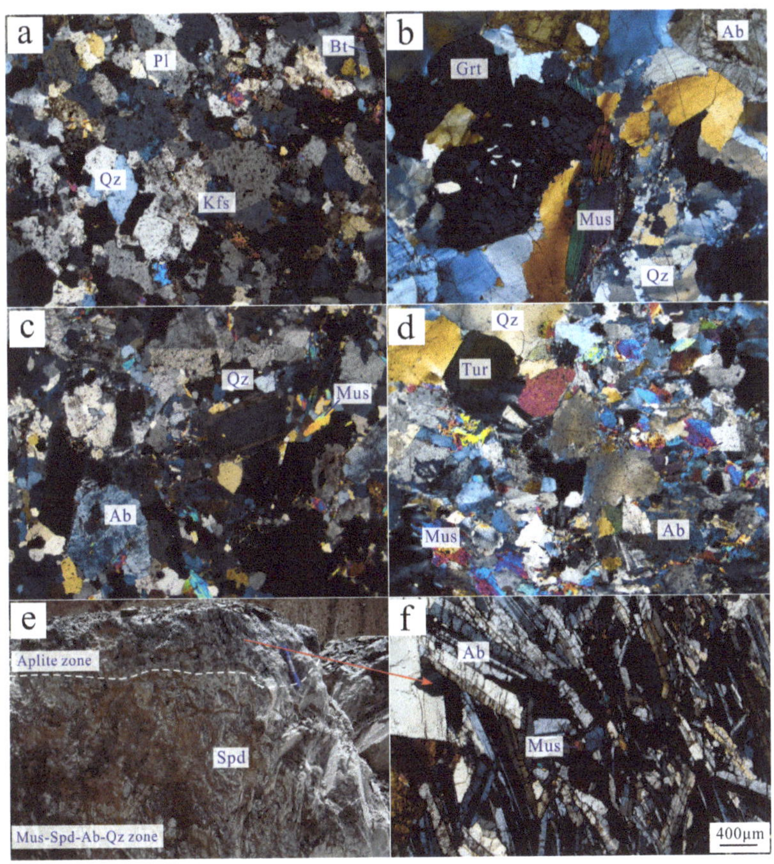

Figure 3. Photographs of the Kumudaban granite facies and pegmatites from the Tamuqie granitic pegmatite Li deposit. (**a**) Photomicrograph of biotite granite from Kumudaban. (**b**) Photomicrograph

of muscovite granite from Kumudaban. (**c**) Photomicrograph of albite granite from Kumudaban. (**d**) Photomicrograph of tourmaline granite from Kumudaban. (**e**) Field photograph of the tρ13 spodumene pegmatite from the Tamuqie granitic pegmatite Li deposit. (**f**) Photomicrograph of the tρ13 spodumene pegmatite from the aplite zone. Abbreviations: Qz: quartz; Pl: plagioclase; Kfs: K-feldspar; Ab: albite; Mus: muscovite; Grt: garnet; Tur: tourmaline; Spd: spodumene.

The Early Paleozoic Suwushijie complex pluton is located in the South of the Tashidaban granitic pegmatite Li deposit and intruded the Jixian System Jinyanshan Formation. The Suwushijie pluton is composed of gabbro–diorite–porphyritic granodiorite fine-grained monzogranite and porphyritic medium-grained monzogranite from early to late [66]. The zircon U-Pb age of monzogranite is 463.6 $\pm$ 9.1 Ma and 474.7 $\pm$ 5.7 Ma [67]. The muscovite ^{40}Ar-^{39}Ar age is 413.8 $\pm$ 7.4 Ma [68], which may represent later hydrothermal events. Several faults cut the metasedimentary rocks in the district with the same direction of the regional deep faults [69].

There are over 50 pegmatite hosted in the Tamuqie granitic pegmatite Li deposit which are located on the northern edge of the district. Among them, more than 40 dykes have a Li_2O grade that meets the standard for industrial mining. The Li-bearing pegmatites represented by spodumene pegmatite contrast with most Li-barren tourmaline-bearing albite pegmatites. The pegmatites concordant with the schist foliation are primarily tabular or lens-shaped intrusions within two-mica schists of the Bashikuergan Group. The main ore-bearing pegmatite dyke, tρ13 spodumene pegmatite, as a whole is approximately 400 m long and is comprised of ten sub-dykes arranged in a left-stepped fashion. CGMs U-Pb dating studies indicated that the tρ13 spodumene pegmatite formed at 448 Ma [64], while ρ8 tourmaline-bearing albite pegmatite formed at 418 Ma [64]. The tρ13 dyke can be classified into the fine-grained albite zone, muscovite–albite–quartz (Mus-Ab-Qz) zone, muscovite–spodumene–albite–quartz (Mus-Spd-Ab-Qz) zone, and Quartz core from outer to inner. The fine-grained albite zone mainly includes muscovite (5~10 vol%), quartz (15~25 vol%), albite (50~60 vol%), and with accessory amounts of apatite, CGMs, and cassiterite (Figure 3e,f).

There are 42 pegmatite dikes hosted in the Tashidaban granitic pegmatite Li deposit, eleven of which have a Li_2O grade that meets the standard for industrial mining (Figure 2). Four types of pegmatite were distinguished in the Tashidaban granitic pegmatite Li deposit, including tourmaline-bearing albite pegmatite, elbaite pegmatite, and spodumene pegmatite. These pegmatites with EW or SE trend intruded into the two-mica schist and marble of the Jixian System Taxidaban Group in the Muzisayi Formation and had a sharp contact with the host rock. For this study, samples were collected from ρ26, ρ36, and ρ40 pegmatite dykes. The pegmatite (ρ40) show a simple internal zoning structure; from the border inward, it can be divided into the elbaite–albite–quartz (Elb-Ab-Qz) zone and aplite zone (Figure 4b). Unlike other zoned pegmatites, aplite zones are not always found in the outer zones of pegmatite [70]. Instead, the aplite zones are dispersed inside pegmatitic dykes, possibly due to multiple episodes of magmatic–hydrothermal developments [71]. The Elb-Ab-Qz zone is characterized by coarse-grained texture (0.3~0.8 cm) and mainly includes quartz (15~20 vol%), albite (30~35 vol%), K-feldspar (5~10 vol%), muscovite (5~10 vol%), elbaite (5~10 vol%), and tourmaline (2~5 vol%), with CGMs and cassiterite as accessory mineral phases (Figures 4e and 5b,c). Coarse-grained quartz with undulose extinction and moderate subgrain orientation displays variable degrees of recrystallization (Figure 5b). The coarse-grained elbaite is replaced by a later fine-grained albite granite (Figure 5c). The aplite zone mainly includes quartz (10~20 vol%), albite (60~70 vol%), apatite (5~10 vol%) and with accessory of CGMs (Figure 5a). The replacement of aplite zone by late-stage fine-grained albite granite can be observed in Figure 5a, and both the aplite zone and albite granite have CGMs intergrown with albite and quartz (Figure 5d,e). The ρ26 tourmaline-bearing albite pegmatite is exposed as a disc-shaped structure on a steep slope with extension in the SE and SW directions with multiple branches. The lithology of the disc-shaped dyke gradually transitions from tourmaline granite to tourmaline-bearing albite pegmatite, a transition marked by a variation in grain size. The tourmaline granite is grey-white and fine- to medium-grained. Quartz (30–35%), plagioclase (30–35%), K-feldspar (5–10%), muscovite (5%), and tourmaline (10%) are the primary components with accessory zircon and CGMs (Figures 4f and 5f). The ρ36 spodumene pegmatite is 5~8 m wide and 70 m long in an intermittent outcrop. Internal zonation of the aplite zone, muscovite–albite–quartz (Mus-Ab-Qz) zone, spodumene–albite–quartz (Spd-Ab-Qz) zone from border to core characterizes the outcrop (Figure 4c,d). The aplite zone mainly includes quartz (25~35 vol%), albite (30~35 vol%), and muscovite (15~25 vol%) with accessory of CGMs (Figure 5g). The Spd-Ab-Qz zone mainly includes quartz (20~30 vol%), albite (25~30 vol%), spodumene (15~20 vol%), microcline (5~10 vol%), muscovite (5 vol%), and apatite (5~10 vol%) with accessory of CGMs (Figure 5g–i). The coarse-grained quartz intergrown with spodumene has undulose extinction. Recrystallization into granoblastic quartz is observed at the edges of the coarse-grained crystals (Figure 5h,i). Moreover, the replacement of the Spd-Ab-Qz zone by late-stage fine-grained muscovite–albite–granite can be observed (Figure 5i). Distinguished from the aplite zone at the edge of the ρ36 spodumene pegmatite, the content of muscovite in fine-grained muscovite–albite–granite is usually less than 10% by volume.

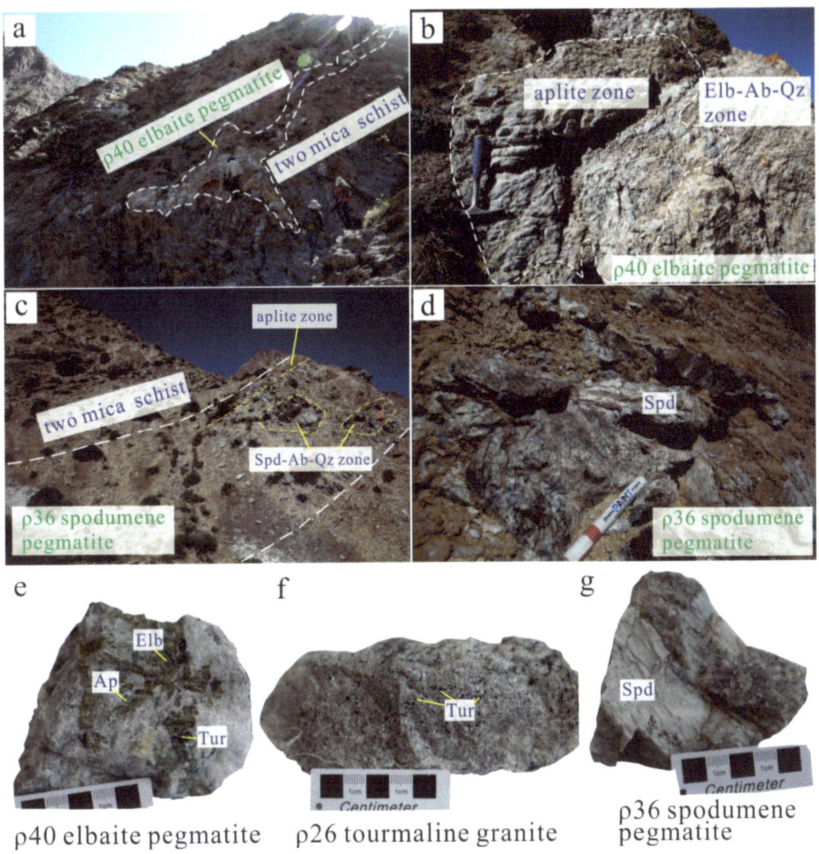

Figure 4. Field photographs of the Tashidaban granitic pegmatite Li deposit and representing hand specimens of the studied pegmatites. (**a**–**d**) Field photographs of the studied pegmatites. (**e**–**g**) Hand specimens of the ρ40 elbaite pegmatite, ρ26 tourmaline granite, and the ρ36 spodumene pegmatite. Abbreviations: Qz: quartz; Ab: albite; Spd: spodumene; Tur: tourmaline; Elb: elbaite.

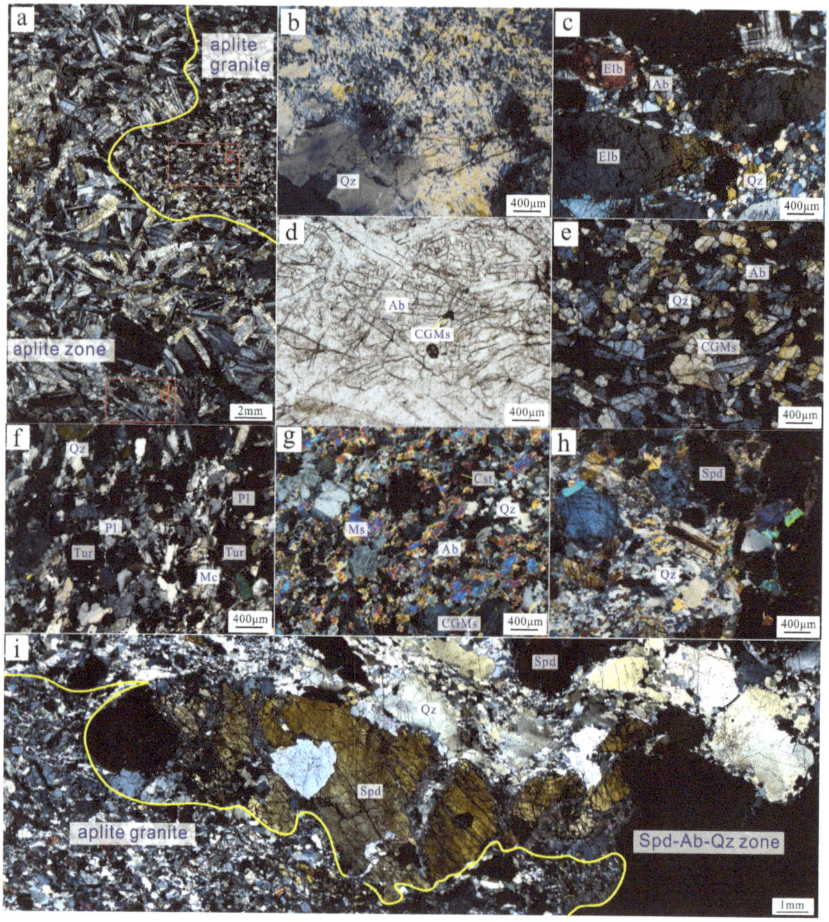

Figure 5. Photographs of the Tashidaban granitic pegmatite Li deposit representing photomicrographs of the studied pegmatites. (**a–e**) Photomicrographs of the ρ40 elbaite pegmatite. (**f**) Photomicrograph of the ρ26 tourmaline granite. (**g**) Photomicrograph of the ρ36 spodumene pegmatite from the aplite zone. (**h**,**i**)Photomicrographs of the ρ36 spodumene pegmatite from the Spd-Ab-Qz zone. Abbreviations: Qz: quartz; Ab: albite; Mc: microcline; Ms: muscovite; Spd: spodumene; Tur: tourmaline; Elb: elbaite; CGMs: columbite group minerals; Cst: cassiterite.

4. Sample Selection and Analytical Methods

The samples selection and description are listed in Table 1. The U-Pb dating samples are from the Tashidaban granitic pegmatite Li deposit (Figure 2). Sample 21TD12-1 was collected from the ρ26 tourmaline granite, sample 21TD10-5 was collected from the fine-grained aplite zone of the ρ40 elbaite pegmatite, and sample 21TD16-1 was collected from the fine-grained aplite zone of the ρ36 spodumene pegmatite. The whole-rock geochemical samples are from the vicinity area of the Kumdaban pluton, the Tamuqie granitic pegmatite Li deposit, and the Tashidaban granitic pegmatite Li deposit area.

Table 1. Sample location, rock type, and analytical method for granites and pegmatites.

No.	Location	Sample	Rock Type	Major Trace Elements	Zircon U-Pb Dating	CGM U-Pb Dating
1	Kumudaban	23TD07-5/6	biotite granite	√ (2)		
2		21TD22-4/5	muscovite granite	√ (2)		
3		23TD06-6/7	albite granite	√ (2)		
4		23TD03-1/2	tourmaline granite	√ (2)		
5	Tamuqie Li granitic pegmatite deposit	23TD01-6-1/2	spodumene pegmatite	√(2)		
6	Tashidaban granitic pegmatite Li deposit	21TD10-3/4/5	ρ40 elbaite pegmatite	√ (2)		√
7		21TD16-1/4/5	ρ36 spodumene pegmatite	√ (2)		√
8		21TD12-1/4/5	ρ26 tourmaline granite	√ (2)	√	

The separation of CGMs and zircon was carried out at the Hebei Regional Geological Survey Institute, China. The samples were crushed to particles with sizes of 40 mesh. Based on differences in mineral physical and chemical properties, zircon was separated using gravity separation, while CGMs were separated using magnetic separation. Subsequently, manual handpicking was performed under a binocular microscope to select appropriate mineral grains.

Textural and mineralogical investigations were conducted using a TM4000plus scanning electron microscopy system at the Institute of Geology and Geophysics, Chinese Academy of Sciences (IGGCAS). Back-scattered electron (BSE) images were obtained at an accelerating voltage of 15 kV to illustrate texture features of CGMs minerals. Cathodoluminenscence (CL) imaging was conducted for zircon grain using a JEOL scanning electron microscope at the Institute of Geology and Geophysics, Chinese Academy of Sciences (IGGCAS).

U-Pb dating of columbite was conducted using an ASI RESOlution S-155 193nm ArF Excimer laser coupled to a Thermo Scientific iCAP Qc quadrupole ICP-MS at the State Key Laboratory for Mineral Deposits Research, Nanjing University, China. The diameter of the laser beam was 43 μm, and the laser energy and the repetition rate of laser pulses were set at ~4 mJ/cm^2 and 4 Hz, respectively. The Coltan 139 standard from Madagascar documented U–Pb ages of 505.4 ± 1.0 Ma (BGR, Hannover, Germany; ID–TIMS), 506.6 ± 2.4 Ma (ID–TIMS), and 506.2 ± 5.0 Ma (LA–ICP–MS; Goethe University Frankfurt, Germany) [72,73]. The obtained concordant U-Pb ages of Coltan 139 are 505.5 ± 2.1 Ma (1σ; MSWD = 0.0078) and 505.1 ± 2.3 Ma (1σ; MSWD = 0.055) for samples 21TD10-5 and 21TD16-1, respectively.

U-Pb dating and trace element analysis of zircon was simultaneously conducted by LA-ICP-MS at the Beijing GeoAnalysis CO., LTD, Beijing, China. The NWR193UC model laser ablation system (Elemental Scientific Lasers LLC, Bozeman, MT, USA) was coupled to an Agilent 7900 ICPMS (Agilent, Santa Clara, CA, USA). Detailed tuning parameters can be seen in [74]. The spot size and frequency of the laser were 30 μm and 6 Hz, respectively. Zircon 91,500 (TIMS ^{207}Pb/^{206}Pb age = 1065.4 ± 0.3 Ma, 1 s, n = 11) was used as external standards for U-Pb dating and trace element calibration, respectively. An Excel-based software ICPMSDataCal was used to perform off-line selection and integration of background and analyzed signals, time-drift correction, and quantitative calibration for trace element analysis and U-Pb dating. Concordia diagrams and weighted mean calculations were performed using Isoplot/Ex_ver3 [75].

Whole-rock major and trace elements were determined at ALS Minerals-ALS Chemx, Guangzhou, China. The rock powder was fused with lithium–lithium metraborate flux which also includes an oxidizing agent (Lithium Nitrate), and then poured into a platinum mold. The concentrations of major elements were analyzed by XRF spectrometry. A prepared sample (0.100 g) was added to lithium metaborate/lithium tetraborate flux and fused in a furnace at 1025 °C. The resulting melt was then cooled and dissolved in an acid mixture containing nitric, hydrochloric, and hydrofluoric acids. Trace element concentrations were analyzed by inductively coupled plasma–mass spectrometry. For lithium analysis, a ~0.4 g sample undergoes an initial digestion process using $HClO_4$, HF, and HNO_3 until complete dryness is achieved. Following this, the residue is further processed through a secondary digestion in concentrated HCl, cooled down, and diluted to the desired volume. Finally, the samples are examined for lithium content using ICPAES spectroscopy.

5. Results

5.1. CL Features and U-Pb Dating Results

The CGM grains in the Tashidaban granitic pegmatite Li deposit from ρ40 elbaite pegmatite are mainly black and subhedral (Figures 5d,e and 6a). Seven analyzed points (Sample number 21TD10-5) varying from 452 to approximately 448 Ma with a weighted mean $^{206}Pb/^{238}U$ age of 450.2 ± 2.4 Ma (1σ; MSWD = 0.37) are shown. Thirteen measured points have a weighted mean $^{206}Pb/^{238}U$ age of 413.0 ± 3.6 Ma (1σ; MSWD = 0.21), varying from 417 to 407 Ma (Figure 7a,b; Table 2).

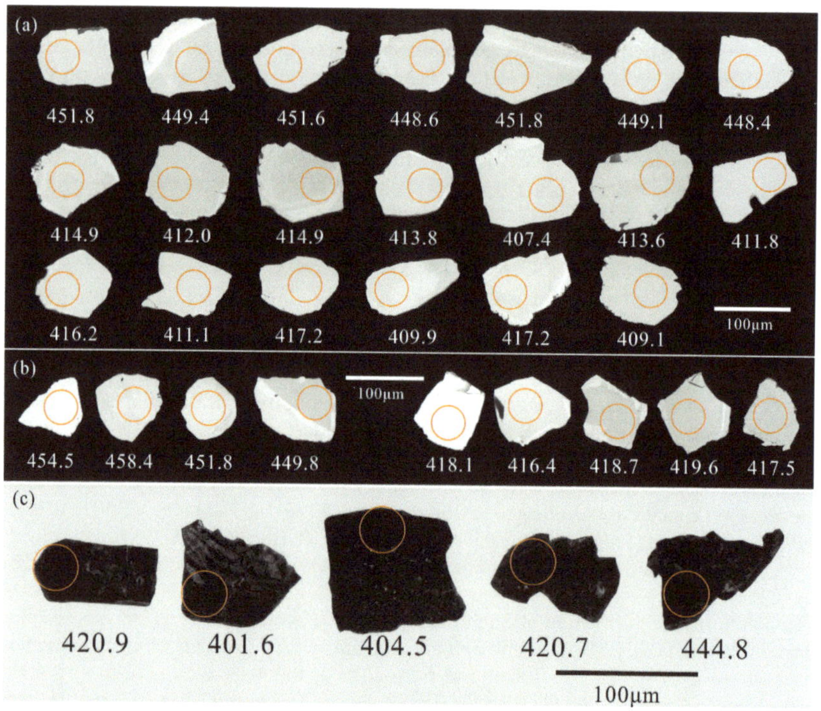

Figure 6. SEM-back-scattered electron (BSE) and Cathodoluminescence (CL) images of CGMs and zircons from the Tashidaban granitic pegmatite Li deposit. (**a**) CGMs from the aplite zone of ρ40 elbaite pegmatite; (**b**) CGMs from the aplite zone of ρ36 spodumene pegmatite; (**c**) Zircons from ρ26 tourmaline granite. The orange circles denote the U-Pb dating sites.

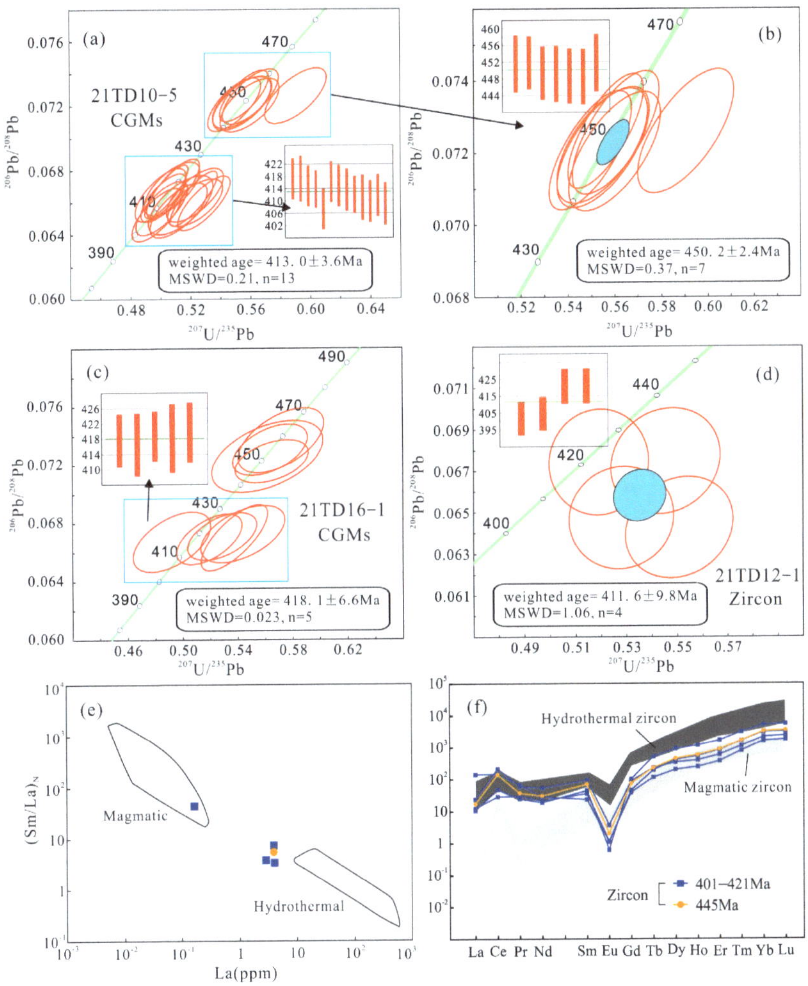

Figure 7. Diagrams of CGMs and zircons from the Tashidaban granitic pegmatite Li deposit. (**a**,**b**) CGMs from the zone of ρ40 elbaite pegmatite; (**c**) CGMs from the zone of ρ36 spodumene pegmatite; (**d**) Zircons from the zone of ρ26 tourmaline granite; (**e**) Chondrite-normalized [76] discrimination plots of $(Sm/La)_N$ vs. La; (**f**) Chondrite-normalized REE concentrations for magmatic and hydrothermal zircon [77].

The CGM grains from ρ36 spodumene pegmatite are mainly black and subhedral (Figures 5g and 6b). Four analyzed points (Sample number 21TD16-1) vary from 458 to 450 Ma, and five analyzed points have a weighted mean $^{206}Pb/^{238}U$ age of 418.1 ± 6.6 Ma (1σ; MSWD = 0.023), varying from 420 to approximately 416 Ma (Figure 7c; Table 2).

The zircon grains from ρ26 tourmaline granite are predominantly black and subhedral to anhedral with a blurred and sponge texture in few inclusions, while the rims have little oscillatory zoning (Figure 6c). One analyzed point has an older $^{206}Pb/^{238}U$ age of 445 Ma, and four points have a weighted mean $^{206}Pb/^{238}U$ age of 411.6 ± 9.8 Ma (1σ; MSWD = 1.06), varying from 421 to 402 Ma (Figure 7d; Table 3).

Table 2. Concentrations of U, Th, and Pb, and U-Pb isotopes of CGMs from the ρ40 elbaite pegmatite and the ρ36 spodumene pegmatite in the Tashidaban granitic pegmatite Li deposit.

No.	Pb (ppm)	Th (ppm)	U (ppm)	Th/U	$^{207}Pb/^{206}Pb$	Error (1σ)	$^{207}Pb/^{235}U$	Error (1σ)	$^{206}Pb/^{238}U$	Error (1σ)	$^{206}Pb/^{238}U$ Age (Ma)	$^{206}Pb/^{238}U$ 1sigma
21TD10-5-1	28	10	367	0.03	0.0556	0.0009	0.5566	0.0108	0.0726	0.0010	451.8	6.3
21TD10-5-2	23	9	302	0.03	0.0557	0.0010	0.5551	0.0129	0.0722	0.0011	449.4	6.3
21TD10-5-3	25	11	332	0.03	0.0562	0.0011	0.5609	0.0123	0.0726	0.0011	451.6	6.8
21TD10-5-4	24	9	323	0.03	0.0555	0.0010	0.5500	0.0117	0.0721	0.0011	448.6	6.6
21TD10-5-5	27	8	381	0.02	0.0589	0.0010	0.5910	0.0137	0.0726	0.0011	451.8	6.8
21TD10-5-6	22	10	306	0.03	0.0560	0.0012	0.5555	0.0131	0.0722	0.0011	449.1	6.6
21TD10-5-7	23	8	322	0.03	0.0565	0.0011	0.5603	0.0130	0.0720	0.0011	448.4	6.6
21TD10-5-8	22	8	323	0.02	0.0548	0.0011	0.5002	0.0115	0.0665	0.0011	414.9	6.6
21TD10-5-9	25	9	372	0.03	0.0576	0.0012	0.5236	0.0122	0.0660	0.0010	412.0	5.9
21TD10-5-10	25	10	378	0.03	0.0573	0.0012	0.5234	0.0127	0.0665	0.0011	414.9	6.7
21TD10-5-11	25	9	371	0.03	0.0551	0.0010	0.5020	0.0101	0.0663	0.0010	413.8	6.0
21TD10-5-12	20	9	307	0.03	0.0554	0.0010	0.4976	0.0112	0.0652	0.0011	407.4	6.6
21TD10-5-13	25	10	373	0.03	0.0575	0.0010	0.5225	0.0105	0.0663	0.0011	413.6	6.8
21TD10-5-14	27	11	390	0.03	0.0583	0.0010	0.5295	0.0116	0.0660	0.0011	411.8	6.8
21TD10-5-15	31	12	444	0.03	0.0550	0.0009	0.5039	0.0097	0.0667	0.0011	416.2	6.6
21TD10-5-16	16	9	235	0.04	0.0563	0.0012	0.5135	0.0144	0.0658	0.0012	411.1	7.3
21TD10-5-17	23	10	320	0.03	0.0551	0.0011	0.5077	0.0124	0.0669	0.0011	417.2	6.6
21TD10-5-18	24	10	348	0.03	0.0556	0.0010	0.5009	0.0115	0.0656	0.0011	409.9	6.7
21TD10-5-19	29	12	429	0.03	0.0552	0.0009	0.5085	0.0118	0.0669	0.0012	417.2	7.4
21TD10-5-20	30	12	454	0.03	0.0565	0.0010	0.5121	0.0127	0.0655	0.0011	409.1	6.8
21TD16-1-1	8	1	108	0.006	0.0571	0.0019	0.5642	0.0179	0.0731	0.0012	454.5	7.0
21TD16-1-2	5	0	63	0.007	0.0577	0.0023	0.5723	0.0207	0.0737	0.0015	458.4	8.8
21TD16-1-3	9	1	117	0.013	0.0572	0.0018	0.5653	0.0176	0.0723	0.0010	449.8	5.9
21TD16-1-4	4	0	52	0.007	0.0564	0.0026	0.5622	0.0268	0.0726	0.0016	451.8	9.8
21TD16-1-5	12	1	170	0.007	0.0586	0.0015	0.5379	0.0160	0.0670	0.0015	418.1	9.0
21TD16-1-6	5	1	81	0.007	0.0539	0.0019	0.4924	0.0187	0.0667	0.0014	416.4	8.2
21TD16-1-7	7	1	106	0.010	0.0584	0.0020	0.5291	0.0166	0.0671	0.0011	418.7	6.5
21TD16-1-8	5	0	73	0.006	0.0595	0.0022	0.5441	0.0202	0.0673	0.0013	419.6	7.8
21TD16-1-9	13	2	182	0.008	0.0565	0.0015	0.5198	0.0157	0.0669	0.0011	417.5	6.9

Table 3. Concentrations of U, Th, and Pb, and U-Pb isotopes of zircons from ρ26 tourmaline granite in the Tashidaban granitic pegmatite Li deposit.

No.	Pb (ppm)	Th (ppm)	U (ppm)	Th/U	$^{207}Pb/^{206}Pb$	Error (1σ)	$^{207}Pb/^{235}U$	Error (1σ)	$^{206}Pb/^{238}U$	Error (1σ)	$^{206}Pb/^{238}U$ Age (Ma)	$^{206}Pb/^{238}U$ 1sigma
21TD12-1-1	745	89	11,102	0.01	0.0587	0.0013	0.5457	0.0143	0.0675	0.0017	420.9	10.2
21TD12-1-2	441	51	6628	0.01	0.0622	0.0013	0.5511	0.0140	0.0643	0.0016	401.6	9.8
21TD12-1-3	690	153	10,479	0.01	0.0592	0.0012	0.5282	0.0136	0.0648	0.0016	404.5	9.8
21TD12-1-4	777	131	11,434	0.01	0.0559	0.0011	0.5194	0.0129	0.0674	0.0017	420.7	10.2
21TD12-1-5	374	27	5140	0.01	0.0585	0.0014	0.5762	0.0163	0.0714	0.0018	444.8	10.9

5.2. Mineral Chemistry of Zircon

The Th/U ratios of both zircons from ρ26 tourmaline granite are 0.01 (Table 3). In the chondrite-normalized REE (rare-earth element) diagrams of tourmaline granite, the samples are distinguished by their enrichment in both LREE (light rare-earth elements) and HREE (heavy rare-earth elements) (Figure 7f, Table 4). These samples exhibit positive Ce anomalies and negative Eu anomalies. In the $(Sm/La)_N$ vs. La magmatic and hydrothermal zircon discrimination plot [77], one zircon grain was plotted in the magmatic zircon field, and three grains were plotted between the magmatic and hydrothermal zircon fields (Figure 7e).

Table 4. Concentrations (ppm) of rare-earth element (REE) in zircons from ρ26 tourmaline granite in the Tashidaban granitic pegmatite Li deposit.

No.	La	Ce	Pr	Nd	Sm	Eu	Gd	Tb	Dy	Ho	Er	Tm	Yb	Lu
21TD12-1-1	43.57	116.75	8.03	16.75	4.47	0.08	11.50	9.04	130.69	38.42	172.96	51.83	629.11	100.41
21TD12-1-2	7.31	170.89	7.22	31.89	18.18	0.27	22.13	10.69	108.32	27.90	121.76	35.29	438.22	74.28
21TD12-1-3	3.28	38.19	2.93	11.07	8.52	0.05	25.66	23.68	283.88	81.51	341.32	97.00	1042.24	174.87
21TD12-1-4	3.77	22.86	3.53	12.54	6.83	0.09	9.90	5.35	63.71	17.40	76.58	25.02	330.04	55.86
21TD12-1-5	5.30	113.99	4.50	18.12	13.49	0.15	18.95	10.68	140.15	40.70	178.51	52.01	656.11	108.85

5.3. Whole-Rock Geochemical Data

Whole-rock geochemical data are listed in Supplementary Table S1. The granite rock facies in the Kumudaban area are composed of biotite granite, muscovite granite, albite granite, and tourmaline granite, with relatively elevated SiO_2 (69.59–75.03%), Al_2O_3 (14.80–15.49%), and Na_2O (3.69–5.75%) contents, and lower TiO_2 content ranging from 0.01% to 0.39%. The A/CNK values range from 1.11 to 1.40 with differentiation index (DI) ranging from 81.87 to 93.62 increasing from biotite granite to tourmaline granite. In the chondrite-normalized diagram (Figure 8a), biotite granite exhibits a rightward-shaped pattern of light rare-earth element (LREE) enrichment ($(La{:}Yb)_N$ = 46.88 to 50.21), while the others show relatively flat distribution trends ($(La{:}Yb)_N$ = 0.20 to 3.76). All samples except biotite granite exhibit significant negative Eu anomalies (δEu = 0.07 to 0.40), and the total rare-earth element (REE) content decreases from biotite granite (average ΣREE = 177.88) to tourmaline granite (average ΣREE = 10.77). In the spider diagram (Figure 8b), all samples except biotite granite show depletions in Ba and Ti.

The fine-grained aplite sample of the spodumene pegmatite (448 Ma) in the Tamuqie granitic pegmatite Li deposit has relatively low SiO_2 content (average 68.58%), higher Al_2O_3 (average 17.89%), and Na_2O content (average 9.39%), with an average A/CNK value of 1.12 and average DI value of 95.70. The total REE content is lower than that of the Kumudaban granite (average ΣREE = 3.02) (Figure 8a).

The tourmaline granite samples compared to the aplite samples of elbaite pegmatite and spodumene pegmatite in the Tashidaban granitic pegmatite Li deposit have relatively high SiO_2 content (average 73.36%) and low Al_2O_3 content (average 15.34%), with an average A/CNK value of 1.20. The fine-grained aplite zone of the ρ36 spodumene pegmatite exhibits the highest A/CNK values (average 1.57) and the lowest total REE content (average ΣREE = 0.54) (Figure 8a). The average DI value of tourmaline granite, elbaite pegmatite, and spodumene pegmatite ranged from 95.09 to 98.13. Furthermore, in the spider diagram, the pegmatitic granite samples exhibit similar distribution patterns (Figure 8b).

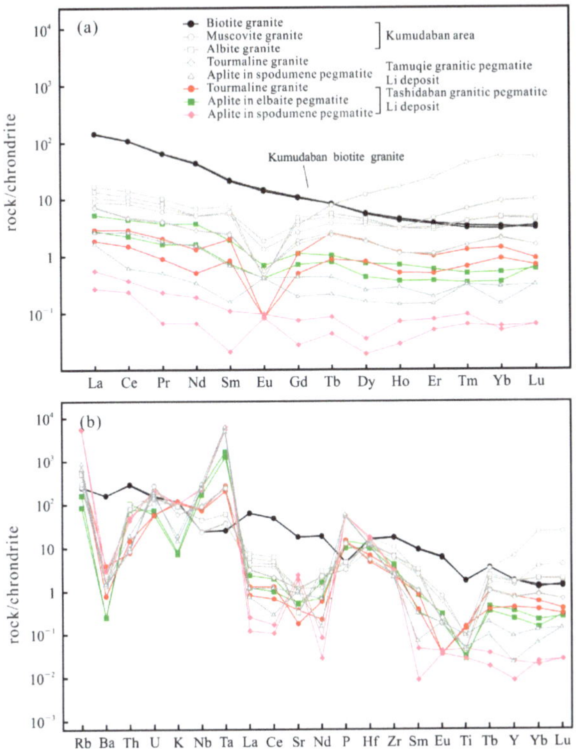

Figure 8. (**a**) Chondrite-normalized REE patterns [78] and (**b**) primitive mantle-normalized trace element diagrams [76]. Kumudaban biotite granite are from [65].

6. Discussion

6.1. Formation Times of the Tashitaban Li Granitic Pegmatites

The CGMs found in the granitic pegmatite contain relatively low common Pb contents and exhibit suitable Th and U contents for U-Pb geochronology [79,80]. Reflected light and BSE images show that most samples of CGMs are homogeneous. Only a few images exhibit bright Ta-Mn oscillatory zoning within dark Mn-rich cores, which indicates that these samples have not undergone hydrothermal metasomatism [81,82]. The ages of CGMs of ρ40 elbaite pegmatite are primarily distributed in two intervals with ranges from 452 to 448 Ma and 417 to 407 Ma. Additionally, those of ρ36 spodumene pegmatite range from 458 to 450 Ma and 420 to 416 Ma. Since the restriction of CGMs stability, the problem of inheritance from precursor rocks can be ruled out [79]. A similar phenomenon showing in a single pegmatite was reported by Melcher et al. [83] and Konzett et al. [84], which have a large age spread of 150–309 Ma in Passeier Valley (Italy) and 70–172 Ma age spread in Eastern Alpine (Italy). Konzett et al. [84] suggested that the oldest CGMs age represents the emplacement age and the age spread due to Cretaceous resetting involves multiple tectono-magmatic events. However, the two distribution interval ages in the single pegmatite of the Tashidaban granitic pegmatite Li deposit possibly represent two independent tectonic thermal events. In the Tashisayi granitic pegmatite Li deposit, three-stage age distribution was found in a composited aplite–pegmatite dike at 472 Ma, 440 Ma, and 416 Ma, with the oldest age originating from the Lep-Ab-Qz pegmatite and the middle age from the Spd-Ab-Qz pegmatite [65]. Additionally, Li-Be mineralization events were found to occur during 472—455 Ma in the North Tugeman granitic pegmatite Li-Be deposit and superimposed by 436–435 Ma and 415 Ma fine-grained granite [46].

In this study, only five age data points are considered reliable, among which the oldest age is 445 Ma, and the other four ages range between 421 and 402 Ma with weighted average age of 411.6 ± 9.8 Ma (MSWD = 1.06). Cores of zircon grains in the ρ26 tourmaline granite are predominantly black and exhibit a blurred and sponge texture with few inclusions, while the rims have little oscillatory zoning (Figure 6c), which indicates that the zircon grains originated from magmatic sources but underwent later recrystallization or metamorphic events [85]. Furthermore, the Th/U ratios of four zircons are 0.1, which are lower than typical magmatic zircons [86]. In the chondrite-normalized REE diagrams, the samples exhibit a similar pattern to hydrothermal zircons (Figure 7f). In the $(Sm/La)_N$ vs. La magmatic and hydrothermal zircon discrimination plot [77], three grains were plotted between the magmatic and hydrothermal zircon fields (Figure 7e). Combined with the CL features, these geochemical characteristics suggest that the zircons have undergone late-stage metamorphic alteration.

The alteration phenomena observed in this study indicate that the pegmatites in the Tashidaban granitic pegmatite Li deposit have undergone a later magmatic–hydrothermal event. The undulose extinction coarse-grained quartz intergrown with spodumene and albite undergoes recrystallization at the edges into granoblastic quartz, which is commonly observed in ρ40 elbaite pegmatite and ρ36 spodumene pegmatite (Figure 5b,h,i). These characteristics represent the synkinematic emplacement of the pegmatites, which are restricted to the shear zones [87]. The Altun region was undergoing simultaneous compression and amalgamation of three micro-continental blocks, resulting in intense compression and the formation of shear zones from the north and south at ~450 Ma (see Section 6.3). Moreover, elbaite was observed being replaced by a later fine-grained albite granite (Figure 5c). The replacement of aplite zone by late-stage fine-grained albite granite can be observed in Figure 5a, and both the aplite zone and albite granite have CGMs intergrown with albite and quartz, while the Spd-Ab-Qz zone is replaced by late-stage fine-grained muscovite albite granite in the ρ36 spodumene pegmatite (Figure 5i). All indicate that the pegmatites in the Tashidaban granitic pegmatite Li deposit have undergone the superimposition of a later magmatic–hydrothermal event. Therefore, the Li-mineralization age of the Tashidaban granitic pegmatite Li deposit is 450.2 ± 2.4 Ma, with superimposing of a late-stage magmatic–hydrothermal event during 418–412 Ma.

Multiple episodes of subduction–collision events occurred on both the northern and southern sides of the CAB due to a complex tectonic evolution in the Early Paleozoic. Statistical results indicate that Early Paleozoic magmatic granitoids in the North Altun Block (NAB) were formed between 514 and 411 Ma, while granitoids in the CAB are concentrated between 522 and 432 Ma. Furthermore, Wang et al. [88] obtained an age of 417 Ma from leucosome in garnet-bearing biotite gneiss of the Altun Group. An age of 413.8 ± 7.4 Ma from muscovite Ar-Ar dating of the Suwushijie monzogranite [68] as well as ages of 406–405 Ma from diorite and granite dykes in the Tashsayi area [89] were obtained. It indicates that contemporaneous magmatic activities exist in the Central Altun during 417–405 Ma. The ages of rare-metal pegmatite in CAB are shown in Figure 9. Ages obtained from the Tashidaban Li granitic pegmatites range in two intervals: 450.2 ± 2.4 Ma and 418–412 Ma, which are consistent with the regional magmatic and mineralization events.

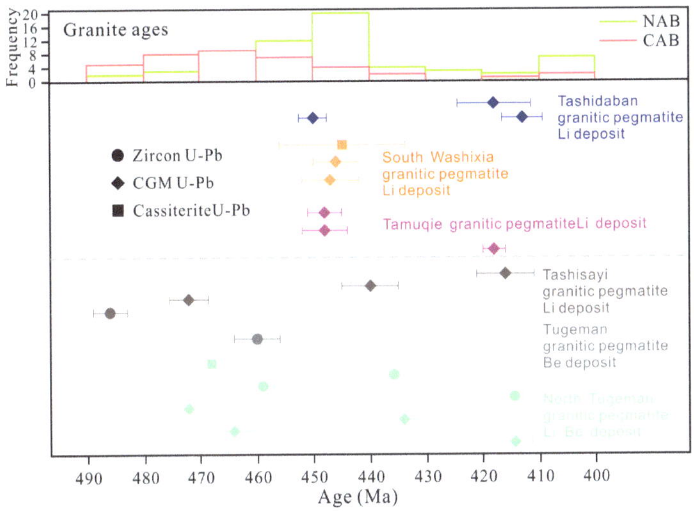

Figure 9. Formation times of granites and granitic pegmatite Li-Be deposits in the CAB. The granite ages of the NAB are from [90–101]; the granite ages of the CAB are from [46–48,60,62,88–90,102–105]. Igneous crystallization and mineralization ages of Li-Be deposits are consistent with the literature in Figure 1.

6.2. Petrogenesis of the Kumudaban Granitoids and Origin of Tashitaban Granitic Pegmatites

6.2.1. Petrogenesis of the Kumudaban Granitoids

An S-type granitoid is generally considered as a peraluminous granitoid, which is mainly formed by the partial melting of crustal materials [106,107]. The Kumudaban granitoid samples, with Alumina saturation index (ASI) values exceeding 1.1, are classified as peraluminous granite (Figure 10a). The content of corundum in granitoid samples in Cross, Iddings, Pirrson, and Washington (CIPW) norm calculation are more than 1%. Muscovite and garnet are identified under the microscope observation (Figure 3b–d). In addition, apatite commonly crystallizes early in I- and A-type granitic magma, resulting in a decrease in the P_2O_5 content of the residual magma. This decrease establishes a strong negative correlation with the SiO_2 content, while, P_2O_5 in S-type granite increases or basically remains unchanged with the increase in SiO_2 [108–110]. P_2O_5 values in these samples are basically unchanged with increased SiO_2 (except for tourmaline granite, with P_2O_5 in the range 0.07~0.09). The Kumudaban granitoids were characterized as S-type high-K to calc-alkaline granite plotting in the area of the high-K to calc-alkaline series (biotite granites to muscovite granites to albite granites, tourmaline granites) (Figure 10b) [111,112].

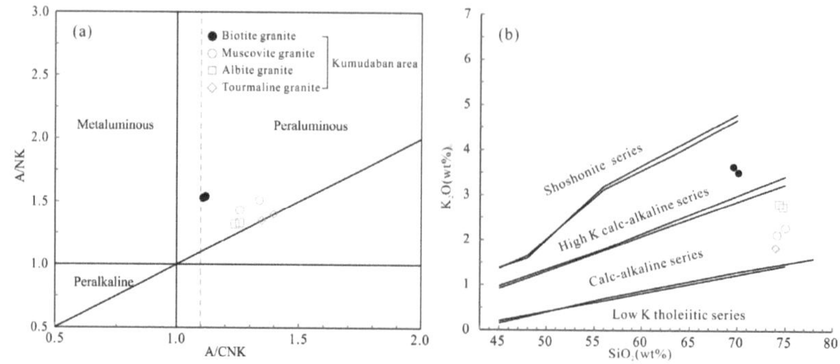

Figure 10. (**a**) ASI shown by A/NK vs. A/CNK diagram [113]; (**b**) the K_2O vs. SiO_2 diagram with dividing curves from [114].

6.2.2. Origin of the Tashidaban Li Granitic Pegmatites

Two petrogenetic processes were proposed to explain the formation of rare-element pegmatites: (1) extensive fractionation of fertile parental granitic magma [115–118], and (2) partial melting of rare-element-rich source rocks [119–121]. The large granite plutons near the Tashidaban granitic pegmatite Li deposit are the Kumudaban granite pluton (with zircon U-Pb ages 445–451 Ma) and the Suwushijie intermediate-acidic pluton (with zircon U-Pb ages 464–475 Ma). In this study, the Li-mineralizition age in the Tashidaban granitic pegmatite Li deposit is 450.2 ± 2.4 Ma. The Tashidaban Li granitic pegmatites should originate from the highly crystallized differentiation of the Kumudaban biotite granite pluton (Suwushijie intermediate-acidic pluton is excluded due to the presence of a >6 Ma age gap). The Tamuqie granitic pegmatite Li deposit is located north of the Tashidaban granitic pegmatite Li deposit and has a Li-mineralization age of 448 Ma with occurrence of a 418 Ma tourmaline-bearing albite pegmatite. Since pegmatites accentuate the trace element signatures of their granitic sources [24,25], the chemical evolution of Tashidaban Li granitic pegmatites should be similar to the Tamuqie spodumene pegmatite.

Zr/Hf ratio is a useful index of the extent of fractional crystallization of a felsic melt [122,123]. The element pairs K/Rb, K/Cs, and Nb/Ta typically exhibit synergistic behavior, and their variation mechanisms are similar [124]. This also suggests that these ratios could be indicators of the degree of differentiation of felsic melts [27,125]. The pairs mentioned above were plotted against Zr/Hf, which shows a well-defined evolutionary trend in the Kumudaban granite rock facies (Figure 11). The biotite granite exhibits the highest K/Rb, K/Cs, and Nb/Ta ratios. The gradual decrease in these ratios from muscovite granite to albite granite and tourmaline granite shows a clear trend in magmatic evolution. With ongoing of magmatic evolution, the ratios of these incompatible and compatible elements pair decrease successively from tourmaline granite to elbaite pegmatite to spodumene pegmatite in the Tashidaban granitic pegmatite Li deposit (Figure 11). Samples from the aplite zone in spodumene pegmatite (448 Ma) in the Tamuqie granitic pegmatite Li deposit and the aplite zone in ρ36 spodumene pegmatite in the Tashidaban granitic pegmatite Li deposit almost overlap (Figure 11). The similar distributions of trace elements (Figure 8) indicate a comparable chemical evolution of the granitic magma. The trace element distribution in the normalized chondrite and primitive mantle-normalized diagram have similar patterns, and total REE content shows a continuous decrease trend (Figure 8). All mentioned above indicate that the Tashidaban Li granitic pegmatites have a close petrogenetic relationship with Kumudaban granite pluton. Another interpretation suggests that the petrogenetic processes of the Li granitic pegmatites originate from the low-degree partial melting of metamorphic rocks. This mode of origin was confirmed to explain some rare-metal granitic pegmatite deposits worldwide [84,126]. However, no evidence of migmatization was found within the Tashidaban granitic pegmatite Li deposit [102], and no metamorphic minerals such as kyanite and chlorite were discovered in the pegmatite [127,128]. Thus, the evidence above suggests that the Tashidaban Li granitic pegmatites originated from extensive fractionation of the Kumudaban granite pluton.

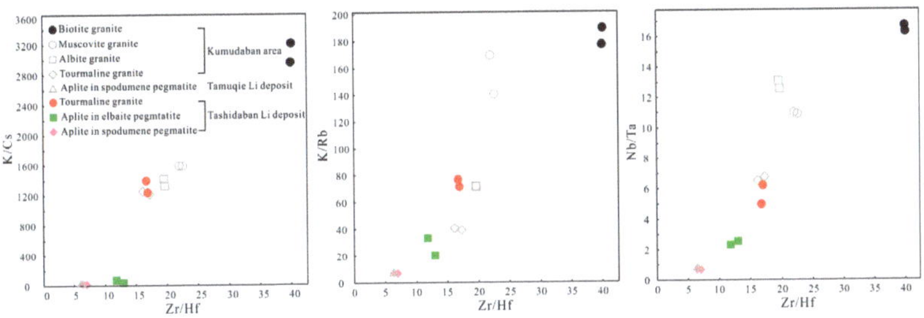

Figure 11. Selected variation diagrams of granites and pegmatites from the Kumudaban area, Tamuqie granitic pegmatite Li deposit, and Tashidaban granitic pegmatite Li deposit.

6.3. Geodynamic Setting of the Tashitaban Li Granitic Pegmatite Deposit

The Central Altun orogenic system is a product of the amalgamation of multiple micro-continental blocks and island arcs [129]. This complex system originated from subduction–accretion–collision processes in the Proto-Tethys Ocean during the Early Paleozoic [130,131].

The existence of supra-subduction zone (SSZ) ophiolite and plagiogranite are evident of the North Altun ocean subducted from north to south under the CAB with the indication of the initial subduction before ~521–512 Ma [56,132,133]. A series of high-temperature–low-pressure (HP–LT) metamorphic rocks (blueschists, eclogites: ages in 491–520 Ma) [134,135] and related magmatism events (arc volcanic rocks and I- and S-type granites) were formed with ongoing subduction [92–94,136–142]. Zircon age reveals these volcanic rocks and granites were formed between 517 and 459 Ma, which is coeval with the age spectra of detrital zircons in the Early Paleozoic sedimentary rocks from the North Altun [143]. A series of syn-collisional granite formed in the thickened lower crust at 458–431 Ma indicated a collisional regime [92–94,96,97,99]. The transition of the tectonic regime has changed to extension at ~425 Ma with the characterizations of low Sr, Sr/Y and $(La/Yb)_N$ granitoids (425–410 Ma) [94,96,99].

The high-pressure (HP)–ultrahigh-pressure (UHP) belt is located south of the Central Altun block, specifically in the Jianggalesayi, Danshuiquan, and Yinggelisayi areas. This belt comprises various rock types, including UHP eclogite, garnet (Grt) pelitic gneisses, Grt clinopyroxenite, Grt granite gneisses, and Grt peridotite. [144–147]. Systematic study of the P-T-t evolution and zircon age in HP-UHP metamorphic rocks inferred three phases of metamorphism corresponding to the continental deep subduction (~509–490 Ma) and rollback of the subducting slab (~490–440 Ma) [148,149]. Extensive magmatism was triggered by this, and consequently granitoids were formed between 522 and 432 Ma, with to the I- and S-type affinities [46–48,59–62,88,89,102–105]. The tectonic regime has changed to extension in South of the Central Altun block after 432 Ma, which is characterized by intraplate granites (432–418 Ma) [66,150], bimodal volcanic rocks (406 Ma) [151], and A-type granites (432–385 Ma) [152–154].

The Li-mineralization age of the Tashidaban granitic pegmatite Li deposit is 450.2 ± 2.4 Ma with superimposition of the late-stage magmatic–hydrothermal event that occurred at 418 to 412 Ma. The Central Altun block experienced intense continental collision between 450 and 445 Ma due to the substantial crustal thickening and regional Barrovian metamorphism [155]. This led to extensive dehydration melting of biotite under granulite facies and generating voluminous granitic magmas, which subsequently crystallized and differentiated to form the Kumudaban to Tatebulake large syn-kinematic biotite granite plutons [60,62,102]. The Tamuqie and Tashidaban Li granitic pegmatites were formed during this period through extreme differentiation by fractional crystallization of granitic magma. The collision ceased on both the northern and southern sides after 432 Ma, and extensional tectonic settings led to lithospheric delamination leading to the lower crust triggered mantle-derived magmatism, resulting in the emplacement of granitic magmas, superimposing the Tashidaban granitic pegmatite Li deposit at 418–412 Ma.

7. Conclusions

Multiple episodes of rare-metal Li-Be mineralization events occurred in the Central Altun during the Early Paleozoic. A newly discovered Tashidaban granitic pegmatite Li deposit, at the northern margin of the Central Altun Block, hosts three distinct types of granitic pegmatities: tourmaline-bearing albite pegmatite, elbaite pegmatite, and spodumene pegmatite. These uniformed pegmatites intruded into the two-mica schist and marble of the Muzisayi Formation of the Tashidaban Group. CGMs and zircon U-Pb dating results reveal the Li-mineralization age of the Tashidaban granitic pegmatite Li deposit is 450.2 ± 2.4 Ma, with a superimposed late-stage magmatic–hydrothermal event occurring at 418 to 412 Ma. The whole-rock geochemical results indicate that the Kumudaban granitoids belongs to the S-type high-K to calc-alkaline granite, and the Tashidaban Li granitic pegmatites have originated from the extreme differentiation by fractional crystallization of the Kumudaban granite pluton.

Supplementary Materials: The following supporting information can be downloaded at: https://www.mdpi.com/article/10.3390/min14060542/s1, Supplementary Table S1: Major element (wt%) and trace element (ppm) geochemistry of the granites and pegmatites from Kumudaban area, Tamuqie granitic pegmatite Li deposit and Tashidaban granitic pegmatite Li deposit.

Author Contributions: Conceptualization, X.X. and K.K.; methodology, X.X. and H.L.; software, Z.L. and K.K.; validation, K.K.; investigation, K.K., X.W. and L.N.; writing—original draft preparation, K.K.; writing—review and editing, Y.M.; visualization, X.X.; supervision, H.L.; project administration, P.Z., J.C. and X.L.; funding acquisition, X.X. All authors have read and agreed to the published version of the manuscript.

Funding: This research was jointly supported by the National Natural Science Foundation of China (Grant Number 42250202, 42272075) and Extreme Geological Events Team (2021ZT09H399), second comprehensive scientific investigation of the Qinghai-Tibet Plateau (Grant Number 2019QZKK0802).

Data Availability Statement: At the request of other researchers, the authors of the article can provide the original data.

Acknowledgments: We are indebted to Xiangjing Yu and Huiwen Wang for sample preparation, as well as to Shuangrong Zhang and Huan Hu for laboratory assistance. We also express our gratitude to the four anonymous reviewers for their constructive suggestions, and Chen Zhao for revising the expression of the article.

Conflicts of Interest: Peng Zhang, Xingzhong Liu and Jianzhong Chen are employees of No. 3 Geological Party, Xinjiang Bureau of Geology and Mineral Exploration and Development, Korla, Xinjiang 841000, China. Lei Niu is employee of Exploration and Development Research Institute, Zhongyuan Oilfield Company, SINOPEC, Puyang 457001, China. The paper reflects the views of the scientists and not the company.

References

1. Schulz, K.J.; DeYoung, J.H.; Seal, R.R.; Bradley, D.C. *Critical Mineral Resources of the United States: Economic and Environmental Geology and Prospects for Future Supply*; U.S. Geological Survey: Reston, VA, USA, 2017; 797p. [CrossRef]
2. Zhai, M.G.; Wu, F.Y.; Hu, R.Z.; Jiang, S.Y.; Li, W.C.; Wang, R.C.; Wang, D.H.; Qi, T.; Qin, K.Z.; Wen, H.J. Critical metal mineral resources: Current research status and scientific issues. *Bull. Natl. Nat. Sci. Found. China* **2019**, 2, 106–111. (In Chinese with English abstract)
3. Hou, Z.Q.; Chen, J.; Zhai, M.G. Current status and frontiers of research on critical mineral resources. *Chin. Sci. Bull.* **2020**, *65*, 3651–3652. (In Chinese with English abstract) [CrossRef]
4. Jaskula, B.W. *Beryllium: U.S. Geological Survey Mineral Commodity Summaries*; U.S. Geological Survey: Reston, VA, USA, 2013; pp. E1–E32. [CrossRef]
5. Linnen, R.L.; Van Lichtervelde, M.; Černý, P. Granitic pegmatites as sources of strategic metals. *Elements* **2012**, *8*, 275–280. [CrossRef]

6. Roda-Robles, E.; Villaseca, C.; Pesquera, A.; Gil-Crespo, P.P.; Vieira, R.; Lima, A.; Garate-Olave, I. Petrogenetic relationships between Variscan granitoids and Li-(FP)-rich aplite-pegmatites in the Central Iberian Zone: Geological and geochemical constraints and implications for other regions from the European Variscides. *Ore Geol. Rev.* **2018**, *95*, 408–430. [CrossRef]
7. Roda-Robles, E.; Vieira, R.; Lima, A.; Errandonea-Martin, J.; Pesquera, A.; Cardoso-Fernandes, J.; Garate-Olave, I. Li-rich pegmatites and related peraluminous granites of the Fregeneda-Almendra field (Spain-Portugal): A case study of magmatic signature for Li enrichment. *Lithos* **2023**, *452*, 107195. [CrossRef]
8. Xu, Z.Q.; Zhu, W.B.; Zheng, B.H.; Shu, L.S.; Li, G.W.; Che, X.D.; Qin, Y.L. New energy strategy for lithium resource and the continental dynamics research: Celebrating the centenary of the School of Earth Sciences and Engineering, Nanjing University. *Acta Geol. Sin.* **2021**, *95*, 2937–2954.
9. Qin, K.Z.; Zhou, Q.F.; Zhao, J.X.; He, C.T.; Liu, X.C.; Shi, R.Z.; Liu, Y.C. Be-rich mineralization features of Himalayan leucogranite belt and prospects for lithium-bearing pegmatites in higher altitudes. *Acta Geol. Sin.* **2021**, *95*, 3146–3162. (In Chinese with English abstract)
10. Zhang, D.; Peng, J.; Coulson, I.M.; Hou, L.H.; Li, S.J. Cassiterite U–Pb and muscovite40Ar–39Ar age constraints on the timing of mineralization in the Xuebaoding Sn–W–Be deposit, western China. *Ore Geol. Rev.* **2014**, *62*, 315–322. [CrossRef]
11. Zhou, Q.F.; Qin, K.Z.; Tang, D.M.; Tian, Y.; Cao, M.J.; Wang, C.L. Formation Age and Evolution Time Span of the Koktokay No. 3 Pegmatite, A ltai, NW China: Evidence from U–Pb Zircon and 40Ar–39Ar Muscovite Ages. *Resour. Geol.* **2015**, *65*, 210–231. [CrossRef]
12. Zhou, Q.F.; Qin, K.Z.; Tang, D.; Wang, C.L.; Asamiah Sakyi, P. LA-ICP-MS U–Pb zircon, columbite-tantalite and 40Ar–39Ar muscovite age constraints for the rare-element pegmatite dykes in the Altai orogenic belt, NW China. *Geol. Mag.* **2018**, *155*, 707–728. [CrossRef]
13. Yan, Q.H.; Qiu, Z.W.; Wang, H.; Wang, M.; Wei, X.P.; Li, P.; Zhang, R.Q.; Li, C.Y.; Liu, J.P. Age of the Dahongliutan rare metal pegmatite deposit, West Kunlun, Xinjiang (NW China): Constraints from LA–ICP–MS U–Pb dating of columbite-(Fe) and cassiterite. *Ore Geol. Rev.* **2018**, *100*, 561–573. [CrossRef]
14. Li, P.; Li, J.; Liu, X.; Li, C.; Huang, Z.B.; Zhou, F.C. Geochronology and source of the rare-metal pegmatite in the Mufushan area of the Jiangnan orogenic belt: A case study of the giant Renli Nb–Ta deposit in Hunan, China. *Ore Geol. Rev.* **2020**, *116*, 103237. [CrossRef]
15. Xiong, Y.Q.; Jiang, S.Y.; Wen, C.H.; Yu, H.Y. Granite–pegmatite connection and mineralization age of the giant Renli TaNb deposit in South China: Constraints from U–Th–Pb geochronology of coltan, monazite, and zircon. *Lithos* **2020**, *358*, 105422. [CrossRef]
16. Zhou, Q.F.; Qin, K.Z.; Tang, D.M. Mineralogy of columbite-group minerals from the rare-element pegmatite dykes in the East-Qinling orogen, central China: Implications for formation times and ore genesis. *J. Asian Earth Sci.* **2021**, *218*, 104879. [CrossRef]
17. Wang, Z.; Li, J.; Chen, Z.; Yan, Q.G.; Xiong, X.; Li, P.; Deng, J.Y. Evolution and Li Mineralization of the No. 134 Pegmatite in the Jiajika Rare-Metal Deposit, Western Sichuan, China: Constrains from Critical Minerals. *Minerals* **2021**, *12*, 45. [CrossRef]
18. Fei, G.C.; Menuge, J.F.; Li, Y.Q.; Yang, J.; Deng, Y.; Chen, C.S.; Yang, Y.F.; Yang, Z.; Qin, L.; Zheng, L.; et al. Petrogenesis of the Lijiagou spodumene pegmatites in Songpan-Garze Fold Belt, West Sichuan, China: Evidence from geochemistry, zircon, cassiterite and coltan U-Pb geochronology and Hf isotopic compositions. *Lithos* **2020**, *364*, 105555. [CrossRef]
19. Zhao, H.; Chen, B.; Huang, C.; Bao, C.; Yang, Q.; Cao, R. Geochemical and Sr–Nd–Li isotopic constraints on the genesis of the Jiajika Li-rich pegmatites, eastern Tibetan Plateau: Implications for Li mineralization. *Contrib. Mineral. Petrol.* **2022**, *177*, 4. [CrossRef]
20. Lv, Z.H.; Chen, J.; Zhang, H.; Tang, Y. Petrogenesis of Neoproterozoic rare metal granite-pegmatite suite in Jiangnan Orogen and its implications for rare metal mineralization of peraluminous rock in South China. *Ore Geol. Rev.* **2021**, *128*, 103923. [CrossRef]
21. Zhu, M.T.; Dong, Z.G.; Zhang, L.C. Volatile accumulation for the mineralization of Li–Be pegmatites in the northeastern Pamir, Western Kunlun, China. *Int. Geol. Rev.* **2022**, *65*, 1354–1371. [CrossRef]
22. Liu, C.Q.; Zhang, H. The lanthanide tetrad effect in apatite from the Altay No. 3 pegmatite, Xingjiang, China: An intrinsic feature of the pegmatite magma. *Chem. Geol.* **2005**, *214*, 61–77. [CrossRef]
23. Tang, Y.; Wang, H.; Zhang, H.; Lv, Z.H. K-feldspar composition as an exploration tool for pegmatite-type rare metal deposits in Altay, NW China. *J. Geochem. Explor.* **2018**, *185*, 130–138. [CrossRef]
24. Yin, R.; Huang, X.L.; Xu, Y.G.; Wang, R.C.; Wang, H.; Yuan, C.; Ma, Q.; Sun, X.M.; Chen, L.L. Mineralogical constraints on the magmatic–hydrothermal evolution of rare-elements deposits in the Bailongshan granitic pegmatites, Xinjiang, NW China. *Lithos* **2020**, *352*, 105208. [CrossRef]
25. Cao, R.; Gao, Y.B.; Chen, B.; Bagas, L.; Yan, S.C.; Huang, C.; Zhao, H. Pegmatite magmatic evolution and rare metal mineralization of the Dahongliutan pegmatite field, Western Kunlun Orogen: Constraints from the B isotopic composition and mineral-chemistry. *Int. Geol. Rev.* **2023**, *65*, 1224–1242.
26. Li, J.; Huang, X.L.; He, P.L.; Li, W.X.; Yu, Y.; Chen, L.L. In situ analyses of micas in the Yashan granite, South China: Constraints on magmatic and hydrothermal evolutions of W and Ta–Nb bearing granites. *Ore Geol. Rev.* **2015**, *65*, 793–810. [CrossRef]
27. Liu, C.; Wang, R.C.; Wu, F.Y.; Xie, L.; Liu, X.C.; Li, X.K.; Yang, L.; Li, X.J. Spodumene pegmatites from the Pusila pluton in the higher Himalaya, South Tibet: Lithium mineralization in a highly fractionated leucogranite batholith. *Lithos* **2020**, *358*, 105421.

28. Li, P.; Li, J.; Chen, Z.; Liu, X.; Huang, Z.B.; Zhou, F.C. Compositional evolution of the muscovite of Renli pegmatite-type rare-metal deposit, northeast Hunan, China: Implications for its petrogenesis and mineralization potential. *Ore Geol. Rev.* **2021**, *138*, 104380. [CrossRef]
29. Yan, Q.H.; Wang, H.; Chi, G.; Wang, Q.; Hu, H.; Zhou, K.L.; Zhang, X.Y. Recognition of A 600-km-Long Late Triassic Rare Metal (Li-Rb-Be-Nb-Ta) Pegmatite Belt in the Western Kunlun Orogenic Belt, Western China. *Econ. Geol.* **2022**, *117*, 213–236.
30. Chen, J.Z.; Zhang, H.; Tang, Y.; Lv, Z.H.; An, Y.; Wang, M.T.; Liu, K.; Xu, Y.S. Lithium mineralization during evolution of a magmatic–hydrothermal system: Mineralogical evidence from Li-mineralized pegmatites in Altai, NW China. *Ore Geol. Rev.* **2022**, *149*, 105058.
31. Han, J.; Chen, H.; Zhang, J.; Zhang, Y.F.; Zhang, R.Q. Age and evolution of the Nanyangshan rare-metal mineralized pegmatite revealed by cassiterite U-Pb geochronology and tourmaline chemistry. *Ore Geol. Rev.* **2022**, *150*, 105121. [CrossRef]
32. Li, J.; Liu, C.Y.; Liu, X.; Li, P.; Huang, Z.B.; Zhou, F.C. Tantalum and niobium mineralization from F-and Cl-rich fluid in the lepidolite-rich pegmatite from the Renli deposit in northern Hunan, China: Constraints of fluid inclusions and lepidolite crystallization experiments. *Ore Geol. Rev.* **2019**, *115*, 103187. [CrossRef]
33. Fei, G.C.; Menuge, J.F.; Chen, C.S.; Yang, Y.L.; Deng, Y.; Li, Y.G.; Zheng, L. Evolution of pegmatite ore-forming fluid: The Lijiagou spodumene pegmatites in the Songpan-Garze Fold Belt, southwestern Sichuan Province, China. *Ore Geol. Rev.* **2021**, *139*, 104441. [CrossRef]
34. Zhang, Z.; Jiang, Y.; Niu, H.; Qu, P. Fluid inclusion and stable isotope constraints on the source and evolution of ore-forming fluids in the Bailongshan pegmatitic Li-Rb deposit, Xinjiang, western China. *Lithos* **2021**, *380*, 105824. [CrossRef]
35. Li, J.K.; Wang, D.H.; Chen, Y.C. The Ore-forming Mechanism of the Jiajika Pegmatite-Type Rare Metal Deposit in Western Sichuan Province: Evidence from Isotope Dating. *Acta Geol. Sin. Engl. Ed.* **2013**, *87*, 91–101.
36. Zhang, X.; Zhang, H.; Ma, Z.L.; Tang, Y.; Lv, Z.H.; Zhao, J.Y.; Liu, Y.L. A new model for the granite–pegmatite genetic relationships in the Kaluan–Azubai–Qiongkuer pegmatite-related ore fields, the Chinese Altay. *J. Asian Earth Sci.* **2016**, *124*, 139–155. [CrossRef]
37. Zhou, J.S.; Wang, Q.; Xu, Y.G.; Cempirek, J.; Wang, H.; Ma, J.L.; Wei, G.J.; Huang, T.Y.; Zhu, G.H.; Zhang, L. Geochronology, petrology, and lithium isotope geochemistry of the Bailongshan granite-pegmatite system, northern Tibet: Implications for the ore-forming potential of pegmatites. *Chem. Geol.* **2021**, *584*, 120484. [CrossRef]
38. Fan, J.J.; Tang, G.J.; Wei, G.J.; Wang, H.; Xu, Y.G.; Wang, Q.; Zhou, J.S.; Zhang, Z.Y.; Huang, T.Y.; Wang, Z.L. Lithium isotope fractionation during fluid exsolution: Implications for Li mineralization of the Bailongshan pegmatites in the West Kunlun, NW Tibet. *Lithos* **2020**, *352*, 105236. [CrossRef]
39. Huang, T.; Fu, X.; Ge, L.; Zou, F.G.; Hao, X.F.; Yang, R.; Xiao, R.Q.; Fan, J.B. The genesis of giant lithium pegmatite veins in Jiajika, Sichuan, China: Insights from geophysical, geochemical as well as structural geology approach. *Ore Geol. Rev.* **2020**, *124*, 103557. [CrossRef]
40. Shen, P.; Pan, H.D.; Li, C.H.; Feng, H.X.; He, L.F.; Bai, Y.X.; Luo, Y.Q.; Suo, Q.Y.; Cap, C. Newly-recognized Triassic highly fractionated leucogranite in the Koktokay deposit (Altai, China): Rare-metal fertility and connection with the No. 3 pegmatite. *Gondwana Res.* **2022**, *112*, 24–51. [CrossRef]
41. Liu, L.; Kang, L.; Cao, Y.T.; Yang, W.Q. Early Paleozoic granitic magmatism related to the processes from subduction to collision in South Altyn, NW China. *Sci. China Earth Sci.* **2015**, *58*, 1513–1522. (In Chinese with English abstract) [CrossRef]
42. Kang, L.; Xiao, P.X.; Gao, X.F.; Xi, R.G.; Yang, Z.C. Early Paleozoic magmatism and collision orogenic process of the South Altyn. *Acta Geol. Sin.* **2016**, *90*, 2527–2550.
43. Romer, R.L.; Smeds, S.A. U-Pb columbite ages of pegmatites from Sveconorwegian terranes in southwestern Sweden. *Precambr. Res* **1996**, *76*, 15–30. [CrossRef]
44. Xu, X.W.; Li, H.; Shi, F.P.; Yao, F.J.; Chen, J.Z.; Yang, Z.Q.; Hong, T.; Ke, Q. Metallogenic characteristics and prospecting of granitic pegmatite-type rare metal deposits in the Tugeman area, Middle Altyn Tagh. *Acta Petrol. Sin.* **2019**, *35*, 3303–3316. (In Chinese with English abstract)
45. Xu, X.W.; Hong, T.; Li, H.; Niu, L.; Ke, Q.; Chen, J.Z.; Liu, S.K.; Zhai, M.G. Concept of high-temperature granite-pegmatite Li-Be metallogenic system with a primary study in the middle Altyn-Tagh. *Acta Petrol. Sin.* **2020**, *36*, 3572–3592. (In Chinese with English abstract)
46. Li, H. Lithium-Beryllium Mineralization of Granite-Pegmatite System in the Tugeman Area in the Middle of Altyn Tagh. Ph.D. Thesis, Chinese Academy of Sciences, Beijing, China, 2021. (In Chinese with English abstract).
47. Hong, T.; Zhai, M.G.; Xu, X.W.; Li, H.; Wu, C.; Ma, Y.C.; Niu, L.; Ke, Q.; Wang, C. Tourmaline and quartz in the igneous and metamorphic rocks of the Tashisayi granitic batholith, Altyn Tagh, northwestern China: Geochemical variability constraints on metallogenesis. *Lithos* **2021**, *400*, 106358. [CrossRef]
48. Gao, Y.B.; Zhao, X.M.; Bagas, L.; Wang, Y.L.; Jin, M.S.; Zhang, J.W.; Lu, L.; Gao, Y.J.; Yan, Z.Q.; Teng, J.X.; et al. Newly discovered Ordovician Li-Be deposits at Tugeman in the Altyn-Tagh Orogen, NW China. *Ore Geol. Rev.* **2021**, *139*, 104515. [CrossRef]
49. Wang, H.; Ma, H.D.; Zhang, S.; Du, X.F.; Huang, L.; Cai, M.Z.; Chen, G.W.; Zhu, B.Y.; Wang, Q.; Wang, K.Y.; et al. Discovery of the Huanglongling giant lithium pegmatite deposit in Altyn Tagh, Xinjiang, China. *Acta Petrol. Sin.* **2023**, *39*, 3307–3318. (In Chinese with English abstract) [CrossRef]
50. Xu, Z.Q.; Yang, J.S.; Zhang, J.X.; Jiang, M.; Li, H.B.; Cui, J.W. A comparison between the tectonic units on the two sides of the Altun sinistral strike-slip fault and the mechanism of lithospheric shearing. *Acta Geol. Sin.* **1999**, *73*, 193–205.

51. Liu, L. Tectonics Significance of High-Pressure Metamorohic Rocks and Ophiolites in Altun area, Xingjiang, China. Ph.D. Thesis, Chinese Academy of Sciences, Beijing, China, 1999. (In Chinese with English abstract).
52. Liu, L.; Wang, C.; Chen, D.L.; Zhang, A.D.; Liou, J.G. Petrology and geochronology of HP-UHP rocks from the South Altyn, northwestern China. *J. Asian Earth Sci.* **2009**, *35*, 232–244. [CrossRef]
53. Liu, L.; Wang, C.; Cao, Y.T.; Chen, D.L.; Kang, L.; Yang, W.Q.; Zhu, X.H. Geochronology of multi-stage metamorphic events: Constraints on episodic zircon growth from the UHP eclogite in the South Altyn, NW China. *Lithos* **2012**, *136*, 10–26. [CrossRef]
54. Wu, C.L.; Gao, Y.H.; Lei, M.; Qin, H.P.; Liu, C.H.; Li, M.Z.; Frost, B.R.; Wooden, J.L. Zircon SHRIMP U-Pb dating, Lu-Hf isotopic characteristics and petrogenesis of the Palaeozoic granites in Mangya area, southern Altun, NW China. *Acta Petrol. Sin.* **2014**, *30*, 2297–2323. (In Chinese with English abstract)
55. Wu, C.L.; Lei, M.; Wu, D.; Zhang, X.; Chen, H.J.; Li, T.X. Zircon U-Pb dating of Paleozoic granite from South Altun and response of the magmatic activity to the tectonic evolution of the Altun Orogenic Belt. *Acta Geol. Sin.* **2016**, *90*, 2276–2315. (In Chinese with English abstract)
56. Zhang, Z.C.; Guo, Z.J. Sm-Nd, Ar Isotopic and Geochemical Characteristics of Plagioclase Amphibolites in the Bashikaogong Area, Altun Tagh Mountains. *Geol. J. China Univ.* **2004**, *10*, 28. (In Chinese with English abstract)
57. Hao, J.B. Composition, Geochronology and Mesoproterozoic-Neoproterozoic Tectonic Evolution of the Central-Southern Altyn Tagh. Ph.D. Thesis, Northwest University, Xi'an, China, 2021. (In Chinese with English abstract).
58. Gehrels, G.E.; Yin, A.; Wang, X.F. Detrital-zircon geochronology of the northeastern Tibetan plateau. *Geol. Soc. Am. Bull.* **2003**, *115*, 881–896. [CrossRef]
59. Chen, B. Characteristics and zircon U-Pb dating of the Kumdaban rock mass swarm on the northern margin of the Azhong block. *Geol. China* **2007**, *34*, 270–275. (In Chinese with English abstract)
60. Cao, Y.T.; Liu, L.; Wang, C.; Yang, W.Q.; Zhu, X.H. Geochemical, zircon U-Pb dating and Hf isotope compositions studies for Tatelekebulake granite in South Altyn Tagh. *Acta Petrol. Sin.* **2010**, *26*, 3259–3271. (In Chinese with English abstract)
61. Kang, L.; Liu, L.; Cao, Y.T.; Wang, C.; Yang, W.Q.; Liang, S. Geochemistry, zircon U-Pb age and its geological significance of the gneissic granite from the eastern segment of the Tatelekebulake composite granite in the south Altyn Tagh. *Acta Petrol. Sin.* **2013**, *29*, 3039–3048. (In Chinese with English abstract)
62. Liu, C.H.; Wu, C.L.; Gao, Y.H.; Lei, M.; Qin, H.P. Age, composition, and tectonic significance of Palaeozoic granites in the Altyn orogenic belt, China. *Int. Geol. Rev.* **2016**, *58*, 131–154. [CrossRef]
63. Li, H.; Hong, T.; Liu, S.K.; Yang, Z.Q.; Ke, Q.; Ma, Y.C.; Wang, X.H.; Niu, L.; Kang, K.; Xu, X.W.; et al. Characteristics of early Paleozoic granite-pegmatite and associated lithium-beryllium mineralization in the Tugeman area, Altun orogenic system, northwestern China. *Ore Geol. Rev.* **2023**, *160*, 105603. [CrossRef]
64. Kang, K.; Hong, T.; Zhang, P.; Liu, S.K.; Yao, F.J.; Li, H.; Ma, Y.C.; Wang, X.H.; Liao, Z.X.; Niu, L.; et al. The large Li pegmatite deposits formed at final collision of a three micro-continent collage in the Central Altun Tagh, Northwest China. *Ore Geol. Rev.* 2024; *Revision Submitted to Journal*.
65. Ma, Y.C.; Xu, X.W.; Hong, T.; Jin, W.K.; Li, H.; Yang, Z.Q.; Liu, S.K.; Kang, K.; Wang, X.H.; Niu, L. Multiphase evolution of a Li-pegmatite field from the Tashisayi area, Altyn Tagh, NW China: Insights from a petrological, geochemical, and geochronological study. *Miner. Depos.* **2023**, *59*, 1–22. [CrossRef]
66. Xiao, P.X. The geological features of Suwushijie rockbody (group) in Altyn Tagh middle-zone. *Northwestern Geol.* **2003**, *36*, 14–23.
67. Liu, C.H. Temporal-Spatial Distribution and Tectonic Evolution of the Granitoids in the Altun Orogenic Belt. Ph.D. Thesis, Chinese Academy of Geological Science, Beijing, China, 2014. (In Chinese with English abstract).
68. Sobel, E.R.; Arnaud, N. A possible middle Paleozoic suture in the Altyn Tagh, NW China. *Tectonics* **1999**, *18*, 64–74. [CrossRef]
69. Gao, Y.B.; Zhao, X.M.; Wang, B.; Zhang, J.W.; Jin, M.S.; Yang, S.F.; Yan, Z.Q.; Teng, J.X.; Zhao, H.B.; Chao, Y.Y. Geological characteristics of fluorite deposits and ore-controlling granites, and prospecting potential of the super-large Kaerqiaer-Kumutashi fluorite mineralization belt in West Altyn-Tagh, China. *China Geol.* **2022**, *50*, 1–32. (In Chinese with English abstract)
70. Zhang, L.; Jiang, S.Y.; Romer, R.L.; Zhang, H.X.; Wan, S.L. Characteristics and formation of rare-metal pegmatites and granites in the Duanfengshan-Guanyuan district of the northern Mufushan granite complex in South China. *Lithos* **2024**, *474*, 107579. [CrossRef]
71. Xu, Z.Q.; Zheng, B.H.; Zhu, W.B.; Chen, Y.J.; Li, G.W.; Gao, J.G.; Che, X.D.; Zhang, R.Q.; Wei, H.Z.; Li, W.Q.; et al. Geologic scenario from granitic sheet to Li-rich pegmatite uncovered by Scientific Drilling at the Jiajika lithium deposit in eastern Tibetan Plateau. *Ore Geol. Rev.* **2023**, *161*, 105636. [CrossRef]
72. Melcher, F.; Sitnikova, M.A.; Graupner, T.; Nicola, M.; Thomas, O.; Friedhelm, H.K.; Eike, G.; Axel, G.; Helene, B.; Don, W.D.; et al. Fingerprinting of conflict minerals: Columbite-tantalite ('coltan') ores. *SGA News* **2008**, *23*, 1–14.
73. Che, X.D.; Wu, F.Y.; Wang, R.C.; Axel, G.; Ji, W.Q.; Zhao, Z.H.; Yang, J.H.; Zhu, Z.Y. In situ U-Pb isotopic dating of columbite-tantalite by LA-ICP-MS. *Ore Geol. Rev.* **2015**, *65*, 979–989. [CrossRef]
74. Thompson, J.; Meffre, S.; Danyushevsky, L. Impact of air, laser pulse width and fluence on U-Pb dating of zircons by LA-ICPMS. *J. Anal. At. Spectrom.* **2018**, *33*, 221–230. [CrossRef]
75. Ludwig, K.R. *ISOPLOT 3.00: A Geochronological Toolkit for Microsoft Excel*; Special Publication; Berkeley Geochronology Center: Berkeley, CA, USA, 2003; Volume 4, pp. 1–71.
76. Sun, S.S.; McDonough, W.F. Chemical and isotopic systematics of oceanic basalts: Implications for mantle composition and processes. *Geol. Soc. Lond. Spec. Publ.* **1989**, *42*, 313–345. [CrossRef]

77. Hoskin, P.W.O. Trace-element composition of hydrothermal zircon and the alteration of Hadean zircon from the Jack Hills, Australia. *Geochim. Cosmochim. Acta* **2005**, *69*, 637–648. [CrossRef]
78. McDonough, W.F.; Sun, S.S. The composition of the Earth. *Chem. Geol.* **1995**, *120*, 223–253. [CrossRef]
79. Romer, R.L.; Wright, J.E. U-Pb dating of columbites: A geochronologic tool to date magmatism and ore deposits. *Geochim. Cosmochim. Acta* **1992**, *56*, 2137–2142. [CrossRef]
80. Baumgartner, R.; Romer, R.L.; Moritz, R.; Sallet, R.; Chiaradia, M. Columbite–tantalite-bearing granitic pegmatites from the Seridó Belt, northeastern Brazil: Genetic constraints from U–Pb dating and Pb isotopes. *Can. Mineral.* **2006**, *44*, 69–86. [CrossRef]
81. Linnen, R.L.; Cuney, M. Granite-related rare-element deposits and experimental constraints on Ta-Nb-W-Sn-Zr-Hf mineralization. In *Rare-Element Geochemistry and Mineral Deposits*; Linnen, R.L., Samson, I.M., Eds.; Geological Association of Canada, GAC (Short Course): St. John's, NF, Canada, 2005.
82. Badanina, E.V.; Sitnikova, M.A.; Gordienko, V.V.; Melcher, F.; Gäbler, H.E.; Lodziak, J.; Syritso, L.F. Mineral chemistry of columbite-tantalite from spodumene pegmatite of Kolmozero, Kola Peninsula (Russia). *Ore Geol. Rev.* **2015**, *64*, 720–735. [CrossRef]
83. Melcher, F.; Graupner, T.; Gäbler, H.E.; Sitnikova, M.; Oberthür, T.; Gerdes, A.; Badanina, E.; Chudy, T. Mineralogical and chemical evolution of tantalum–(niobium–tin) mineralisation in pegmatites and granites. Part 2: Worldwide examples (excluding Africa) and an overview of global metallogenetic patterns. *Ore Geol. Rev.* **2017**, *89*, 946–987. [CrossRef]
84. Konzett, J.; Schneider, T.; Nedyalkova, L.; Hauzenberger, C.; Melcher, F.; Gerdes, A.; Whitehouse, M. Anatectic granitic pegmatites from the Eastern Alps: A case of variable rare-metal enrichment during high-grade regional metamorphism–I: Mineral assemblages, geochemical characteristics, and emplacement ages. *Can. Mineral.* **2018**, *56*, 555–602. [CrossRef]
85. Geisler, T.; Schaltegger, U.; Tomaschek, F. Re-equilibration of zircon in aqueous fluids and melts. *Elements* **2007**, *3*, 43–50. [CrossRef]
86. Sun, W.B.; Li, H. Research progress on zircon from pegmatites and insights into rare-metal mineralization—A review. *Earth Sci. Front.* **2023**, *30*, 171–184. (In Chinese with English abstract)
87. Fabricio-Silva, W.; Rosière, C.A.; Bühn, B. The shear zone-related gold mineralization at the Turmalina deposit, Quadrilátero Ferrífero, Brazil: Structural evolution and the two stages of mineralization. *Miner. Depos.* **2019**, *54*, 347–368. [CrossRef]
88. Wang, T.; Tong, Y.; Jahn, B.; Zou, T.R.; Wang, Y.B.; Hong, D.W.; Han, B.F. SHRIMP U–Pb Zircon geochronology of the Altai No. 3 Pegmatite, NW China, and its implications for the origin and tectonic setting of the pegmatite. *Ore Geol. Rev.* **2007**, *32*, 325–336. [CrossRef]
89. Wang, C.; Liu, L.; Yang, W.Q.; Zhu, X.H.; Cao, Y.T.; Kang, L.; Chen, S.F.; Li, R.S.; He, S.P. Provenance and ages of the Altyn Complex in Altyn Tagh: Implications for the early Neoproterozoic evolution of northwestern China. *Precambrian Res.* **2013**, *230*, 193–208. [CrossRef]
90. Zhang, K. Geochronology, Geochemistry of Early Paleozoic Igneous Rocks in Tashisayi Area, South Altyn Tagh and Its Geological Signification. Ph.D. Thesis, Northwest University, Xi'an, China, 2018. (In Chinese with English abstract).
91. Cowgill, E.; Yin, A.N.; Harrison, T.M. Reconstruction of the Altyn Tagh fault based on U-Pb geochronology: Role of back thrusts, mantle sutures, and heterogeneous crustal strength in forming the Tibetan Plateau. *J. Geophys. Res. Solid Earth* **2003**, *108*, 7.
92. Qi, X.X.; Li, H.B.; Wu, C.L.; Yang, J.S.; Zhang, J.X.; Meng, F.C.; Shi, R.D.; Chen, S.Y. SHRIMP U-Pb zircon dating for Qiashikansayi granodiorite, the northern Altyn Tagh mountains and its geological implications. *Chin. Sci. Bull.* **2005**, *50*, 440–445. [CrossRef]
93. Hao, J.; Wang, E.Q.; Liu, X.H.; Sang, H.Q. Jinyanshan collisional orogenic belt of the early Paleozoic in the Altun mountains: Evidence from single zircon U-Pb and 40Ar/39Ar isotopic dating for the arc magmatite and ophiolitic mélange. *Acta Petrol. Sin.* **2006**, *22*, 10. (In Chinese with English abstract)
94. Wu, C.L.; Yao, S.Z.; Zeng, L.S.; Yang, J.S.; Wooden, J.L.; Chen, S.Y.; Frank, K.M. Bashikaogong-Shimierbulake granitic complex, north Altun, NW China: Geochemistry and zircon SHRIMP ages. *Sci. China Ser. D Earth Sci.* **2006**, *49*, 1233–1251. [CrossRef]
95. Wu, C.L.; Yang, J.S.; Robinson, P.T.; Wooden, J.L.; Frank, K.M.; Gao, Y.H.; Wu, S.P.; Chen, Q.L. Geochemistry, age and tectonic significance of granitic rocks in north Altun, northwest China. *Lithos* **2009**, *113*, 423–436. [CrossRef]
96. Yang, Z.J. *Early Palaeozoic Tectonic Evolution in HongLiuGou, Altyn, Xinjiang*; Chinese Academy of Geological Science: Beijing, China, 2012; (In Chinese with English abstract).
97. Zhang, Z.W.; Huang, G.; Li, H.M.; Zhang, W.F. LA-ICP-MS zircon U-Pb geochronology and geochemistry of gabbro and diorite from Qilesayi pluton in Lapeiquan area of northern Altun Mountains and their tectonic implications. *Acta Petrol. Mineral.* **2012**, *31*, 13–27. (In Chinese with English abstract)
98. Han, F.B.; Chen, B.L.; Cui, L.L.; Wang, S.X.; Chen, Z.L.; Jiang, R.B.; Li, L.; Qi, W.X. Zircon SHRIMP U-Pb age of intermediate-acid intrusive rocks in Kaladawan area, eastern Altun Mountains, NW China, and its implications. *Acta Petrol. Sin.* **2012**, *28*, 2277–2291. (In Chinese with English abstract)
99. Wu, D.; Zheng, K.; Wu, C.L.; Chen, H.J.; Gao, D.; Zou, F.H. Geochronology and geochemistry of the Bashikaogong S-type granitic rocks: A record of Early Paleozoic subduction and collision in North Altun, Northwestern China. *Arab. J. Geosci.* **2021**, *14*, 1327. [CrossRef]
100. Meng, L.T.; Chen, B.L.; Zhao, N.N.; Wu, Y.; Zhang, W.G.; He, J.T.; Wang, B.; Han, M.M. The distribution, geochronology and geochemistry of early Paleozoic granitoid plutons in the North Altun orogenic belt, NW China: Implications for the petrogenesis and tectonic evolution. *Lithos* **2017**, *268*, 399–417. [CrossRef]

101. Yu, S.Y.; Zhang, J.X.; Li, S.Z.; Sun, D.Y.; Peng, Y.B.; Zhao, X.L. Continuity of the North Qilian and North Altun orogenic belts of NW China: Evidence from newly discovered Palaeozoic low-Mg and high-Mg adakitic rocks. *Geol. Mag.* **2018**, *155*, 1684–1704. [CrossRef]
102. Xinjiang Bureau of Geology and Mineral Resources Exploration and Development. 1:250,000 Geological Map of the Suwushijie Area. 2003; 1–275, unpublished. (In Chinese)
103. Kang, L. Early Paleozoic Multi-Stage Granitic Magmatism and the Geological Significance in the South Altyn Tagh HP-UHP Metamorphic Belt. Ph.D. Thesis, Northwest University, Xi'an, China, 2014. (In Chinese with English abstract).
104. Zhang, R.Y.; Zeng, Z.C.; Zhu, W.P.; Chen, N.; Zhao, J.L.; Li, Q.; Wang, Q.W.; Rao, J.W. LA-ICP-MS zircon U-Pb dating, geochemical features and their geological implications of Paxialayidang plutons on the southern margin of Altyn Tagh. *Geol. Rev.* **2016**, *62*, 1283–1299. (In Chinese with English abstract)
105. Huang, Y.Q.; Wu, M.Q.; Germain, B.; Yu, H.C.; Qiao, B.X.; Zhao, Z.G.; Qiu, K.F. Geodynamic setting and ore formation of the Younusisayi thorium deposit in the Altyn orogenic belt, NW China. *Ore Geol. Rev.* **2022**, *140*, 104552. [CrossRef]
106. Patiño Douce, A.E.; Johnston, A.D. Phase equilibria and melt productivity in the pelitic system: Implications for the origin of peraluminous granitoids and aluminous granulites. *Contrib. Mineral. Pet.* **1991**, *107*, 202–218. [CrossRef]
107. Castro, A.; Patiño Douce, A.E.; Corretgé, L.G.; Rosa, J.D.; El Biad, M.; El-Hmidi, H. Origin of peraluminous granites and granodiorites, Iberian massif, Spain: An experimental test of granite petrogenesis. *Contrib. Mineral. Petrol.* **1999**, *135*, 255–276. [CrossRef]
108. Watson, E.B.; Capobianco, C.J. Phosphorus and the rare earth elements in felsic magmas: An assessment of the role of apatite. *Geochim. Cosmochim. Acta* **1981**, *45*, 2349–2358. [CrossRef]
109. Wolf, M.B.; London, D. Apatite dissolution into peraluminous haplogranitic melts: An experimental study of solubilities and mechanisms. *Geochim. Cosmochim. Acta* **1994**, *58*, 4127–4145. [CrossRef]
110. Li, X.H.; Li, W.X.; Li, Z.X. On the genetic classification and tectonic implication of the Early Yanshanian granitoids in the Nanling rang. South China. *Chin. Sci. Bull.* **2007**, *52*, 981–992. (In Chinese with English abstract)
111. Chappell, B.W.; White, A.J.R. I-and S-type granites in the Lachlan Fold Belt. *Earth Environ. Sci. Trans. R. Soc. Edinb.* **1992**, *83*, 1–26.
112. Clemens, J.D. S-type granitic magmas-petrogenetic issues, models and evidence. *Earth-Sci. Rev.* **2003**, *61*, 1–18. [CrossRef]
113. Maniar, P.D.; Piccoli, P.M. Tectonic discrimination of granitoids. *Geol. Soc. Am. Bull.* **1989**, *101*, 635–643. [CrossRef]
114. Peccerillo, A.; Taylor, S.R. Geochemistry of Eocene Calc-alkaline volcanic rocks from the Kastamonu Area, Northern Turkey. *Contrib. Mineral. Petrol.* **1976**, *56*, 63–81. [CrossRef]
115. Jahns, R.H.; Burnham, C.W. Experimental studies of pegmatite genesis; I, A model for the derivation and crystallization of granitic pegmatites. *Econ. Geol.* **1969**, *64*, 843–864. [CrossRef]
116. Shearer, C.K.; Papike, J.J.; Redden, J.A.; Simon, S.B.; Walker, R.J.; Laul, J.C. Origin of pegmatitic granite segregations, Willow Creek, Black Hills, South Dakota. *Can. Mineral.* **1987**, *25*, 159–171.
117. London, D. Internal differentiation of rare-element pegmatites: A synthesis of recent research. *Geol. Soc. Am. Spec. Pap.* **1990**, *246*, 35–50.
118. Webster, J.D.; Thomas, R.; Rhede, D.; Förster, H.J.; Seltmann, R. Melt inclusions in quartz from an evolved peraluminous pegmatite: Geochemical evidence for strong tin enrichment in fluorine-rich and phosphorus-rich residual liquids. *Geochim. Cosmochim. Acta* **1997**, *61*, 2589–2604. [CrossRef]
119. Simmons, W.B.; Foord, E.E.; Falster, A.U.; King, V.T. Evidence for an anatectic origin of granitic pegmatites, western Maine, USA. *Geol. Soc. Am. Abstr. Programs* **1995**, *27*, A411.
120. Shaw, R.A.; Goodenough, K.M.; Roberts, N.M.W.; Horstwood, M.S.A.; Chenery, S.R.; Gunn, A.G. Petrogenesis of rare-metal pegmatites in high-grade metamorphic terranes: A case study from the Lewisian Gneiss Complex of north-west Scotland. *Precambrian Res.* **2016**, *281*, 338–362. [CrossRef]
121. Müller, A.; Romer, R.L.; Pedersen, R.B. The Sveconorwegian Pegmatite Province—Thousands of pegmatites without parental granites. *Can. Mineral.* **2017**, *55*, 283–315. [CrossRef]
122. Linnen, R.L.; Keppler, H. Melt composition control of Zr/Hf fractionation in magmatic processes. *Geochim. Cosmochim. Acta* **2002**, *66*, 3293–3301. [CrossRef]
123. Zaraisky, G.P.; Aksyuk, A.M.; Devyatova, V.N.; Udoratina, O.V.; Chevychelov, V.Y. Zr/Hf ratio as an indicator of fractionation of rare-metal granites by the example of the Kukulbei complex, eastern Transbaikalia. *Petrology* **2008**, *16*, 7710–7736. [CrossRef]
124. Černý, P.; Meintzer, R.E.; Anderson, A.J. Extreme fractionation in rare-element granitic pegmatites: Selected examples of data and mechanisms. *Can. Mineral.* **1985**, *23*, 381–421.
125. Ballouard, C.; Poujol, M.; Boulvais, P.; Branquet, Y.; Tartèse, R.; Vigneresse, J.L. Nb-Ta fractionation in peraluminous granites: A marker of the magmatic hydrothermal transition. *Geology* **2016**, *44*, 231–234. [CrossRef]
126. Knoll, T.; Huet, B.; Schuster, R.; Mali, H.; Ntaflos, T.; Hauzenberger, C. Lithium pegmatite of anatectic origin-A case study from the Austroalpine Unit Pegmatite Province (Eastern European Alps): Geological data and geochemical model. *Ore Geol. Rev.* **2023**, *154*, 105298. [CrossRef]
127. Dill, H.G. Geology and chemistry of Variscan-type pegmatite systems (SE Germany)–with special reference to structural and chemical pattern recognition of felsic mobile components in the crust. *Ore Geol. Rev.* **2018**, *92*, 205–239.
128. Zhang, H.; Lv, Z.H.; Tang, Y. Metallogeny and prospecting model as well as prospecting direction of pegmatite-type rare metal ore deposits in Altay orogenic belt, Xinjiang. *Miner. Depos.* **2019**, *38*, 792–814. (In Chinese with English abstract)

129. Xu, Z.Q.; Li, S.T.; Zhang, J.X.; Yang, J.S.; He, B.Z.; Li, H.B.; Lin, C.S.; Cai, Z.H. Paleo-Asian and Tethyan tectonic systems with docking the Tarim block. *Acta Petrol. Sin.* **2011**, *27*, 1–22. (In Chinese with English abstract)
130. Şengör, A.M.C.; Natal'In, B.A.; Burtman, V.S. Evolution of the Altaid tectonic collage and Palaeozoic crustal growth in Eurasia. *Nature* **1993**, *364*, 299–307.
131. Pan, G.; Wang, L.; Li, R.; Yuan, S.H.; Ji, W.H.; Yin, F.G.; Zhang, W.P.; Wang, B.D. Tectonic evolution of the Qinghai-Tibet plateau. *J. Asian Earth Sci.* **2012**, *53*, 3–14.
132. Gao, X.F.; Xiao, P.X.; Guo, L.; Dong, Z.C.; Xi, R.G. Opening of an early Paleozoic limited oceanic basin in the northern Altyn area: Constraints from plagiogranites in the Hongliugou-Lapeiquan ophiolitic mélange. *Sci. China Earth Sci.* **2011**, *54*, 1871–1879. (In Chinese with English abstract) [CrossRef]
133. Gai, Y.S.; Liu, L.; Kang, L.; Yang, W.Q.; Liao, X.Y.; Wang, Y.W. The origin and geologic significance of plagiogranite in ophiolite belt at North Altyn Tagh. *Acta Petrol. Sin.* **2015**, *31*, 2549–2565. (In Chinese with English abstract)
134. Zhang, J.X.; Meng, F.C.; Yu, S.Y.; Chen, W.; Chen, S.Y. ^{39}Ar-^{40}Ar geochronology of high-pressure/low-temperature blueschist and eclogite in the North Altyn Tagh and their tectonic implications. *Geol. China* **2007**, *34*, 558–564. (In Chinese with English abstract)
135. Zhang, J.X.; Meng, F.C.; Yu, S.Y. Two contrasting HP/LT and UHP metamorphic belts: Constraint on Early Paleozoic orogeny in Qilian-Altun orogen. *Acta Petrol. Sin.* **2010**, *26*, 1967–1992. (In Chinese with English abstract)
136. Chen, B.L.; Li, S.B.; Jiang, R.B.; Chen, Z.L.; Han, F.B.; Cui, L.L.; Li, L.; Zhao, S.M.; Qi, W.X.; Yang, Y.; et al. Zircon SHRIMP U-Pb dating of intermediate-felsic volcanic rocks from the Kaladawan area, Altun Mountains and its tectonic environment. *Acta Geol. Sin.* **2016**, *90*, 708–727. (In Chinese with English abstract)
137. Meng, L.T.; Chen, B.L.; Luo, D.K.; Wang, Y.; Sun, Y.; Wu, Y.; Zhang, H.; Wang, T. SHRIMP Zircon U-Pb Geochronology of Northern 4337Highland Granodiorite in Kaladawan Area of Northern Altun Mountains and Its Tectonic Implications. *J. Jilin Univ. Earth Sci. Ed.* **2015**, *45*, 1757–1771. (In Chinese with English abstract)
138. Liu, J.H. The Associations, Geochronology and Geochemistry of Igneous Rocks in Lapeiquan, North Altyn Tagh and Its Tectonic Significance. Ph.D. Thesis, Northwest University, Xi'an, China, 2017. (In Chinese with English abstract).
139. Wang, C.M.; Tang, H.S.; Zheng, Y. Early Paleozoic magmatism and metallogeny related to Proto-Tethys subduction: Insights from volcanic rocks in the northeastern Altyn Mountains, NW China. *Gondwana Res.* **2019**, *75*, 134–153. [CrossRef]
140. Zheng, K.; Wu, C.; Lei, M.; Zhang, X.; Chen, H.J.; Wu, D.; Gao, D. Petrogenesis and tectonic implications of granitoids from western North Altun, Northwest China. *Lithos* **2019**, *340*, 255–269. [CrossRef]
141. Ye, X.T.; Zhang, C.L.; Wang, Q.; Wang, G.D. Subduction initiation of Proto-Tethys Ocean and back-arc extension in the northern Altun Mountains, northwestern China: Evidence from high-Mg diorites and A-type rhyolites. *Lithos* **2020**, *376*, 105748. [CrossRef]
142. Wu, Y.; Chen, Z.L.; Chen, B.L.; Wang, Y.; Sun, Y.; Meng, L.T.; He, J.T.; Wang, B. Early Palaeozoic syn-collisional granitic magmatism in the North Altun and its implication for accretionary orogenic processes. *Acta Petrol. Sin.* **2021**, *37*, 1321–1346. (In Chinese with English abstract)
143. Liu, Q.; Tsunogae, T.; Zhao, G.C.; Li, J.H.; Yao, J.L.; Han, Y.G.; Wang, P. Multiphase ophiolite formation in the Northern Altyn Tagh Orogen, southeastern Tarim. *Am. J. Sci.* **2021**, *321*, 788–821. [CrossRef] [PubMed]
144. Liu, L.; Sun, Y.; Luo, J.H.; Wang, Y.; Chen, D.; Zhang, A.D. Ultra-high pressure metamorphism of granitic gneiss in the Yinggelisayi area, Altun Mountains, NW China. *Sci. China (Ser. D Earth Sci.)* **2004**, *47*, 338–346. [CrossRef]
145. Liu, L.; Zhang, J.; Green, H.W., II; Jin, Z.; Bozhilov, K.N. Evidence of former stishovite in metamorphosed sediments, implying subduction N350 km. *Earth Planet. Sci. Lett.* **2007**, *263*, 180–191. [CrossRef]
146. Wang, C.; Liu, L.; Chen, D.; Cao, Y.T. Petrology, geochemistry, geochronology, and metamorphic evolution of garnet peridotites from South Altyn Tagh UHP terrane, northwestern China: Records related to crustal slab subduction and exhumation history. In *Ultrahigh-Pressure Metamorphism*; Elsevier: Amsterdam, The Netherlands, 2011; pp. 541–577.
147. Cao, Y.T. HP-UHP Metamorphic Evolution and Fluid Activity from the South Altyn Tagh and Shenglikou Area. Ph.D. Thesis, Northwest University, Xi'an, China, 2013. (In Chinese with English abstract).
148. Dong, J.; Wei, C.; Chen, J.; Zhang, J.X. P–T–t path of garnetites in South Altyn Tagh, West China: A complete record of the ultradeep subduction and exhumation of continental crust. *J. Geophys. Res. Solid Earth* **2020**, *125*, e2019JB018881. [CrossRef]
149. Dong, J.; Wei, C. Multi-stage metamorphism of the South Altyn ultrahigh-pressure metamorphic belt, West China: Insights into tectonic evolution from continental subduction to arc–backarc extension. *J. Petrol.* **2021**, *62*, egab082. [CrossRef]
150. Gao, D. Study on the Petrogenesis and Tectonic Geological Significance of the Yusupuleke Tagh Granitic Composite Intrusion, Southern Altun. Ph.D. Thesis, Chang'an University, Xi'an, China, 2017. (In Chinese with English abstract).
151. Kang, L.; Xiao, P.X.; Gao, X.F.; Gao, X.F.; Xi, R.G.; Yang, Z.C. Age, petrogenesis and tectonic implications of Early Devonian bimodal volcanic rocks in the South Altyn, NW China. *J. Asian Earth Sci.* **2015**, *111*, 733–750. [CrossRef]
152. Wu, S.P.; Wu, C.L.; Chen, Q.L. Characteristics and tectonic setting of the Tula aluminous A- type gr anite at the south side of the Altyn Tagh fault, NW China. *Geol. Bull. China* **2007**, *26*, 1385–1392. (In Chinese with English abstract)
153. Wang, C.; Liu, L.; Zhang, A.D.; Yang, W.Q.; Cao, Y.T. Geochemistry and petrography of Early Paleozoic Yusupuleke Tagh rapakivi-textured granite complex, South Altyn: An example for magma mixing. *Acta Petrol. Sin.* **2008**, *24*, 2809–2819. (In Chinese with English abstract)

154. Wang, C.; Liu, L.; Xiao, P.X.; Cao, Y.T.; Yu, H.Y.; Meert, J.G.; Liang, W.T. Geochemical and geochronologic constraints for Paleozoic magmatism related to the orogenic collapse in the Qimantagh–South Altyn region, northwestern China. *Lithos* **2014**, *202*, 1–20. [CrossRef]
155. Zhang, J.X.; Yu, S.Y.; Mattinson, C.G. Early Paleozoic polyphase metamorphism in northern Tibet, China. *Gondwana Res.* **2017**, *41*, 267–289. [CrossRef]

minerals

MDPI

Article

Characteristics and Metallogenic Significance of Fe-Mn Carbonate Minerals in the Erdaokan Ag Deposit, Heilongjiang Province, Northeast China: Constraints from Sm-Nd Geochronology and Trace Elements

Yuanjiang Yang [1], Chenglu Li [1], Zeyu Wang [2,3,*], Huajuan Gu [1], Wenpeng Yang [1], Maowen Yuan [4], Anzong Fu [1], Bo Zheng [1], Zhaoxun Cheng [1] and Baoshan Liu [2,3]

1 Natural Resources Survey Institute of Heilongjiang Province, Harbin 150036, China; geo1@foxmail.com (Y.Y.); lcl230881@163.com (C.L.); guhuajuan02@163.com (H.G.); orcidme@163.com (W.Y.); tpp_0009@163.com (A.F.); hljdcyzb1987@163.com (B.Z.); czx158840@163.com (Z.C.)

2 Shenyang Geological Survey Center, China Geological Survey, Shenyang 110034, China; liubaoshan1111@163.com

3 Northeast Geological S&T Innovation Center of China Geological Survey, Shenyang 110034, China

4 School of Earth Science and Resources, China University of Geosciences, Beijing 100083, China; 2023020003@cugb.edu.cn

* Correspondence: wangyuzhe@mail.cgs.gov.cn

Abstract: Fe-Mn carbonate is the dominant mineral in the Erdaokan Ag deposit, which represents the first large independent silver deposit during the Late Triassic Period in the Duobaoshan Cu-Mo-Au mineralization concentrated area of Heilongjiang Province, NE China. The Fe-Mn carbonates in the deposit frequently co-exist with Ag minerals. Thus, the presence of Fe-Mn carbonates plays a crucial role in the ore-formation process, making their analysis essential for obtaining valuable metallogenic information about the Erdaokan deposit. Through microexamination, SEM and EDS analysis, a clear relationship between Fe-Mn carbonate minerals and Ag minerals was established. Furthermore, electron probe microanalysis, LA-ICP-MS, and Sr-Nd isotope tests were conducted to analyze Fe-Mn carbonates for significant metallogenic insights. The distribution pattern of trace elements and rare-earth elements in Fe-Mn carbonates is similar, characterized by Zr depletion (below 0.131 ppm), enrichment of light rare-earth elements, a noticeable deficit of Eu (δEu = 0.06–0.63), and an average Y/Ho value of 34.29, indicating the involvement of upper mantle-derived deep magma in the formation of ore-forming materials. The samples had a Sm-Nd isochron age of 233.7 $\pm$ 1.2 Ma, suggesting that the Erdaokan Ag deposit was formed during the Late Triassic Period. This study highlights the significance of Fe-Mn carbonate as a valuable mineral indicator for regional silver prospecting purposes, and confirms the Late Triassic Period as another important metallogenic stage in the Duobaoshan Cu-Mo-Au mineralization concentrated area.

Keywords: Fe-Mn carbonate minerals; Erdaokan Ag deposit; Sm-Nd isotopes; Duobaoshan Cu-Mo-Au mineralization concentrated area; ore-forming fluid evolution

1. Introduction

The Duobaoshan Cu-Mo-Au mineralization concentrated area, situated in the eastern part of the Xing'an–Mongolia orogenic belt, represents a region that has undergone superimposed transformations associated with the Paleo-Asian Ocean, Mongol-Okhotsk Ocean, and Pacific Ocean tectonic domains. This area is characterized by numerous ore deposits exhibiting rich mineral species and complex types, making it a focal point for research on ore prospecting and metallogenic theory in Northeast China. During the Caledonian period, the region was influenced by the subduction of the ancient Asian Ocean (480–470 Ma), resulting in the formation of porphyry-type deposits such as the

Duobaoshan copper deposit [1–3] and the Tongshan copper deposit [4]. During the Indonesian and early Yanshanian periods, the region was influenced by the subduction of the Mongolian Okhotsk Ocean, resulting in the formation of a skarn-type Xiaodubaoshan copper deposit [5] and a hydrothermal Erdaokan silver deposit [6,7], both of which were formed between 240 and 230 Ma, and the skarn-type Sankuanggou iron deposit [8], which was formed between 180 and 170 Ma. Since the middle–late Yanshan period, the region has been mainly affected by the subduction of the Pacific Plate. Epithermal-type deposits such as Zhengguang gold deposits [9], Yongxin gold deposits [10], Sandaowanzi gold deposits [11], and Shangmachang gold deposits [12] were generated between 150 and 100 Ma. Among these deposits, the Erdaokan Ag deposit stands out as a recently discovered large-scale independent silver deposit from the Late Triassic Period [6,7]. Its unique formation age renders it an important research site for understanding the metallogenic mechanisms within both the Duobaoshan Cu-Mo-Au mineralization concentrated area and the CAOB (Central Asian Orogenic Belt). Although some mineralogical studies have been conducted on this deposit, there remains uncertainty regarding its precise origin and the formation processes of the ore-forming fluid [13–15].

Fe-Mn carbonates, commonly encountered in deposits, serve as indicators for the nature, environment, and evolution of ore-forming fluids [16–25]. In our study, Fe-Mn carbonate is a pervasive mineral within the Erdaokan deposit and occurs abundantly in the ore. Microscopically, there is a conspicuous symbiotic relationship between Ag and Fe-Mn carbonate. Therefore, it is imperative to conduct a comprehensive analysis of the mineral composition and elemental distribution characteristics of Fe-Mn carbonates to gain profound insights into the specific ore-forming processes associated with the Erdaokan deposit.

Based on a comprehensive geological investigation of the Erdaokan deposit, this study employed various analytical techniques, including electron probe microanalysis (EPMA), scanning electron microscopy (SEM), energy-dispersive spectrometry (EDS), and laser ablation–inductively coupled plasma–mass spectrometry (LA-ICP-MS), to trace the origin of ore-forming fluids. Additionally, Sm-Nd isotope testing was utilized to obtain the metallogenic age of the Erdaokan deposit. Building upon these experimental findings, we have refined the metallogenic model for the Erdaokan silver deposit to better reflect its actual formation process. The findings confirm that the Late Triassic Period was a significant period for ore-formation in the Duobaoshan Cu-Mo-Au mineralized area, and highlight Fe-Mn carbonate as a promising exploration indicator in this region. Fe-Mn carbonate and Ag minerals such as pyrargyrite and argentite, have an intimate symbiotic relationship, and these findings provide valuable insights into ore-forming element migration, and offer new guidance for prospecting strategies in the Duobaoshan Cu-Mo-Au mineralization concentrated area.

2. Geological Setting

2.1. Regional Geology

The Erdaokan silver deposit belongs to the Duobaoshan Cu-Mo-Au mineralization concentrated area, which is one of the most significant metallogenic regions in Northeast China. It is situated in the eastern part of the Central Asian orogenic belt, and is northeast of the Greater Khingan mountains (Figure 1a). The Greater Khingan mountains, an extensive accretionary orogenic belt connecting Siberia and North China Craton [26,27], are linked to the Pacific Plate in the east and the Mongol-Okhotsk plate in the west. From northwest to southeast, there are four blocks: the Erguna block, the Xing'an block, the Songnen–Zhangguangcailing block, and the Jiamusi block (Figure 1b). This region exhibits a Paleozoic (Ordovician–Silurian, Carboniferous, Permian) arc basin system, along with magmatic arcs and Mesozoic volcano-sedimentary superposition basins. The geological evolution of this area can be divided into two stages: during the Paleozoic period, it was influenced by the Paleo–Asian Marine tectonic system; from the late Paleozoic to the early Mesozoic period, it experienced microcontinent amalgamation, leading to the final closure of the Paleo–Asian Ocean [28]. In addition to these influences, during the Mesozoic, NE–SW trending fault systems formed due to the Circum–Pacific tectonic system, and the Okhotsk tectonic system in Mongolia resulted in the formation of various epithermal and porphyry deposits, such as the Duobaoshan Cu deposit [1–3], Tongshan Cu deposit [4], Sankuanggou Fe deposit [8], Zhengguang Au deposit [9], Yongxin Au deposit [10], Sandaowanzi Au deposit [11], Shangmachang Au deposit [12], and Erdaokan Ag deposit [6,7].

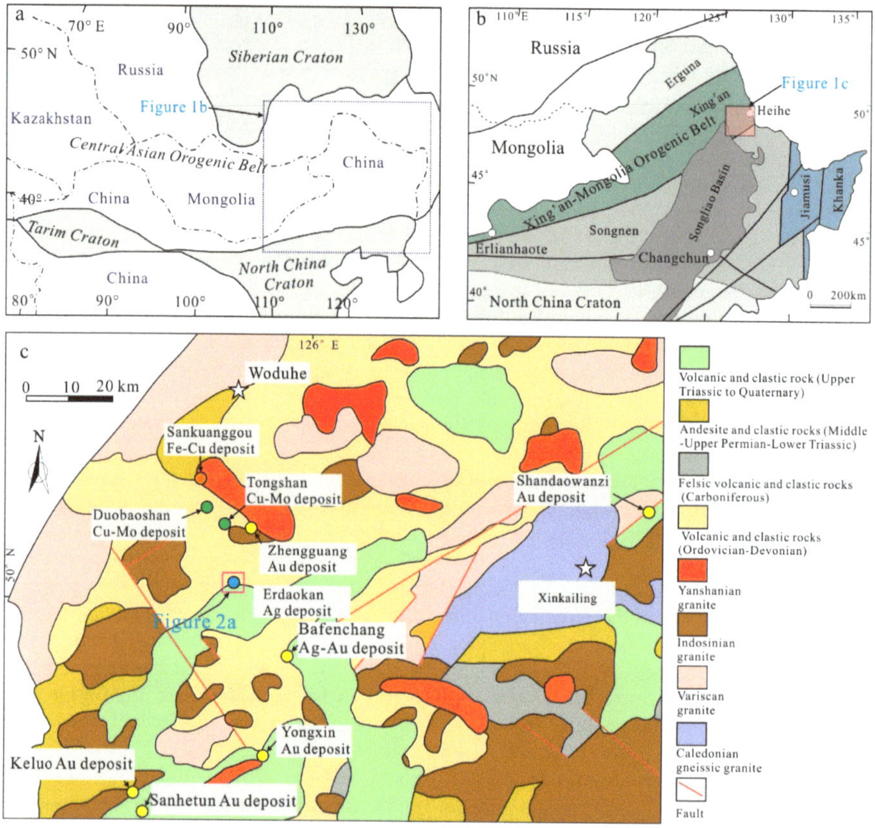

Figure 1. Geological map of Duobaoshan Cu-Mo-Au mineralization concentrated area: (**a**) schematic map of the Central Asian orogenic belt; (**b**) the tectonic map of Northeast China; (**c**) geology and mineral map of Duobaoshan Cu-Mo-Au mineralization concentrated area (modified after reference [29]).

2.2. Deposit Geology

The Erdaokan deposit represents a recent discovery of a large-scale silver (Ag) deposit. The geological units in the area predominantly comprise the Lower Devonian Niqiuhe Formation, which is characterized by the deposition of clastic sedimentary rocks ranging from fine to coarse in shallow marine sedimentary facies during the transgression period. These formations exhibit distinct sericitization and chloritization, highlighting their significance as host rocks for the Ag deposit. Moreover, the mining area encompasses several volcanic formations, including rhyolite and andesite from the Upper Triassic Qingshuihe Formation. Intrusive rocks, such as Middle to Upper Triassic of porphyritic diabase, diorite, and dacite occur as veins, closely associated with the mineralization process.

In the mining area, two Ag ore bodies (Figure 2a) were discovered, exhibiting a steep occurrence of ore (Figure 2b). The first ore body, named I AgB, has a controlling length of 556.27 m and a maximum dip controlling depth of 273.35 m. It possesses a maximum true thickness of 26.75 m, an average true thickness of 6.23 m. The second ore body, referred to as II AgB, spans across a controlling length of 689.39 m with a maximum dip controlling depth of up to 231.64 m. It has a maximum thickness of 40.29 m and an average thickness of approximately 10.20 m. Furthermore, its inferred economic resource quantity of the deposit is estimated to be 1777 t of silver metal content. The highest recorded grade for silver was found to be 25,516.00 g/t, with an average grade of 431.10 g/t. Additionally, the associated Mn ore reserve in Erdaokan deposit amounts to 90.0 × 10^4 t with an average grade of 11.75%, while the associated Pb metal quantity in Erdaokan deposit is 1.68 × 10^4 t with an average grade of 0.75%. Moreover, the associated Zn metal quantity in Erdaokan deposit is 0.16 × 10^4 t with an average grade of 1.15%.

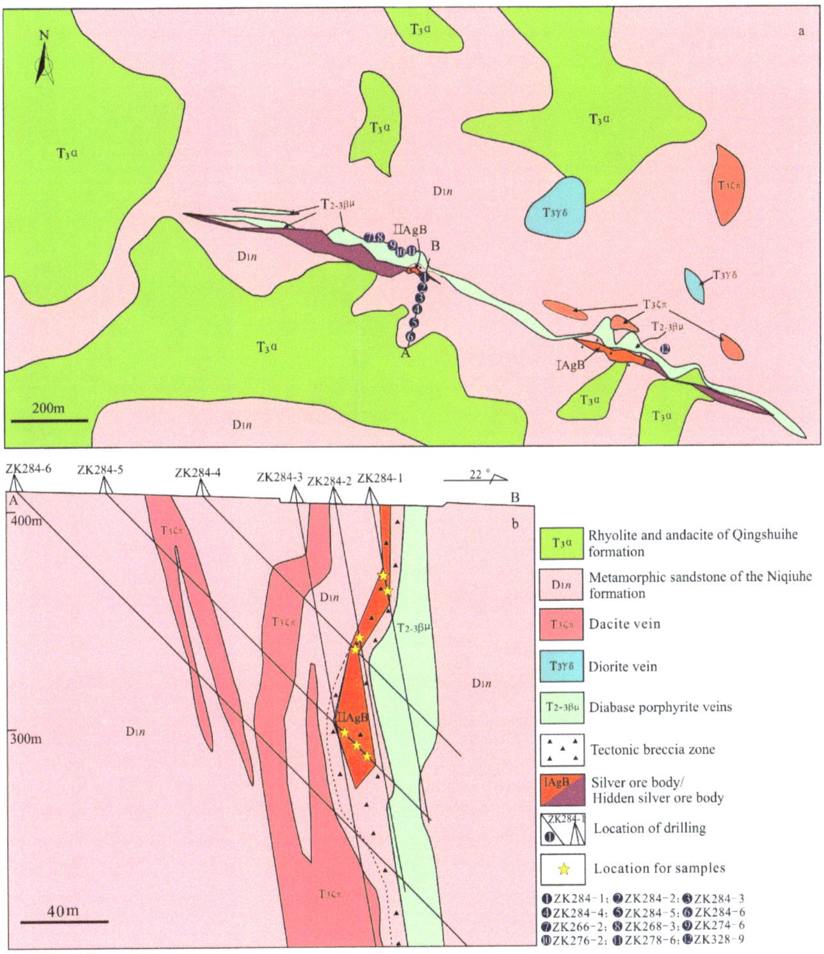

Figure 2. Geological map and prospecting line profile map of the Erdaokan deposit: (**a**) geological map depicting the Erdaokan deposit; (**b**) prospecting line profile map illustrating line 284 (The section in Figure 2b corresponds to the A–B section in Figure 2a).

The Erdaokan Ag deposit predominantly contains pyrargyrite, stephanite, and argentite as its primary silver minerals. Additional silver-bearing minerals include galenite, pyrite, sphalerite, and a small amount of tetrahedrite [13–15]. The total content of silver minerals accounts for 0.074%. Carbonate minerals within the deposit mainly consist of Mn siderite, Fe rhodochrosite, and ankerite. The Fe-Mn carbonates make up 59.06% of the total carbonate present in the deposit. They are distributed in a banded manner (Figure 3a) or nodular manner in the ore (Figure 3b,c), often intersected by quartz veins (Figure 3d) and replaced by siliceous material (Figure 3b–d). Under microscopic examination, rhombic twin structures can be observed (Figure 3e), along with oscillatory zoning patterns (Figure 3e–h), semi-automorphic rhombic aggregates (Figure 3f,g), and residual structures (Figure 3h); Ag and Fe-Mn carbonate minerals have a close symbiotic relationship (Figure 3i). The deposit also contains trace amounts of arsenopyrite and covellite. Additionally, magnetic oxides, including magnetite and hematite (Figure 3a), along with minor quantities of pyrolusite and ferromanganese, can be found. Quartz is the main gangue mineral, followed by feldspar and minor amounts of pyroxene, calcite, muscovite, and hematite.

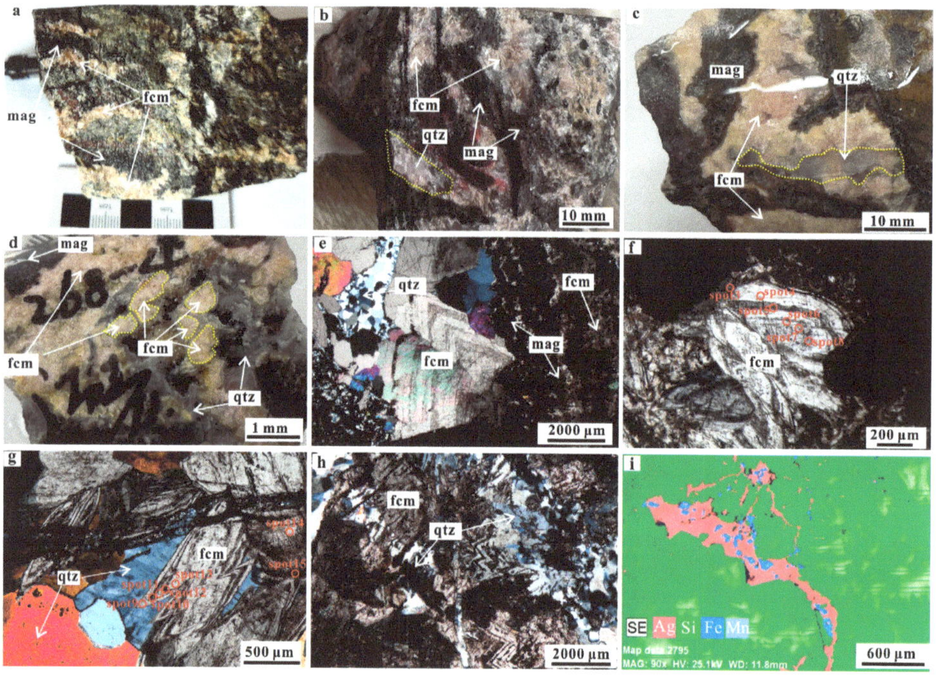

Figure 3. Ore and microscopic images: (**a**) Fe-Mn carbonate minerals (fcm) are observed to be distributed in a banded manner within the ore, which is subsequently replaced by magnetite (mag); (**b**,**c**) nodular distribution of Fe-Mn carbonate minerals can be observed in the ore; (**d**) Quartz (qtz) veins intersect with Fe-Mn carbonate minerals; (**e**) rhombohedral lattice structure of Fe-Mn carbonate minerals is evident; (**f**,**g**) hemidiomorphic rhomboid assembly can be observed; (**h**) residual texture of Fe-Mn carbonate minerals is present; (**i**) the presence of Ag is closely associated with the occurrence of Fe-Mn carbonate minerals. (The red circles indicate the location of in-situ laser testing points, (**e**–**h**) are under crossed polars images).

The alteration in the Erdaokan deposit exhibits distinct spatial zoning, which can be divided into three stages (Figure 4): the sheet-like precipitation of quartz–sulfide minerals, the formation of galenite–sphalerite–Fe–Mn carbonate assemblages, and the occurrence of pyrite–arsenopyrite–carbonate aggregates. In the first stage, well-formed quartz, pyrite, and a small amount of arsenopyrite precipitated with some quartz-filled voids within the pyrite and arsenopyrite. The second stage represents the primary mineralization period characterized by the development of galenite, sphalerite, and copper sulfides. Silver acanthite and pyrargyrite were also observed within the lattice structure of the galenite, sphalerite, and pyrite. Additionally, Fe-Mn carbonates, along with magnetite asphalt calcite hematite, appeared later in this stage. In the third stage, disseminated or vein-like aggregates consisting predominantly of pyrite and arsenopyrite were observed along with abundant calcite precipitation.

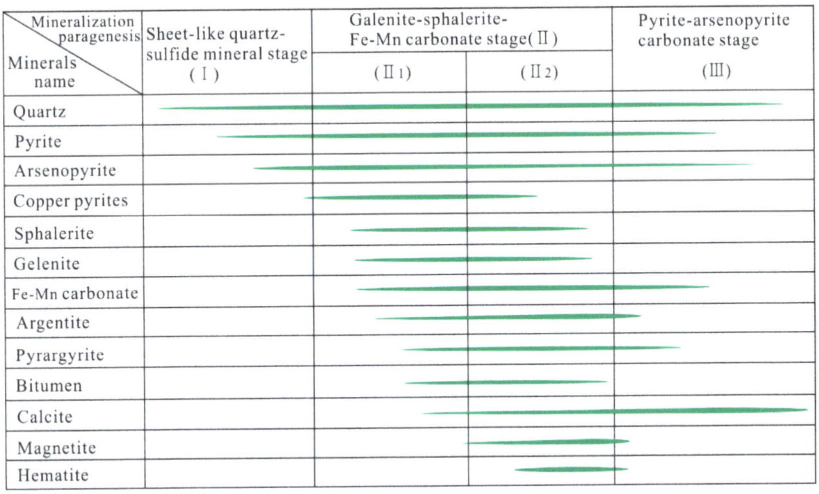

Figure 4. The sequence diagram depicting the hydrothermal mineralization period of the Erdaokan Ag deposit.

3. Materials and Methods

Based on the observation and analysis of rock specimens and microscope samples in this area, this study employed electron probe microanalysis (EPMA), scanning electron microscopy (SEM), and LA-ICP-MS testing to investigate the morphology, distribution patterns, coexistence relationships, elemental distribution, and elemental composition characteristics of minerals at a micro scale. Subsequently, the origin of Fe-Mn carbonate minerals was inferred. Additionally, Sm-Nd isotope analysis was conducted to determine the formation time of Fe-Mn carbonates and to compare it with the deposit's age. These findings will contribute to establishing the ore-forming process of the Erdaokan Ag deposit. Detailed sampling and experimental information are provided below.

The A-B exploration line controls the main ore body, and the drill holes were sequentially numbered as 1, 2, 3, 4, 5, and 6, as shown in Figure 2a. The test samples for EPMA and LA-ICP-MS analysis were collected from four specific drill holes (ZK284-1, ZK284-2, Zk284-4, ZK284-5), with their sampling locations indicated by yellow stars in Figure 2b. A total of thirty testing points were examined using EPMA, while nineteen testing points were analyzed using LA-ICP-MS. Additionally, Sm-Nd isotope samples were obtained from six distinct drill holes (ZK266-2, ZK268-3, ZK274-6, ZK276-2, ZK278-6, and ZK328-9), corresponding to numbers 7, 8, 9, 10, 11, and 12 in Figure 2a.

The EPMA and LA-ICP-MS were conducted at Wuhan Spectrum Analysis Technology Co., Ltd. The EPMA was performed using a JXA-8230 instrument (JEOL Ltd., Tokyo, Japan) with an acceleration voltage of 15 kV, current of 20 nA, and a beam spot diameter of 5 μm. Prior to the experiment, the sample underwent carbon film coating. A total of 30 Fe-Mn carbonate samples were measured using EPMA (Figure 5a). The LA-ICP-MS analyses were carried out using GeoLas HD and Agilent 7900 instruments. The laser energy used was 80 mJ, with a frequency of 5 Hz and a laser beam diameter of 44 μm. For specific analysis conditions and procedures, please refer to the literature [30]. NIST610, BHVO-2G, BIR-1G, and BCR-2G international standard substances were used as trace element calibration standards for this study. ICP MS DATACAL 10.8 software was employed for data processing. A total of 19 measurement points were utilized, and the test minerals and their corresponding locations can be observed in Figures 3f,g and 5a,b. For microscopic morphology observation and elemental composition analysis of minerals, we used an FEI Quanta650 scanning electron microscope (SEM) operating at an acceleration voltage range between 200 and 300 Kv, with magnification of up to 6×10^6 in combination with a Bruker QUANTAX EDS X-ray diffractometer. Sm-Nd isotope analysis was conducted at Nantai Geological Testing and Research Institute in Nanjing using an ISOPROBE-T thermal ionization mass spectrometer for Sm-Nd isochron age measurement. The Nd standard sample ($^{143}Nd/^{144}Nd = 0.511860 \pm 0.000008$, 2σ, $n = 8$) complied with the US La Jolla Nd isotope standard. A correction factor based on $^{146}Nd/^{144}Nd = 0.7219$ was applied to account for mass fractionation. Details regarding the procedure and methodology can be found in references [31,32].

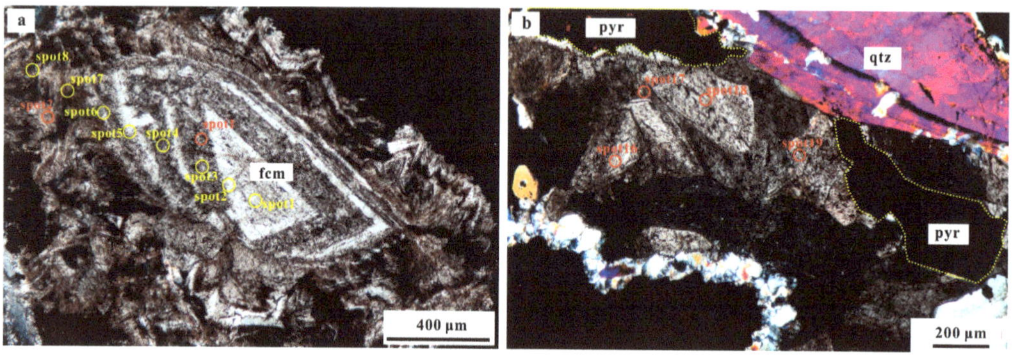

Figure 5. Microscopic image of Fe-Mn carbonate minerals: (**a**) hemidiomorphic ring band structure; (**b**) pyrargyrite (pyr) coexisting with Fe-Mn carbonate minerals. (The yellow circle indicates a portion of the location for in-situ electron probe testing, while the red circle represents the location of in-situ laser testing; Figure 5a,b are under crossed polars images).

4. Results

4.1. Microscopic Analysis of Fe-Mn Carbonate Minerals

The Fe-Mn carbonates constitute 59.06% of the total carbonate content in the deposit, which is consistent with EPMA results indicating that Fe and Mn are the primary components of the carbonate minerals (Figure 6a–d). Atomic absorption spectroscopy analysis of 774 whole rock samples from the Erdaokan Ag deposit revealed a positive correlation between Ag and Mn concentrations (Figure 7). Microscopic observations demonstrated that pyrargyrite was distributed within the Fe-Mn carbonate minerals (Figures 5b and 8a,b), while SEM analysis identified argentite as crystals < 20 μm distribution within these minerals (Figure 8c). Furthermore, EDS mapping confirmed widespread distribution of Ag in areas where Fe and Mn were present, highlighting a close relationship between Ag and Fe-Mn carbonate minerals (Figure 8d).

The EDS image analysis reveals the presence of zoning structures in the Fe-Mn carbonate (Figure 9a), characterized by a gradual transition in the concentrations of Fe and Mn (Figure 9b,c). In the Fe-Mn carbonates, Ca content surpasses that of Mg (Figure 9d,e), while Si is predominantly absent (Figure 9f). Notably, there is a decrease in Mn content from core to rim, accompanied by an increase in both Fe and Mg content. Furthermore, higher concentrations of Ca are observed at the core compared to its outer regions, exhibiting a rhythmic pattern.

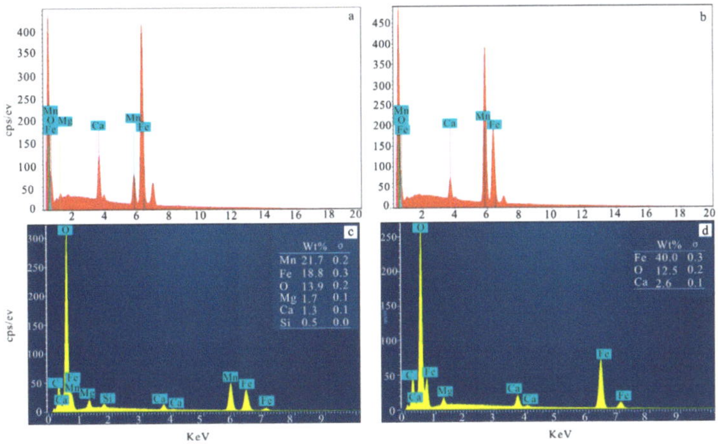

Figure 6. X-ray spectrum of Fe-Mn carbonate: (**a**,**b**) samples from ZK284-1; (**c**) samples from ZK284-4; (**d**) samples from ZK284-5.

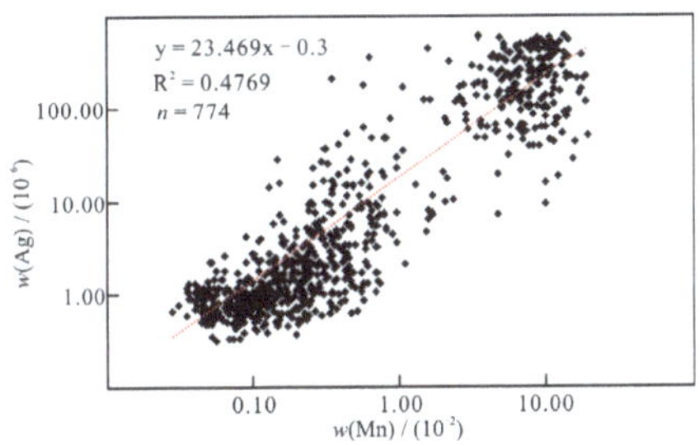

Figure 7. The content relating to Ag and Mn in whole rock samples.

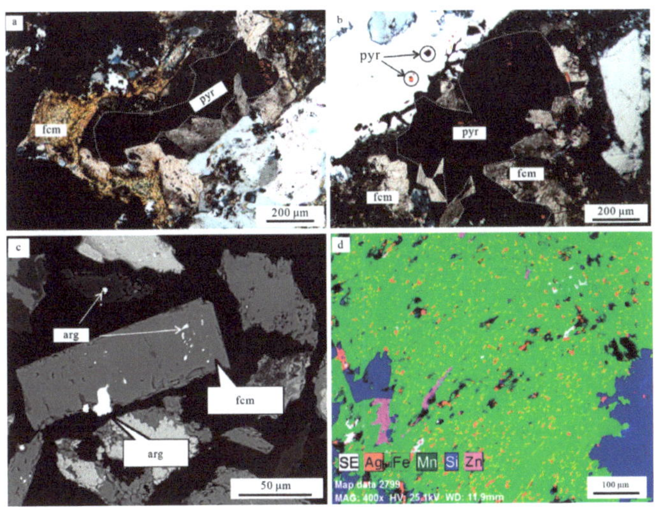

Figure 8. Microscopic images depicting Fe-Mn carbonate minerals: (**a**,**b**) Images illustrating the presence of pyrargyrite (pyr) within Fe-Mn carbonate minerals; (**c**) Image demonstrating the distribution of argentite (arg) in Fe-Mn carbonate; (**d**) Elemental mapping showcasing the distribution of Fe, Mn, Ag, Si, and Zn ((**a**,**b**) are under crossed polars images).

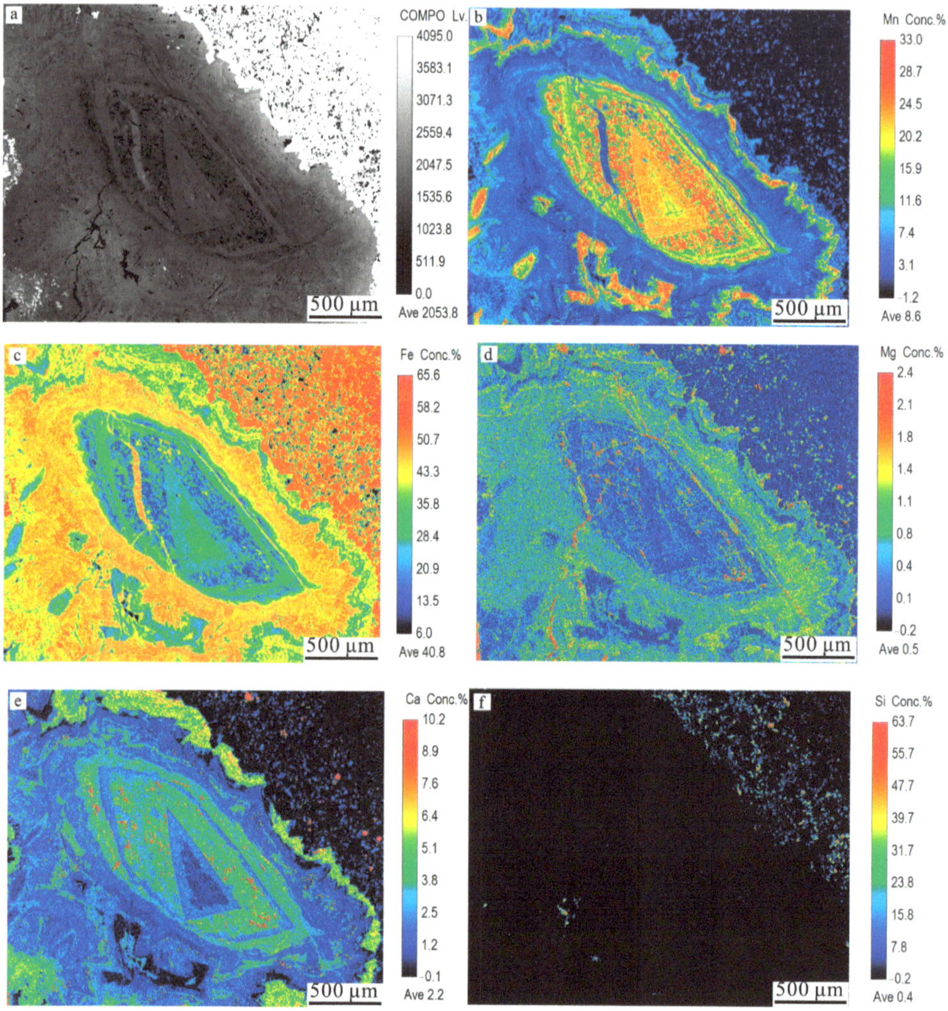

Figure 9. Characterization of Fe-Mn carbonate: (**a**) high-angle annular dark-field (HAADF) image of Fe-Mn carbonate mineral; (**b**–**f**) elemental mapping depicting the spatial distribution of Mn, Fe, Mg, Ca, and Si within this particular sample of Fe-Mn carbonate mineral.

4.2. Analysis of Major and Trace Elements

The major element data obtained by EPMA are presented in Table 1. The composition of Fe-Mn carbonates primarily consists of Fe, Mn, and O, with minor amounts of Ca and Mg, as well as trace amounts of Si (Figure 6a–d, Figure 10a), furthermore, the FeO and MnO content in Fe-Mn carbonate shows continuous transitional changes (Figure 10b). The content of FeO + MnO exhibits a negative correlation with CaO (Figure 11a), while no relationship is observed between CaO and SrO (Figure 11b). The carbonate equivalents for these elements are as follows: $MnCO_3$ = 2.88%–97.53%, $FeCO_3$ = 1.34%–92.39%, $CaCO_3$ = 0.30%–11.26%, $MgCO_3$ = 0.02%–8.03%. Notably, there is a negative correlation between MnO and FeO (Figure 11c), suggesting the possibility of isomorphic replacement between Mn and Fe.

Table 1. Key constituents of Fe-Mn carbonate minerals with a weight percentage of 10^{-2}.

		MnO	FeO	MgO	CaO	BaO	SrO	SiO_2	Al_2O_3	ZnO	Total	$MnCO_3$	$FeCO_3$	$MgCO_3$	$CaCO_3$
Spot1	ZK284-1	25.70	28.43	1.11	2.88	0.05	0.14	1.43	0.11	0.07	59.91	42.90	47.45	1.85	4.81
Spot2	ZK284-1	27.25	27.90	0.40	4.27	0.04	0.05	0.00	0.01	0.07	59.98	45.42	46.51	0.67	7.12
Spot3	ZK284-1	37.05	19.30	0.55	2.01	-	0.10	0.05	0.04	0.01	59.11	62.69	32.66	0.93	3.40
Spot4	ZK284-1	34.53	21.50	0.50	2.14	0.01	0.11	0.00	-	-	58.79	58.72	36.57	0.85	3.64
Spot5	ZK284-1	23.57	32.10	0.57	2.52	-	-	0.02	-	-	58.77	40.10	54.61	0.98	4.28
Spot6	ZK284-1	21.41	36.88	0.57	1.10	0.03	0.02	-	-	-	60.02	35.68	61.45	0.95	1.84
Spot7	ZK284-1	11.14	40.39	4.73	2.65	-	-	-	-	0.04	58.95	18.90	68.52	8.03	4.49
Spot8	ZK284-1	9.09	40.73	3.06	5.36	-	-	-	0.06	-	58.30	15.59	69.86	5.25	9.20
Spot9	ZK284-2	58.36	0.80	0.04	0.52	-	0.01	-	-	0.11	59.84	97.53	1.34	0.06	0.87
Spot10	ZK284-2	55.25	3.25	0.01	0.18	0.05	-	-	-	0.03	58.76	94.02	5.52	0.02	0.30
Spot11	ZK284-2	51.42	4.99	0.28	2.42	-	0.10	-	0.02	-	59.24	86.81	8.43	0.47	4.08
Spot12	ZK284-2	50.08	5.98	0.18	3.45	-	0.05	-	0.01	-	59.75	83.81	10.01	0.30	5.78
Spot13	ZK284-2	48.44	5.88	0.13	5.37	0.02	0.04	0.05	-	0.08	60.00	80.74	9.80	0.22	8.94
Spot14	ZK284-2	47.43	9.54	0.17	3.43	0.04	0.01	0.00	-	0.12	60.74	78.09	15.70	0.28	5.64
Spot15	ZK284-4	46.72	9.75	0.08	3.48	-	-	-	-	0.15	60.18	77.64	16.20	0.13	5.79
Spot16	ZK284-4	44.46	7.51	0.25	6.65	-	-	0.09	-	0.04	58.99	75.37	12.74	0.42	11.26
Spot17	ZK284-4	43.21	11.55	0.32	3.59	0.04	-	0.05	0.05	-	58.79	73.49	19.64	0.55	6.10
Spot18	ZK284-4	37.81	20.26	0.44	1.69	-	-	0.03	-	0.07	60.29	62.71	33.60	0.73	2.80
Spot19	ZK284-4	32.14	24.64	0.24	1.43	0.10	0.03	0.06	-	-	58.65	54.80	42.02	0.42	2.44
Spot20	ZK284-4	30.36	22.62	3.01	2.95	-	0.07	0.06	0.04	-	59.10	51.37	38.27	5.09	4.99
Spot21	ZK284-4	21.48	34.09	1.71	2.10	-	0.04	-	0.00	0.18	59.60	36.03	57.20	2.88	3.52
Spot22	ZK284-4	16.16	38.92	1.04	2.86	0.02	-	0.80	0.08	0.20	60.06	26.90	64.80	1.72	4.76
Spot23	ZK284-4	15.91	40.28	1.42	2.10	0.02	0.03	0.54	0.06	-	60.36	26.36	66.74	2.35	3.49
Spot24	ZK284-4	12.10	45.80	0.91	1.31	0.09	-	0.34	0.09	-	60.63	19.96	75.54	1.50	2.15
Spot25	ZK284-5	5.30	48.60	1.52	4.18	-	-	-	0.02	0.13	59.74	8.87	81.35	2.54	7.00
Spot26	ZK284-5	4.39	49.64	2.04	2.90	0.23	-	0.46	0.06	-	59.71	7.35	83.13	3.41	4.85
Spot27	ZK284-5	3.46	53.85	1.79	0.49	0.05	-	-	-	0.13	59.77	5.79	90.10	2.99	0.83
Spot28	ZK284-5	3.06	56.28	0.35	0.85	0.20	-	0.06	0.09	0.03	60.91	5.02	92.39	0.57	1.40
Spot29	ZK284-5	1.73	53.47	0.71	1.69	-	-	0.06	0.04	0.49	58.19	2.97	91.90	1.22	2.90
Spot30	ZK284-5	1.70	51.40	0.16	5.64	0.07	-	0.02	0.10	0.04	59.12	2.88	86.93	0.27	9.54

The contents of $MnCO_3$, $FeCO_3$, $MgCO_3$, and $CaCO_3$ were determined by extrapolating from the measured contents of MnO, FeO, MgO, and CaO. "-" denotes values below the detection limit of the instrument.

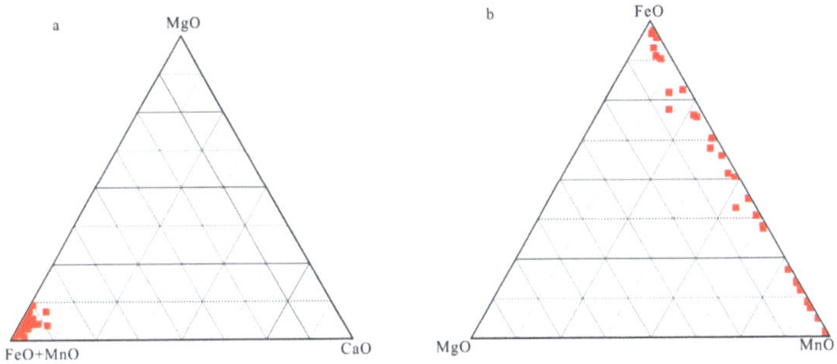

Figure 10. The correlation between the concentrations of MnO, FeO, MgO, and CaO in Fe-Mn carbonate: (**a**) The correlation between the concentration of (FeO + MnO) − CaO − MgO; (**b**) The correlation between the concentrations of MgO − MnO − FeO.

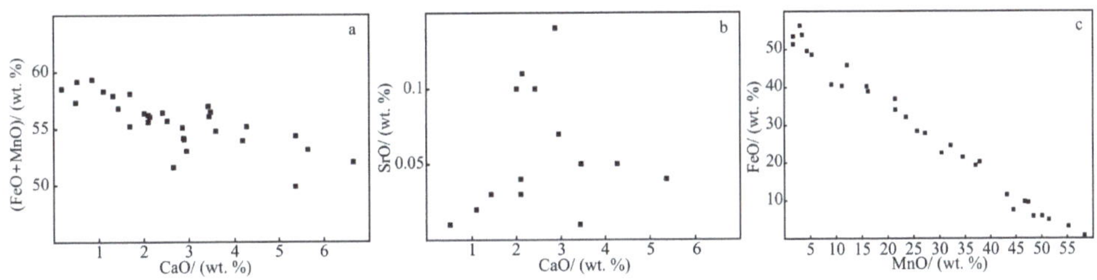

Figure 11. The content correlation of MnO, FeO, CaO, and SrO in Fe-Mn carbonate: (**a**) FeO + MnO vs. CaO; (**b**) SrO vs. CaO; (**c**) FeO vs. MnO.

The trace elements data obtained by LA-ICP-MS are presented in Table 2. Phosphorus ranges from 15.22 to 46.14 ppm, while K ranges from 0.75 to 14.56 ppm, indicating an overall deficit in these elements. Sr content varies from 0.02 to 12.20 ppm, whereas Ti content ranges from 0.03 to 8.95 ppm. It should be noted that the values for Zr and Nb fall below the detection limit of the instrument used in this study. The similar radii of Mn^{2+},

Fe^{2+}, and Mg^{2+} (0.083 nm, 0.083 nm, and 0.072 nm, respectively) facilitate replacement reactions among them, contributing to their comparable distribution patterns observed in siderite, Mn siderite, Fe rhodochrosite, and rhodochrosite samples as shown by the spidergram of trace elements normalized to the primitive mantle (Figure 12a–c). These Fe-Mn carbonate samples exhibit deficits in large ion lithophile elements, such as K and Sr, as well as high field strength elements including Zr, Ti, and Nb.

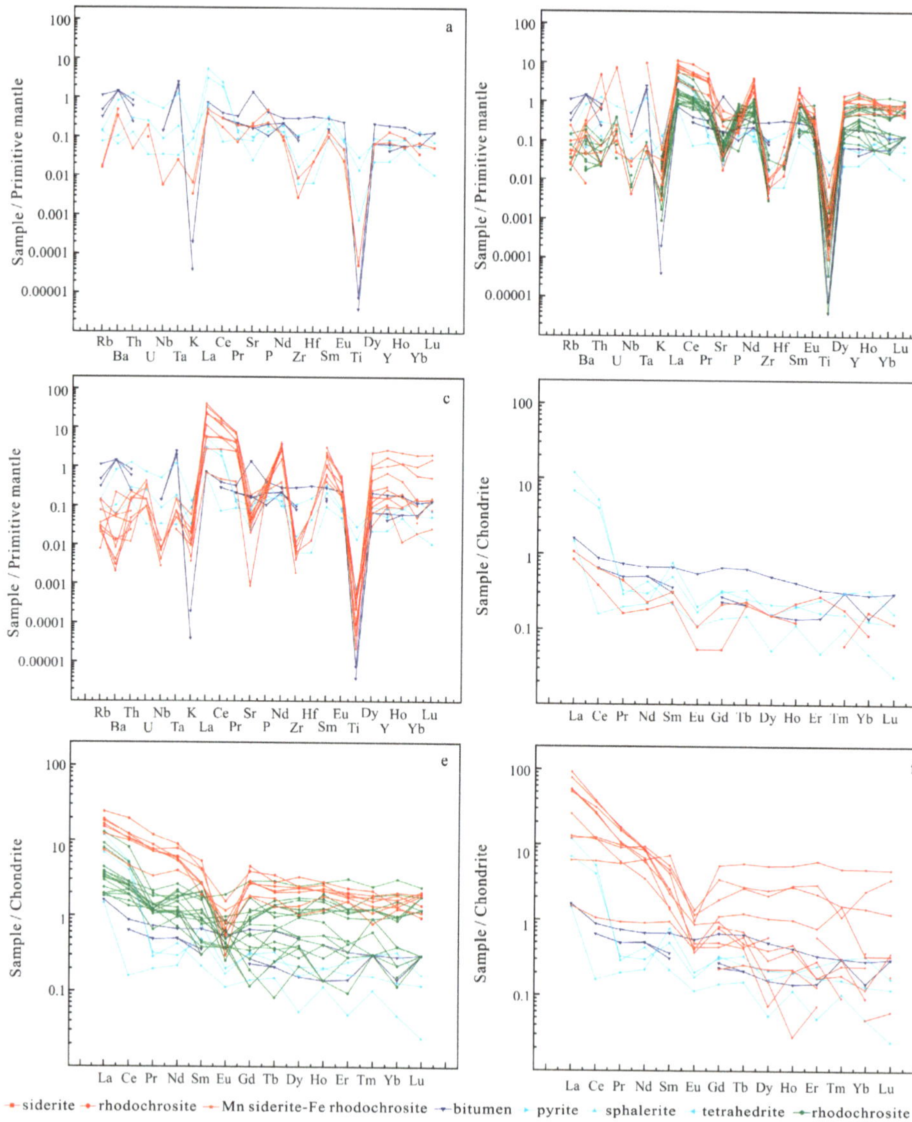

Figure 12. Partitioning of trace elements and rare earth elements is as follows: (**a**) siderite trace element; (**b**) rhodochrosite trace element; (**c**) Fe rhodochrosite and Mn siderite trace element; (**d**) siderite rare earth element; (**e**) rhodochrosite rare earth element; (**f**) Mn siderite-Fe rhodochrosite rare earth element (The data on bitumen is cited [14], the data on pyrite, sphalerite, and tetrahedrite have been cited [15], the rhombosite data, represented by the green lines, are referenced in citation [33]).

Table 2. The LA-ICP-MS data obtained from Fe-Mn carbonate samples.

	Spot1	Spot2	Spot3	Spot4	Spot5	Spot6	Spot7	Spot8	Spot9	Spot10	Spot11	Spot12	Spot13	Spot14	Spot15	Spot16	Spot17	Spot18	Spot19
	ZK284-1	ZK284-1	ZK284-2	ZK284-2	ZK284-2	ZK284-2	ZK284-2	ZK284-2	ZK284-4	ZK284-4	ZK284-4	ZK284-4	ZK284-4	ZK284-4	ZK284-4	ZK284-5	ZK284-5	ZK284-5	ZK284-5
P	45.13	21.54	32.90	15.86	18.60	30.53	15.22	24.73	22.75	32.03	24.23	30.08	24.04	28.66	40.70	43.85	37.80	19.29	46.14
K	0.84	1.61	7.86	5.39	14.36	0.75	4.06	1.80	1.24	14.56	0.95	3.23	4.82	6.28	2.54	5.37	4.16	2.37	2.56
Ti	0.07	0.07	0.77	0.33	0.24	0.45	1.14	0.13	0.07	0.13	0.42	0.39	0.58	0.12	0.72	0.30	8.95	0.03	2.72
Zn	271.19	316.24	39.06	10.30	4.70	7.46	69.61	3.92	7.79	36.02	33.01	40.89	33.19	36.08	35.53	32.97	333.43	43.40	136.20
Rb	0.01	0.01	0.06	0.02	0.15	0.03	0.02	0.02	0.02	0.05	0.01	0.01	0.02	0.02	0.08	0.09	0.02	0.02	0.03
Sr	4.54	3.83	7.36	0.43	0.84	0.96	5.07	0.37	0.02	1.05	0.81	3.92	1.32	0.60	2.11	0.89	12.20	0.78	7.26
Y	0.35	0.62	4.28	8.00	8.28	6.51	5.81	3.66	3.18	0.77	0.50	1.21	1.44	0.45	0.74	12.32	2.33	4.78	4.15
Zr	0.03	0.10	0.12	0.07	-	0.11	0.07	-	0.02	0.09	0.07	0.13	0.05	0.08	0.08	0.11	0.04	0.07	0.05
Nb	-	-	-	-	-	-	-	0.02	0.01	0.01	-	0.01	0.01	-	-	0.01	0.09	-	0.02
Ba	3.34	2.33	0.82	0.10	0.33	0.30	1.14	0.05	0.02	0.34	0.03	1.51	0.39	0.01	0.06	0.40	1.75	0.09	2.16
La	0.26	0.33	2.32	4.05	4.75	5.16	5.74	6.01	0.47	7.92	16.50	23.92	15.45	16.91	28.43	1.90	3.83	3.75	7.55
Ce	0.31	0.51	3.71	9.42	8.71	8.46	9.92	9.78	0.85	9.62	20.60	29.20	25.10	21.47	30.74	4.84	8.06	10.02	15.89
Pr	0.02	0.05	0.41	1.10	0.93	0.93	1.07	0.86	0.11	0.71	1.30	2.08	1.87	1.27	1.98	0.67	0.86	1.19	1.45
Nd	0.11	0.14	2.38	5.62	4.71	3.22	3.44	3.49	0.55	2.10	3.89	5.18	5.07	3.48	4.91	3.75	3.67	4.98	5.35
Sm	0.05	0.06	0.51	1.01	1.04	0.51	0.53	0.37	0.19	0.28	0.27	0.80	0.49	0.49	0.43	1.39	0.80	0.89	0.82
Eu	-	0.01	0.04	0.10	0.02	0.06	0.05	0.09	0.03	-	0.03	0.03	-	0.04	0.06	0.07	0.04	0.09	0.11
Gd	0.01	0.06	0.71	0.89	1.18	0.75	0.69	0.47	0.29	0.06	0.11	0.21	0.20	0.13	0.24	1.35	0.47	0.48	1.00
Tb	0.01	0.01	0.10	0.13	0.13	0.10	0.12	0.06	0.06	0.01	0.03	0.03	0.02	0.02	0.02	0.26	0.08	0.12	0.17
Dy	0.05	0.05	0.75	0.79	1.06	0.67	0.77	0.43	0.34	0.07	0.05	0.09	0.19	0.02	0.13	1.67	0.33	0.68	0.79
Ho	0.01	0.02	0.16	0.20	0.20	0.17	0.19	0.13	0.07	0.02	-	0.03	-	0.02	0.03	0.38	0.09	0.21	0.15
Er	-	0.06	0.37	0.47	0.49	0.43	0.38	0.32	0.16	0.03	0.02	0.04	0.13	0.06	0.04	1.27	0.33	0.62	0.41
Tm	-	0.01	0.04	0.05	0.07	0.06	0.06	0.04	0.05	-	-	0.01	0.01	-	0.01	0.16	0.03	0.04	0.05
Yb	0.04	0.02	0.28	0.30	0.28	0.43	0.34	0.30	0.07	0.08	0.01	0.05	0.02	-	0.03	1.00	0.26	0.51	0.40
Lu	-	-	0.03	0.04	0.06	0.06	0.07	0.07	0.01	-	-	-	-	0.01	0.01	0.15	0.04	0.12	0.05
Hf	0.01	0.01	0.01	-	0.01	0.01	-	0.01	-	0.02	-	0.02	0.02	-	-	-	-	-	0.02
Ta	-	-	-	-	0.39	-	-	-	-	0.01	-	0.01	-	-	-	-	-	-	-
Th	-	-	-	-	-	-	0.01	-	-	-	-	0.01	0.01	0.01	0.01	0.02	0.06	-	0.40
U	-	-	-	0.01	0.01	-	-	-	-	-	-	0.01	-	0.01	-	-	0.15	-	-
Y/Ho	38.93	39.24	27.51	40.58	40.84	39.55	31.31	28.13	44.20	48.19	201.88	42.73	-	29.96	21.63	32.45	27.09	23.28	27.28
ΣREE	0.88	1.32	11.81	24.15	23.63	20.99	23.35	22.41	3.25	20.90	42.80	61.66	48.54	43.92	67.06	18.85	18.89	23.68	34.19
LREE	0.75	1.11	9.37	21.29	20.15	18.34	20.74	20.59	2.20	20.64	42.58	61.20	47.97	43.66	66.55	12.61	17.26	20.92	31.18
HREE	0.12	0.22	2.45	2.86	3.47	2.66	2.61	1.82	1.05	0.26	0.22	0.46	0.57	0.26	0.51	6.24	1.62	2.76	3.02
LREE:HREE	6.06	5.14	3.83	7.45	5.80	6.90	7.94	11.31	2.10	78.60	194.47	132.23	84.77	171.44	131.20	2.02	10.63	7.58	10.34
δEu	0.37	0.40	0.21	0.31	0.06	0.29	0.25	0.63	0.40	-	0.48	0.15	-	0.32	0.56	0.16	0.19	0.36	0.39
(La:Sm)N	3.67	3.43	2.85	2.51	2.87	6.34	6.86	10.19	1.61	17.71	38.89	18.90	20.02	21.51	41.68	0.86	3.02	2.65	5.82
(La:Yb)N	5.01	12.38	5.61	9.24	11.51	8.17	11.24	13.32	4.57	68.87	1112.90	317.34	552.57	-	755.28	1.28	9.97	4.98	12.86
(Sm:Nd)N	1.25	1.36	0.66	0.56	0.68	0.49	0.47	0.33	1.04	0.41	0.21	0.47	0.30	0.44	0.27	1.14	0.67	0.55	0.47
(Gd:Yb)N	0.31	2.49	2.07	2.43	3.42	1.42	1.61	1.24	3.33	0.61	9.10	3.37	8.63	-	7.76	1.10	1.48	0.76	2.04

The element content unit is (ppm), "-" indicates that the measured value falls below the detection limit of the instrument.

Fe-Mn carbonate minerals exhibit a ∑REE ranging from 0.88 to 67.06 ppm, with LREE content ranging from 0.75 to 66.55 ppm and HREE content ranging from 0.12 to 6.24 ppm for rare-earth elements (REE). The LREE/HREE ratio ranges from 1.72 to 194.47, the $(La{:}Yb)_N$ ranges from 1.24 to 1112.90, the $(Sm{:}Nd)_N$ is mostly <1 and ranges from 0.21 to 1.59, the $(La{:}Sm)_N$ ranges from 0.86 to 41.68, and them$(Gd{:}Yb)_N$ ranges from 0.31 to 9.10. When standardized using chondritic values, rhodochrosite exhibits significantly higher ΣREE than siderite (Figure 12d–f), while siderite, Mn siderite, and Fe rhodochrosite show similar distribution patterns. Fe-Mn carbonates display a right-skewed curve for REE enrichment, indicating preferential enrichment of LREE over HREE. The δEu values range between 0.06 and 0.63, suggesting the presence of a notable negative Eu anomaly. The substitution of Eu^{2+} for Ca^{2+} is impeded by the lower mass concentration of Ca^{2+} in the Fe-Mn carbonates of the Erdaokan Ag deposit, thereby resulting in the observed negative Eu anomaly.

4.3. Dating Analysis of Fe-Mn Carbonate Minerals

The Sm and Nd contents, as well as the isotopic compositions in the Fe-Mn carbonate mineral from the Erdaokan Ag deposit, are presented in Table 3. The results indicate that the Sm and Nd contents meet the requirements for isotope dating. Specifically, the Sm content ranges from 0.2115 to 0.9604 ppm, while the Nd content ranges from 0.7859 to 0.9827 ppm. Moreover, the values of $^{147}Sm/^{144}Nd$ vary between 0.0815 and 0.7169, with corresponding values of $^{143}Nd/^{144}Nd$ ranging from 0.512374 to 0.513346 being observed as well (Table 3). Importantly, a reliable Sm-Nd isochron age of approximately 233.7 ± 1.2 Ma has been determined for these samples (Figure 13a), which is further supported by a small weighted mean square deviation (MSWD = 1.5).

Table 3. Isotopic analysis of Sm-Nd in Fe-Mn carbonate minerals.

		Sm (ppm)	Nd (ppm)	Testing Error	$^{147}Sm/^{144}Nd$	$^{143}Nd/^{144}Nd$
Sample1	ZK266-2	0.7815	0.9827	0.000009	0.4802	0.512988
Sample2	ZK268-3	0.9604	0.8103	0.000008	0.7169	0.513346
Sample3	ZK274-6	0.4683	0.7859	0.000007	0.3647	0.512805
Sample4	ZK276-2	0.1327	0.9716	0.000009	0.0815	0.512374
Sample5	ZK278-6	0.3624	0.8512	0.000006	0.2562	0.512641
Sample6	ZK328-9	0.2115	0.9761	0.000007	0.1308	0.512452

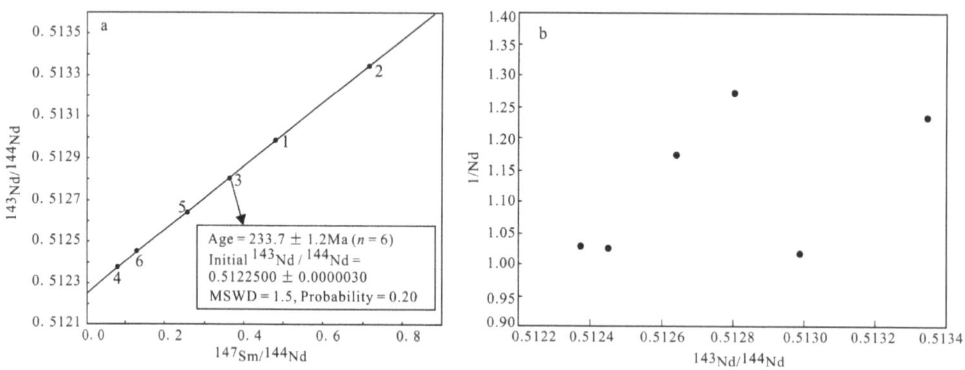

Figure 13. Sm-Nd isochron diagrams for the Erdaokan deposit: (**a**) Sm-Nd isotopic diagram of Fe-Mn carbonate minerals (The numbers in Figure 13 are consistent with the samples in Table 3); (**b**) 1/Nd-$^{143}Nd/^{144}Nd$ diagram of Fe-Mn carbonate.

The high concentrations of Sm and Nd in Fe-Mn carbonates ensure precise and reliable determination of Sm-Nd isochron ages. Figure 13b demonstrates no correlation between 1/Nd and $^{143}Nd/^{144}Nd$ in Fe-Mn carbonate samples, excluding the possibility of an isotope mixing line [34]. This Sm-Nd isochron age of the Fe-Mn carbonate mineral confirms that the main period of Erdaokan Ag deposit mineralization occurred during the Late Triassic Period (233.7 ± 1.2 Ma).

5. Discussion

5.1. Analysis of Fluid Source in Erdaokan Deposit

The main carbonates in the Erdaokan deposit are Fe-Mn carbonates (Figure 10a). Fe-Mn carbonate commonly occurs in the later stage of carbonate magma evolution, being closely associated with low-temperature hydrothermal deposits. Fe-Mn carbonate can be classified into two forms of isomorphism: Fe siderite and Mn rhodochrosite. The Fe and Mn content in the Fe-Mn carbonate of the Erdaokan deposit exhibits continuous transitional changes (Figure 10b), indicating that the formation of Fe-Mn carbonates in this deposit is a continuous process. Therefore, it can be inferred that these Fe-Mn carbonates originate from the same fluid source. A negative correlation between CaO and (FeO + MnO) is observed (Figure 11a), suggesting that Ca may replace Fe and Mn within the lattice structure of the Fe-Mn carbonate due to water–rock reactions involving Ca during deep-seated hydrothermal activity interacting with limestone from the Niqiuhe Formation [35]. The distribution of trace elements in Fe-Mn carbonate is comparable to that of limestone in the Niqiuhe formation. Both exhibit a shortage of Zr, Nb, and Ti, indicating that some of the limestone from the Niqiuhe formation contributed to the creation of Fe-Mn carbonates [33].

The distribution of trace elements in Fe-Mn carbonates exhibits a strong correlation with the main mineralization sulfides, such as pyrite, chalcopyrite, sphalerite, and bitumen in the mining area (Figure 12a–c). The $\delta^{34}S$ values of galena range from −0.97‰ to +3.8‰, indicating the presence of deep magmas in the ore-forming fluid [13]. Furthermore, the LREE/HREE ratio is between 1.72 and 194.47, while (La:Yb)N is between 1.24 and 1112.90, indicating that LREE has a higher fractionation degree than HREE. Note that the material source has the primary characteristics of a deep source [33]. The patterns of trace elements (Figure 12a,c) and rare-earth elements (Figure 12d,f) suggest a close relationship between Fe-Mn carbonate and deep magmatic–hydrothermal processes [14]. Furthermore, the depletion of Zr in Fe-Mn carbonates suggests the involvement of upper-mantle material in the hydrothermal system. Therefore, it can be inferred that the ore-forming fluid consists primarily of deep-source magma with a minor contribution from upper-mantle materials. The absence of barite in the Erdaokan Ag ore indicates that sulfur content is not significantly high within the ore-forming system, and that redox conditions do not play a key controlling role in ore-formation [36].

The formation of magnetite is observed as a result of late thermal action on Fe-Mn carbonate minerals [37]. The correlation between CaO and SrO in Fe-Mn carbonates was found to be insignificant (Figure 11b), indicating that the displacement of Ca by Sr mainly occurs during the high-temperature stage of magma crystallization [38]. This suggests that there is no apparent substitution between these two elements, and mineralization took place during the medium- to low-temperature stage. The negative Eu anomaly (0.06–0.63) is indicative of carbonate crystallization from a low-temperature fluid (<200 °C) [39]. Fluid inclusions within quartz veins indicate an average deposit formation temperature of 152.1 °C [13], implying characteristics of low-temperature mineralization. Overall, it can be stated that the Erdaokan Ag deposit exhibits medium- to low-temperature mineralization.

The distribution pattern of rare-earth elements (REE) in Fe-Mn carbonates reflects their formation by the same fluid, indicating a close relationship between Fe-Mn carbonate and deep magmatic–hydrothermal processes [13,15]. This is consistent with the distribution patterns observed in metal sulfides and bitumen (Figure 12d,f). Additionally, it should be noted that rhodochrosite exhibits higher REE content than siderite (Figure 12d,e).

Postma (1981) [40] demonstrated that Mn has a higher affinity for carbonate, leading to the formation of carbonate rocks and the enrichment of REE. Consequently, it is hypothesized that the sequential occurrence of fluid phases involves the initial formation of Mn siderite followed by Fe rhodochrosite. In EPMA's EDS mapping (Figure 9b,c), Mn is observed in the core region of Fe-Mn carbonate, while Fe predominantly resides in its rim. This observation further supports the notion that Mn siderite forms prior to Fe rhodochrosite. Previous investigations have indicated that modern seawater exhibits a Y/Ho ratio greater than 44, whereas hydrothermal seabed and chondrite samples typically fall within the range of 24–34 [16,20,41]. The average Y/Ho ratio in Fe-Mn carbonates from the Erdaokan deposit mining area was determined as 34.29, which aligns with values observed in hydrothermal vents and chondritic meteorites. These findings suggest a deep magmatic–hydrothermal origin for these Fe-Mn carbonates.

5.2. Sm-Nd Isotopes and Metallogenic Age Indication

The Sm-Nd isotopic system has been widely employed to determine the metallogenic geological age of hydrothermal deposits. Commonly used minerals for Sm-Nd isotope dating include fluorite, calcite, tourmaline, scheelite, and sulfide metal minerals [42–48]. However, the Sm-Nd systematics of Fe-Mn carbonate minerals formed through hydrothermal processes have received less attention [49]. In the Erdaokan Ag deposit, Fe-Mn carbonate is abundant throughout the entire mineralization stage and occurs mainly in the galena–sphalerite–Fe–Mn carbonate stage, which represents the primary silver metallogenic phase. Therefore, determining its formation age is crucial for constraining the overall mineralization age. Due to their high Sm/Nd ratios, Fe-Mn carbonate minerals in the Erdaokan deposit are suitable for Sm-Nd isotope dating. The obtained results indicate a mineralization age of 233.7 ± 1.2 Ma for Fe-Mn carbonate minerals, which aligns well with both bitumen (234.6 ± 1.2 Ma) [14] and sulfide Rb-Sr (232.9 ± 2.3 Ma) [15] data from the same deposit area. These findings further confirm that the ore-forming event at Erdaokan occurred during the Late Triassic Period.

A significant number of Cu-Au deposits have been discovered in the Duobaoshan Cu-Mo-Au mineralization concentrated area, where the Erdaokan silver deposit is located. Unlike the Triassic formations of Duobaoshan and Tongshan deposits, which primarily formed during the Ordovician [4,8,50,51], the Sankuanggou Fe deposit mainly originated in the Jurassic Period [8,52,53]. Similarly, the Sandaowanzi gold deposit and Yongxin gold deposit were predominantly formed during the Cretaceous Period [10,12,54–56]. Recent studies have revealed some Triassic diagenetic events in the Duobaoshan Cu-Mo-Au mineralization concentrated area [2,5,57], reporting that magmatic rocks with ages ranging from 244 ± 2 Ma to 214 ± 3 Ma are present in this region. These rock types include tonalite and granodiorite, as well as porphyritic-like granodiorite. This evidence suggests that Triassic magmatic activity within this concentration area aligns closely with the metallogenic age of Erdaokan deposit. The Mengdehe gold deposit situated south of this region was formed during the Late Triassic Period (209.6 ± 3.1 Ma) [58], coinciding with metallogenesis within the Erdaokan silver deposit's time frame. Collectively, these findings establish a newly recognized metallogenic age for the Duobaoshan Cu-Mo-Au mineralization concentrated area involving Triassic diagenetic and metallogenic events. It should be noted that there is a distinct difference between Erdaokan's metallogenic age and that of large Ag-Pb-Zn deposits such as Shuangjianzishan (135.0 ± 0.6 Ma) [59], Erentaolegai (139 ± 2.3 Ma) [60], Jiawula (133 ± 0.66 Ma) [61], and Chagan Buragen (138 ± 1 Ma) [62] found southwest of the Greater Khingan metallogenic belt, thus emphasizing Erdaokan's crucial indicative significance within the Duobaoshan Cu-Mo-Au mineralization concentrated area.

5.3. Metallogenic Model of Erdaokan Silver Deposit and Favorable Conditions for Ag Enrichment

In summary, the Duobaoshan Cu-Mo-Au mineralization concentrated area experienced significant magmatic activity during the Middle to Late Triassic Period due to the superposition of the Okhotsk Ocean tectonic domain [5,63–66]. The upward migration of deep-seated magma carrying Ag and other ore-forming materials in the Erdaokan area resulted in the formation of a medium- to low-temperature magmatic hydrothermal deposits with weak structures (Figure 14). Moreover, interaction between the ascending fluid and surrounding rocks was observed. Late-stage magmatic hydrothermal activity caused the heating of limestone from the Niqiuhe Formation, which was exposed in the Erdaokan mining area, and this led to its transformation into structures with high specific surface area [67,68]. This transformation facilitated adsorption and enrichment of Ag, Mn, and other ore minerals within the hydrothermal system. Additionally, magnetite and hematite were commonly found in peripheral areas of Fe-Mn carbonate minerals in this mining region, indicating significant thermal alteration subsequent to mineralization. Such thermal alteration likely contributed to further enhancement of Ag deposition. These findings provide an explanation for the supernormal enrichment of Ag (highest grade: 25,516.00 g/t) at the Erdaokan deposit, while offering new insights into prospecting strategies within this ore cluster area.

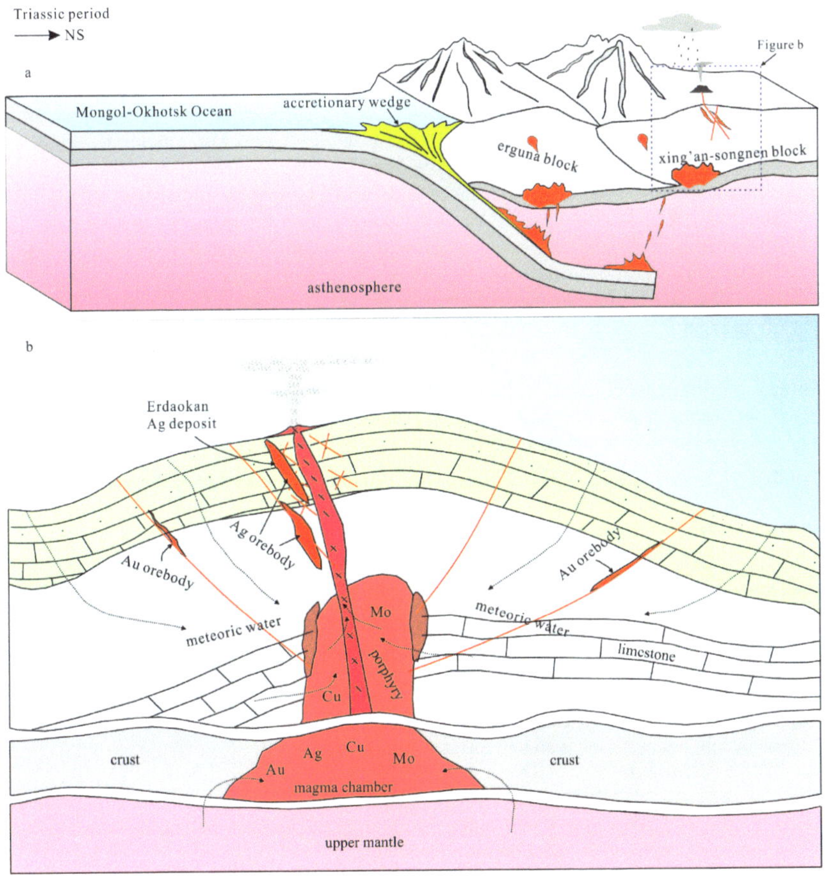

Figure 14. Geotectonic location and model illustration of the Erdaokan deposit: (**a**) schematic diagram depicting the geotectonic location of the Duobaoshan Cu-Mo-Au mineralization concentrated area; (**b**) schematic model illustrating the metallogenic process of the Erdaokan deposit.

6. Conclusions

- Fe-Mn carbonate is a pervasive mineral in the Erdaokan Ag deposit, primarily formed during the galenite–sphalerite–Fe–Mn carbonate stage. Through microscopic analysis and EPMA, pyrargyrite was found to coexist with Fe-Mn carbonate, while argentite is finely disseminated within Fe-Mn carbonates. Consequently, the presence of Fe-Mn carbonation has great potential as an indicator for prospecting regional Ag deposits, and provides a novel avenue for comprehending mineral deposit information.
- The Fe-Mn carbonates can be classified into manganese siderite and iron rhodochrosite, representing the two extremes of isomorphism. These Fe-Mn carbonates are primarily distributed in banded or nodular forms within the ore, often intersected by quartz veins. Microscopically, the Fe-Mn carbonates exhibit semi-idiomorphic rhomboid morphology, semi-idiomorphic rhomboid assemblage, rhomboid twinning, and residual structures resulting from quartz metasomatism. The concentrations of Fe and Mn show continuous variations, and display a similar distribution pattern to that of rare-earth trace elements.
- The Fe-Mn carbonates in the Erdaokan Ag deposit exhibit a similar distribution pattern of trace elements and REE, indicating a pronounced resemblance to the pyrite pattern observed in the deposit. Furthermore, these carbonates display a depletion of Zr, an enrichment of light rare-earth elements, a noticeable deficit of Eu, and an average Y/Ho value of 34.29. These findings suggest that the ore-forming materials originated from deep magma sources with involvement from upper-mantle materials.
- The Sm-Nd isotopic isochron age of the Fe-Mn carbonate monomineral has been determined to be 233.7 ± 1.2 Ma (MSWD = 1.5), indicating that the formation of the Erdaokan Ag deposit occurred during the Late Triassic Period in a magmatic hydrothermal environment characterized by moderate to low temperatures. This study further reinforces the significance of the Late Triassic Period as an additional metallogenic stage in the Duobaoshan Cu-Mo-Au mineralization concentrated area.

Author Contributions: Conceptualization, Y.Y. and C.L.; methodology, Z.W.; validation, H.G. and W.Y.; formal analysis, M.Y.; investigation, A.F., B.Z. and B.L.; software, Z.C. and B.L. All authors have read and agreed to the published version of the manuscript.

Funding: This research was Supported by Heilongjiang Provincial Key R&D Program Project (No. GA21A204), Heilongjiang Provincial Natural Science Foundation of China (No. LH2023D025), Scientific Research Project of Heilongjiang Provincial Bureau of Geology and Mineral Resources (No. HKY202301) and National Natural Science Foundation of China (No. 41872038).

Data Availability Statement: The data were used for the research described in this article.

Acknowledgments: The authors wish to acknowledge Zhenjiang Liu, Xuefei Liu and Yi Cao of China University of Geosciences (Beijing) for their contribution to improving the quality of this article.

Conflicts of Interest: The authors declare no conflicts of interest. The funders had no role in the design of this study, nor in the collection, analyses, or interpretation of the data, nor in the writing of the manuscript, nor in the decision to publish the results.

References

1. Liu, J.; Wu, G.; Li, Y.; Zhu, M.T.; Zhong, W. Re-Os sulfide (chalcopyrite, pyrite and molybdenite) systematicand fluid inclusion study of the Duobaoshan porphyry Cu (Mo) deposit, Heilongjiang Province, China. *J. Asian Earth Sci.* **2012**, *49*, 300–312. [CrossRef]
2. Zeng, Q.D.; Liu, J.M.; Chu, S.X.; Wang, Y.B.; Sun, Y.; Duan, X.X.; Zhou, L.L.; Qu, W.J. Re-Os and U-Pb geochronology of the Duobaoshan porphyry Cu-Mo-(Au) deposit, northeast China, and its geological significance. *J. Asian Earth Sci.* **2014**, *79*, 895–909. [CrossRef]
3. Gao, R.Z.; Xue, C.J.; Lu, X.B.; Zhao, X.B.; Yang, Y.S.; Li, C.C. Genesis of the Zhengguang gold deposit in the Duobaoshan ore field, Heilongjiang Province, NE China: Constraints from geology, geochronology and S-Pb isotopic Compositions. *Ore Geol. Rev.* **2017**, *84*, 202–217. [CrossRef]
4. Hao, Y.J.; Ren, Y.S.; Duan, M.X.; Tong, K.Y.; Chen, C.; Li, C. Re-Os Isotopic Dating of the Molybdenite from the Tongshan Porphyry Cu-Mo Deposit in Heilongjiang Province, NE China. *Acta Geol. Sin. (Engl. Ed.)* **2014**, *88* (Suppl. S2), 522–523.

5. Hao, Y.J.; Ren, Y.S.; Duan, M.X.; Zhao, H.L.; Tong, K.Y.; Sun, Z.M. Tectonic setting of Triassic magmatic and metallogenic event in the Duobaoshan mineralization area of Heilongjiang Province, NE China. *Geol. J.* **2017**, *52*, 67–91. [CrossRef]
6. Bao, X.B.; Yin, G.L.; Yu, X.Z. Geological characteristics of Erdaokancun Silver Polymetallic Deposit, Nenjiang County, Heilongjiang Province and its prospecting indicators. *Gold* **2019**, *7*, 20–23. (In Chinese with English abstract)
7. Xu, W.X.; Li, C.L.; Bao, X.B.; Yuan, M.W. Geological characteristics and genesis analysis of the first Triassic silver deposit discovered in Northeast of Da Hinggan Mountains. *Miner. Resour. Geol.* **2019**, *33*, 434–441. (In Chinese with English abstract)
8. Chu, S.X.; Zeng, Q.D.; Liu, J.M.; Wang, Y.B. Early–Middle Jurassic magmatism and skarn-porphyry mineralization in NE China: Geochronological and geochemical constraints from the Sankuanggou skarn Fe-Cu-(Mo) deposit, and tectonic implications. *J. Geochem. Explor.* **2019**, *200*, 84–103. [CrossRef]
9. Wang, L.; Qin, K.Z.; Cao, M.J.; Danišík, M.; Evans, N.J.; Li, G.M.; Song, G.X.; Pang, X.Y. Thermal history of an Early Paleozoic epithermal deposit: Constraints from 40Ar/39Ar and (U-Th)/He thermochronology at Zhengguang, eastern Central Asian Orogenic Belt. *Ore Geol. Rev.* **2020**, *126*, 103791. [CrossRef]
10. Li, C.L.; Li, L.; Yuan, M.W.; Alam, M.; Li, S.R.; Santosh, M.; Deng, C.Z.; Liu, H.; Xu, G.Z. Study on pyrite thermoelectricity, ore-forming fluiAll authors have been informed and agreed.ds and H-O-Rb-Sr isotopes of the Yongxin gold deposit, Central Asian Orogenic Belt: Implications for ore genesis and exploration. *Ore Geol. Rev.* **2020**, *121*, 103568. [CrossRef]
11. Gao, S.; Xu, H.; Zang, Y.Q.; Yang, L.J.; Yang, B.; Wang, T. Late Mesozoic magmatism and metallogeny in NE China: The Sandaowanzi-Beidagou example. *Int. Geol. Rev.* **2017**, *59*, 1413–1438. [CrossRef]
12. Gao, S.; Xu, H.; Zang, Y.Q.; Wang, T. Mineralogy, ore-forming fluids and geochronology of the Shangmachang and Beidagou gold deposits, Heilongjiang province, NE China. *J. Geochem. Explor.* **2018**, *188*, 137–155. [CrossRef]
13. Yuan, M.W.; Li, L.; Li, S.R. Mineralogy, fluid inclusions and S-Pb-H-O isotopes of the Erdaokan Ag-Pb-Zn deposit, Duobaoshan metallogenic belt, NE China: Implications for ore genesis. *Ore Geol. Rev.* **2019**, *113*, 103074. [CrossRef]
14. Yuan, M.W.; Li, L.; Li, S.R.; Santosh, M.; Li, C.L.; Alam, M.; Hou, Z.Q. Bitumen Sm-Nd, pyrite Rb-Sr and zircon U-Pb isotopes constrain timing of ore formation and hydrocarbon deposition in the Erdaokan Ag-Pb-Zn deposit, NE China. *Ore Geol. Rev.* **2021**, *134*, 104161. [CrossRef]
15. Yuan, M.W.; Li, L.; Li, C.L.; Li, S.R.; Santosh, M.; Alam, M.; Hou, Z.Q. The genesis of bitumen and its relationship with mineralization in the Erdaokan Ag-Pb-Zn deposit from the Great Xing'an Range, northeastern China. *Ore Geol. Rev.* **2021**, *139*, 104464. [CrossRef]
16. Yang, Z.; Jiang, M.J.; Zhao, S.R.; Ding, Z.J.; He, M.C. Stable isotopes and halogen geochemistry of the Huayuan carbonate-hosted Pb-Zn ore district, South China: Implications for the salt source of ore-forming fluids. *Acta Geol. Sin. (Engl. Ed.)* **2022**, *96*, 506–516. [CrossRef]
17. Xiong, S.F.; Gong, Y.J.; Yao, S.Z.; Shen, C.B.; Ge, X.; Jiang, S.Y. Nature and evolution of the ore-forming fluids from Nanmushu carbonate-hosted Zn-Pb deposit in the Mayuan district, Shaanxi Province, Southwest China. *Geofluids* **2017**, *2017*, 2410504. [CrossRef]
18. Xu, Y.W.; Mao, G.Z.; Geng, H.Y.; He, T.L.; Xu, Q.L.; Meng, Y.K.; Cao, M.P.; Yang, F.J.; An, P.R.; Song, L.G.; et al. Ore-forming materials and fluids and ore-controlling factors of the Liaoshang gold deposit in Jiaodong Peninsula, NE China. *Ore Geol. Rev.* **2023**, *154*, 105330. [CrossRef]
19. Myint, A.Z.; Wagner, T.; Fusswinkel, T. Calcite trace element geochemistry of Au deposits in the Singu-Tabeikkyin Gold District, Myanmar: Implications for the sources of ore-forming fluids. *Ore Geol. Rev.* **2022**, *145*, 104892. [CrossRef]
20. Bau, M.; Dulski, P. Distribution of yttrium and rare–earth elements in the Penge and Kuruman iron-formations, Transvaal Supergroup, South Africa. *Precambrian Res.* **1996**, *79*, 37–55. [CrossRef]
21. Bau, M. Controls on the fractionation of isovalent trace elements in magmatic and aqueous systems: Evidence form Y/Ho, Zr/Hf, and lanthanide tetrad effect. *Contrib. Mineral. Petrol.* **1996**, *123*, 323–333. [CrossRef]
22. Jiang, S.Y.; Chen, X.; Chen, Y.Q.; Jiang, Y.H.; Dai, B.Z.; Ni, P. Geochemistry and genetic model for the giant magnesite deposits in the eastern Liaoning Province, China. *Acta Petrol. Sin.* **2004**, *20*, 765–772.
23. Planavsky, N.; Bekker, A.; Rouxel, O.J.; Kamber, B.; Hofmann, A.; Knudsen, A.; Lyons, T.W. Rare Earth Element and yttrium compositions of Archean and Paleoproterozoic Fe formations revisited: New perspectives on the significance and mechanisms of deposition. *Geochim. Cosmochim. Acta* **2010**, *74*, 6387–6405. [CrossRef]
24. Presley, B.J.; Kaplan, I.R. Changes in dissolved sulfate, calcium and carbonate from interstitial water of near-shore sediments. *Geochim. Cosmochim. Acta* **1968**, *32*, 1037–1048. [CrossRef]
25. Chen, F.G.; Pufahl, K.P.; Wang, Q.F.; Matheson, E.J.; Shabaga, B.M.; Zhang, Q.Z.; Zeng, Y.S.; Le, X.W.; Ruan, D.; Zhao, Y.T. A new model for the genesis of Carboniferous Mn Ores, Longtou deposit, South China Block. *Econ. Geol.* **2022**, *117*, 107–125. [CrossRef]
26. Wang, F.; Zhou, X.H.; Zhang, L.C.; Ying, J.F.; Zhang, Y.T.; Wu, F.Y.; Zhu, R.X. Late Mesozoic volcanism in the Great Xing'an Range (NE China): Timing and implications for the dynamic setting of NE Asia. Earth Planet. *Sci. Lett.* **2006**, *251*, 179–198. [CrossRef]
27. Fan, W.M.; Guo, F.; Wang, Y.J.; Lin, G. Late Mesozoic calc-alkaline volcanism of post-orogenic extension in the northern Da Hinggan Mountains, northeastern China. *J. Volcanol. Geotherm. Res.* **2003**, *121*, 115–135. [CrossRef]
28. Wu, H.Y.; Zhang, L.C.; Wan, B.; Chen, Z.G.; Xiang, P.; Pirajno, F.; Du, A.D.; Qu, W.J. Re-Os $^{40}Ar/^{39}Ar$ ages of the Jiguanshan porphyry Mo deposit, Xilamulun metallogenic belt, NE China, constraints on mineralization events. *Miner. Depos.* **2011**, *46*, 171–185. [CrossRef]

29. Deng, C.Z.; Li, C.L.; Rong, Y.M.; Chen, D.; Zhou, T.; Wang, X.Y.; Chen, H.Y.; Lehmann, B.; Yin, R.S. Different metal sources in the evolution of an epithermal ore system: Evidence from mercury isotopes associated with the Erdaokan epithermal Ag-Pb-Zn deposit, NE China. *Gondwana Res.* **2021**, *95*, 1–9. [CrossRef]
30. Liu, Y.S.; Hu, Z.C.; GAO, S.; Günther, D.; Xu, J.; Gao, C.G.; Chen, H.H. In situ analysis of major and trace elements of anhydrous minerals by LA-ICP-MS without applying an internal standard. *Chem. Geol.* **2008**, *257*, 34–43. [CrossRef]
31. Yang, Y.H.; Wu, F.Y.; Yang, J.H.; Chew, D.M.; Xie, L.W.; Chu, Z.Y.; Zhang, Y.B.; Huang, C. Sr and Nd isotopic compositions of apatite reference materials used in U-Th-Pb geochronology. *Chem. Geol.* **2014**, *385*, 35–55. [CrossRef]
32. Yang, Y.H.; Wu, F.Y.; Li, Q.L.; Rojas-Agramonte, Y.; Yang, J.H.; Li, Y.; Ma, Q.; Xie, L.W.; Huang, C.; Fan, H.R.; et al. In situ U–Th–Pb dating and Sr-Nd isotope analysis of bastnäsite by LA-(MC)-ICP-MS. *Geostand. Geoanal. Res.* **2019**, *43*, 543–565. [CrossRef]
33. Dong, J.; Yuan, M.W.; Li, C.L. Geochemical characteristics of hydrothermal rhodochrosite in Erdaokan Ag-Pb-Zn deposit Duobaoshan metallogenic belt, Heilongjiang province and its indications. *J. Jilin Univ. Earth Sci. Ed.* **2022**, *52*, 855–865, (In Chinese with English abstract).
34. Faure, G. *Principle of Isotope Geology*; John Wiley and Sons Inc.: Hoboken, NJ, USA, 1986.
35. Peter, S.M. Relation between depositional environment and the elemental composition of early diagenetic siderite. *Geology* **1989**, *17*, 704–706.
36. Kelley, K.D.; Leach, D.L.; Johnson, C.A.; Clark, J.L.; Fayek, M.; Slack, J.F.; Anderson, V.M.; Ayuso, R.A.; Ridley, W.I. Textural, compositional, and sulfur isotope variations of sulfide minerals in the Red Dog Zn-Pb-Ag deposits, Brooks Range, Alaska: Implications for ore formation. *Econ. Geol.* **2004**, *99*, 1509–1532. [CrossRef]
37. Kang, N.; Schmidt, M.W.; Poli, S.; Franzolin, E.; Connolly, J.A.D. Melting of siderite to 20 GPa and thermodynamic properties of $FeCO_3$-melt. *Chem. Geol.* **2015**, *400*, 34–43. [CrossRef]
38. Wall, F.; Le Bas, M.J.; Srivastave, R.K. Calcite and carbocernaite exsolution and cotectic textures in a Sr, REE–rich carbonatite dyke from Rajasthan, India. *Mineral. Mag.* **1993**, *57*, 495–513. [CrossRef]
39. Gamyanin, G.N.; Vikent'eva, O.V.; Prokof'ev, V.Y.; Bortnikov, N.S. Arkachan: A new Gold–Bismuth–Siderite–Sulfide type of deposits in the West Verkhoyansky Tin District, Yakutia. *Geol. Ore Depos.* **2015**, *57*, 465–495. [CrossRef]
40. Postma, D. Formation of siderite and vivianite and the pore-water composition of a recent bog sediment in Denmark. *Chem. Geol.* **1981**, *31*, 225–244. [CrossRef]
41. Bolhar, R.; Kamber, B.S.; Moorbath, S.; Fedo, C.N.; Whitehouse, M.J. Characterisation of early Archaean chemical sediments by trace element signatures. *Earth Planet. Sci. Lett.* **2004**, *222*, 43–60. [CrossRef]
42. Chesley, J.T.; Halliday, A.N.; Scrivener, R.C. Samarium–neodymium direct dating of fluorite mineralization. *Science* **1991**, *252*, 949–995. [CrossRef]
43. Chesley, J.T.; Halliday, A.N.; Kyser, T.K.; Spry, P.G. Direct dating of Mississippi Valley-type mineralizations: Use of Sm-Nd in fluorite. *Econ. Geol.* **1994**, *89*, 1192–1199. [CrossRef]
44. Jiang, S.Y.; John, F.S.; Martin, R.P. Sm-Nd dating of the giant Sullivan Pb-Zn-Ag deposit, British Columbia. *Geology* **2000**, *28*, 751–754. [CrossRef]
45. Peng, J.T.; Hu, R.Z.; Burnard, P.G. Samarium-neodymium isotope systematics of hydrothermal calcites from the Xikuangshan antimony deposit (Hunan, China): The potential of calcite as a geochronometer. *Chem. Geol.* **2003**, *200*, 129–136. [CrossRef]
46. Oberthür, T.; Melcher, F.; Henjes, K.F.; Gerdes, A.; Stein, H.; Zimmerman, A.E.; Ghorfi, M. Hercynian age of the cobalt-nickel-arsenide–(gold) ores, Bou Azzer, Anti-Atlas, Morocco: Re-Os, Sm-Nd, and U-Pb age determinations. *Econ. Geol.* **2009**, *104*, 1065–1079. [CrossRef]
47. Su, W.; Hu, R.; Xia, B.; Xia, Y.; Liu, Y. Calcite Sm-Nd isochron age of the Shuiyindong Carlin-type gold deposit, Guizhou, China. *Chem. Geol.* **2009**, *258*, 269–274. [CrossRef]
48. Henjes, K.F.; Prochaska, W.; Niedermayr, A.; Sullivan, N.; Baxter, E. Sm-Nd dating of hydrothermal carbonate formation: An example from the Breitenau magnesite deposit (Styria, Austria). *Chem. Geol.* **2014**, *387*, 184–201. [CrossRef]
49. Krupenin, M.T.; Kuznetsov, A.B.; Chervyakovskaya, M.V.; Gulyaeva, T.Y.; Konstantinova, G.V. The Source of Ore Fluids and Sm-Nd Age of Siderite from the Largest Bakal Deposit, Southern Urals. *Geol. Ore Depos.* **2021**, *63*, 324–340. [CrossRef]
50. Zeng, H.; Zhao, Y.Y.; Fu, J.J.; Li, Y. Features of sulfide and plumbum isotopes of the copper deposits in Duobaoshan deposit cluster in Heilongjiang Province, China. *Acta Geol. Sin. (Engl. Ed.)* **2014**, *88*, 643–644. [CrossRef]
51. Hu, X.L.; Yao, S.Z.; Ding, Z.J.; He, M.C. Early Paleozoic magmatism and metallogeny in Northeast China: A record from the Tongshan porphyry Cu deposit. *Miner. Depos.* **2016**, *52*, 85–103. [CrossRef]
52. Hao, Y.J.; Ren, Y.S.; Duan, M.X.; Tong, K.Y.; Chen, C.; Yang, Q.; Li, C. Metallogenic events and tectonic setting of the Duobaoshan ore field in Heilongjiang Province, NE China. *J. Asian Earth Sci.* **2015**, *97*, 442–458. [CrossRef]
53. Deng, K.; Li, Q.; Chen, Y.J.; Zhang, C.; Zhu, X.F.; Xu, Q.W. Geochronology, geochemistry and Sr-Nd-Pb-Hf isotopes of the Early Jurassic granodiorite from the Sankuanggou intrusion, Heilongjiang Province, Northeastern China: Petrogenesis and geodynamic implications. *Lithos* **2018**, *296*, 113–128. [CrossRef]
54. Yu, R.T.; Li, B.L.; Sun, F.Y.; Li, Z.H.; Li, H.W.; Shi, Y.F. Geochronology, geochemistry and Hf isotopes of andesites in the Sandaowanzi gold deposit (Great Xing'an Range, NE China): Implications for petrogenesis, tectonic setting, and mineralization. *Acta Geochim.* **2021**, *40*, 251–270. [CrossRef]
55. Zhai, D.G.; Liu, J.J.; Ripley, E.M.; Wang, J.P. Geochronological and He-Ar-S isotopic constraints on the origin of the Sandaowanzi gold-telluride deposit, northeastern China. *Lithos* **2015**, *212*, 338–352. [CrossRef]

56. Zhao, Z.H.; Sun, J.G.; Li, G.H.; Xu, W.X.; Lu, C.L.; Guo, Y.; Liu, J.; Zhang, X. Zircon U-Pb geochronology and Sr-Nd-Pb-Hf isotopic constraints on the timing and origin of the Early Cretaceous igneous rocks in the Yongxin gold deposit in the Lesser Xing'an Range, NE China. *Geol. J.* **2020**, *55*, 2684–2703. [CrossRef]
57. Zhao, C.; Qin, K.Z.; Song, G.X.; Li, G.M. Switch of geodynamic setting from the Paleo-Asian Ocean to the Mongol-Okhotsk Ocean: Evidence from granitoids in the Duobaoshan ore field, Heilongjiang Province, Northeast China. *Lithos* **2019**, *336*, 202–220. [CrossRef]
58. Li, C.L.; Deng, C.Z.; Li, S.R.; Yuan, M.W.; Yang, Y.J. Geochronology and genesis of the newly discovered Mengdehe orogenic-type Au deposit in the Xing'an-Mongolia orogenic Belt, NE China. *Ore Geol. Rev.* **2021**, *133*, 104083. [CrossRef]
59. Zhai, D.G.; Williams–Jones, A.E.; Liu, J.; David, S.; Voudouris, P.C.; Tombros, S.; Li, K.; Li, P.; Sun, H. The genesis of the giant Shuangjianzishan epithermal Ag-Pb-Zn deposit, Inner Mongolia, Northeastern China. *Econ. Geol.* **2020**, *115*, 101–128. [CrossRef]
60. Xu, L.Q.; Liu, C.; Deng, J.F.; Li, N.; Dai, M.; Bai, L.B. Geochemical characteristics and ziron U-Pb SHRIMP age of igneous rocks in Erentaolegai silver deposit, Inner Mongolia. *Acta Petrol. Sin.* **2014**, *30*, 3203–3212. (In Chinese with English abstract)
61. Niu, S.D.; Li, S.R.; Huizenga, J.M.; Santosh, M.; Zhang, D.H.; Li, Z.D.; Tang, L. 40Ar/39Ar geochronology fluid inclusions and ore-grade distribution of the Jiawula Ag-Pb-Zn deposit NE China: Implications for deposit genesis and exploration. *Geol. J.* **2019**, *55*, 1115–1127. [CrossRef]
62. Li, T.G.; Wu, G.; Liu, J.; Wang, G.R.; Hu, Y.Q.; Zhang, Y.F.; Luo, D.F.; Mao, Z.H.; Xu, B. Geochronology fluid inclusions and isotopic characteristics of the Chaganbulagen Pb-Zn-Ag deposit, Inner Mongolia, China. *Lithos* **2016**, *261*, 340–355. [CrossRef]
63. Li, Y.; Xu, W.L.; Wang, F.; Pei, F.P.; Tang, J.; Zhao, S. Triassic volcanism along the eastern margin of the Xing'an Massig, NE China: Constraints on the spatial-emporal extent of the Mongol-khotsk tectonic regime. *Gondwana Res.* **2017**, *48*, 205–223. [CrossRef]
64. Li, Y.; Xu, W.L.; Tang, J.; Pei, F.P.; Wang, F.; Sun, C.Y. Geochronology and geochemistry of Mesozoic intrusive rocks in the Xing'an Massif of NE China: Implications for the evolution and spatial extent of the Mongol-khotsk tectonic regime. *Lithos* **2018**, *304*, 57–73. [CrossRef]
65. Zhou, J.B.; Cao, J.L.; Wilde, S.A.; Zhao, G.C.; Zhang, J.J.; Wang, B. Paleo-Pacific subduction-accretion: Evidence from Geochemical and U-Pb zircon dating of the Nadanhada accretionary complex, NE China. *Tectonics* **2014**, *33*, 2444–2466. [CrossRef]
66. Aouizerat, A.; Xiao, W.J.; Schulmann, K.; Jerabek, P.; Monie, P.; Zhou, J.B.; Zhang, J.J.; Ao, S.J.; Li, R.; Li, Y.C. Structures, Strain Analyses, and 40Ar/39Ar Ages of Blueschist-Bearing Heilongjiang Complex (NE China): Implications for the Mesozoic Tectonic Evolution of NE China. *Geol. J.* **2018**, *160*, 1–30. [CrossRef]
67. Ptáček, P.; Bartoníčková, E.; Švec, J. The kinetics and mechanism of thermal decomposition of $SrCO_3$ polymorphs. *Ceram. Int.* **2015**, *41*, 115–126. [CrossRef]
68. Xie, J.J.; Chen, T.; Xing, B.B.; Liu, H.B.; Xie, Q.Q.; Li, H.W.; Wu, Y.C. The thermochemical activity of dolomite occurred in dolomite-palygorskite. *Appl. Clay Sci.* **2016**, *119*, 42–48. [CrossRef]

minerals

MDPI

Article

Crystallization Sequence of the Spodumene-Rich Alijó Pegmatite (Northern Portugal) and Related Metasomatism on Its Host Rock

Idoia Garate-Olave [1,*], Encarnación Roda-Robles [1], Nora Santos-Loyola [1], Tania Martins [2], Alexandre Lima [3,4] and Jon Errandonea-Martin [1]

[1] Geology Department, University of the Basque Country (UPV/EHU), Sarriena Street n/n, 48940 Leioa, Spain; encar.roda@ehu.eus (E.R.-R.); nora.santos@ehu.eus (N.S.-L.); jon.errandonea@ehu.eus (J.E.-M.)

[2] Manitoba Geological Survey, 360-1395 Ellice Avenue, Winnipeg, MB R3G 3P2, Canada; tania.martins@gov.mb.ca

[3] Department of Geosciences, Environment and Spatial Planning, Faculty of Sciences, University of Porto, Rua Campo Alegre, 4169-007 Porto, Portugal; allima@fc.up.pt

[4] ICT (Institute of Earth Sciences), Porto Pole, Rua Campo Alegre, 4169-007 Porto, Portugal

* Correspondence: idoia.garate@ehu.eus

Abstract: The Barroso–Alvão region is an excellent setting for studying Li mineralization associated with granitic pegmatites and developing Li exploration techniques. Among the distinguished pegmatite types in this pegmatite field, the spodumene-bearing dyke from Alijó is a representative example of an Iberian Li–Cs–Ta (LCT) pegmatite currently under exploitation. In this work, we examine the internal evolution of the Alijó dyke and its external metasomatic effect on the surrounding metasediments, contributing to lithium exploration techniques. Electron microprobe analyses provided clues about the crystallization conditions and the degree of differentiation of the pegmatitic melt, whereas the external metasomatism induced by the spodumene-bearing pegmatite was studied through whole-rock geochemistry. The obtained results indicate that the primary crystallization of the studied dyke likely occurred at temperatures between 450–500 °C, with emplacement at shallow crustal levels of about 2–3 kbar. The high concentrations of trace elements such as Li, Cs, Rb, Be, Sn, Nb, Ta, Ge, U, and Tl in the pegmatitic melt suggests high availability of these elements, allowing their partitioning into an early exsolved fluid phase. The exsolution of this fluid phase, subtracting components such as F and B, from the pegmatitic melt would cause a significant undercooling of the melt. Moreover, the interaction of this expelled fluid with the country rock generated a metasomatic overprint in the surrounding metasedimentary host rocks. The metasomatic effect in Alijó is strongly influenced by the nature of the host metasediments, with a significantly higher grade of metasomatism observed in pelitic (mica-rich) samples compared to psammitic (mica-poor) samples collected at same distances from the dyke. The greisen developed close to the pegmatite contact reflects this metasomatic signature, characterized by the mobilization of at least B, F, Li, Rb, Cs, Sn, Be, Nb, Ta, and Tl. We cautiously suggest that whole rock Li concentrations greater than 300 ppm, combined with a minimum value of 1000 ppm for the sum of B, F, Li, Rb, Cs, and Sn in pelitic metasediments of Barroso–Alvão, may be indicative of a mineralized pegmatite in this region.

Keywords: spodumene; pegmatite; lithium exploration; metasomatism; Barroso–Alvão; Iberian Massif

1. Introduction

Lithium demand as a high-technology metal has increased considerably in the last decades [1–3]. The use of this element in rechargeable batteries, high-capacity energy storage, and capacitors for wireless technology has made it a strategic metal for the current society [4–6]. As a logical consequence, scientific knowledge about the location of Li-bearing mineralization has become a priority for the European Union [7,8].

Among the different types of existing lithium mineralization, pegmatites are considered a feasible source of many critical raw materials [5,9]. Rare element pegmatites, especially those included in the Li–Cs–Ta (LCT) family, show marked enrichment in a variety of chemical elements such as Li, Rb, Cs, Be, Sn, Ta, P, B, and F [10,11]. The chemical behavior of these incompatible elements in granite-pegmatite systems is of increasing interest among the scientific community. The distribution of incompatible elements in pegmatite-forming minerals during pegmatite crystallization has been proven to be an excellent tool for estimating the fractionation degree of a pegmatite body (e.g., [11,12]). As a result, the chemical variation in micas, quartz, and feldspars, among others, has become a principal objective in the exploration of pegmatite-related critical raw materials [8,13,14]. However, the magmatic–hydrothermal transition in pegmatitic systems is still a matter of debate, and the behavior of the aforementioned elements in the presence of coexisting fluids is not fully understood [15–18].

In this work, we present a mineralogical and whole-rock geochemical characterization of a spodumene-bearing pegmatite and its metasedimentary host rocks. This contribution provides new insights not only into the internal magmatic–hydrothermal evolution of this Li-bearing pegmatite but also into the metasomatic processes developed in the adjacent host rocks. The studied pegmatite, located in northern Portugal, is currently under exploitation and represents a good example of the late-Variscan LCT pegmatites that occur in NW Iberia, so its study would contribute to the understanding of the formation history of pegmatitic rocks. Moreover, the inferences about the geochemical halos generated around the studied pegmatite could be extended to other mineralized LCT pegmatites in the European Variscan Belt, providing new data to consider in less-invasive mineral exploration techniques.

2. Geological Setting

The Barroso–Alvão (BA) region is known for its vast Li mineralization associated with pegmatites (e.g., [19–21]). This proliferous pegmatite field is located in northern Portugal, in the most westerly portion of the European Variscan Belt, the Iberian Massif (Figure 1). Of the several geotectonic zones that compose the Iberian Massif, the study area belongs to the Galicia-Trás-os-Montes Zone (GTMZ) as defined by [22], specifically to the Schistose Domain or Parautochthon, and is close to the basal thrust that represents the southern boundary with the Central Iberian Zone (CIZ) [22–24].

The Schistose Domain or Parautochthon comprises preorogenic Cambro–Silurian sedimentary successions and Devono–Carboniferous synorogenic sequences, with minor Ordovician felsic volcanics [24,25]. This sedimentary record has been interpreted as distal deposits of the Gondwanan continental margin, showing affinities with the autochthonous CIZ (e.g., [24,26]). The main hosts for the pegmatites of the BA Pegmatite Field include mica schists, quartz-feldspathic schists, phyllites, quartz phyllites, quartzites, metagreywackes, cherts, and minor calcsilicate rocks [27].

The BA region was affected by three deformation phases of the Variscan Orogeny (D1 to D3) recognized by S1–S3 foliations, resulting in an imbricated internal structure [28–30]. Intermediate-pressure Barrovian metamorphism reached up to medium-grade conditions in the internal areas of the GTMZ [26], with pressure–temperature peak estimations for the BA region established at T = 500–550 °C and P = ~3 kbar, within the amphibolite facies [30]. The studied areas belong to the biotite and andalusite zones, displaying isogrades with a general WNW–ESE direction, roughly parallel to the nearby Variscan granitoids located to the south of the pegmatite field [28,30] (Figure 1). Another remarkable structural feature of the area is its proximity to the Regua-Verin fault, one of the major sets of NNE–SSW faults in the region.

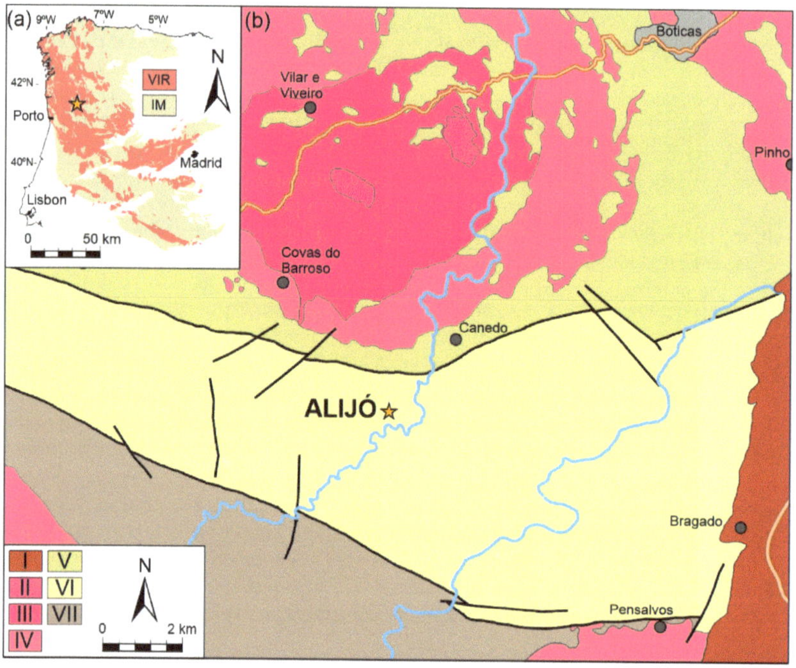

Figure 1. (**a**) Geological map of the Iberian Massif (IM; excluding the Ossa-Morena and South Portuguese zones), highlighting the Variscan Igneous Rocks (VIR). (**b**) Regional geological map of the Barroso-Alvão region: I: post-D3 biotite granites; II: syn-D3 biotite granites; III: syn-D3 biotite > muscovite granites; IV: syn-D3 two-mica granites; V: mica-schist and metagreywackes (lower parautochthon); VI: mica-schist/phyllites and quartzites (upper parautochthon); VII: dark schist and quartzites (upper parautochthon): Star: location of studied dyke. Based on [27,31–33].

Different types of late Variscan granitoids surround the BA Pegmatite Field, ranging from syn-D3 two mica granites to late/post-D3 biotite granites [34–38]. The syn-D3 plutons are represented in the area by the Barroso and Vila da Ponte granites, the Cabeceiras de Basto Pluton, and the Boticas/Leiranco granite, whereas late- to post-D3 massifs correspond to the Vila Pouca de Aguiar and Peneda-Gerês plutons [27,35,38–40]. The syn-D3 two-mica granites are peraluminous, and their origin has been proposed to be mainly from a crustal derivation, involving the melting of distinct metasedimentary and/or metaigneous sources [35,41]. By contrast, the genesis of the biotite dominant late- to post-D3 granitoids has been attributed to mantle-derived magmas in diverse ways, from hybridization of mantle-derived juvenile magmas with crustal melts/rocks to the melting of metabasic lower crustal protoliths [35,42,43] (Figure 1).

Pegmatites from Barroso–Alvão

The pegmatites from the BA Pegmatite Field are unevenly distributed in the region. They form local swarms of various sizes, up to 300 m in outcrop length. Their width is variable, from less than a few meters to up to 30 m across. These pegmatites display varied structures in terms of emplacement, some are flat-lying, whereas others occur gently or steeply dipping. Typically, these pegmatites cut discordantly through the host rock, showing sharp contacts. According to [44,45], five different groups of granitic pegmatites are identified in BA:

(1) Intragranitic pegmatites with quartz, feldspar, muscovite, biotite, minor tourmaline, beryl, garnet, fluorapatite, chlorite, and zircon.

(2) Barren pegmatites with quartz, feldspar, muscovite, minor biotite, fluorapatite, beryl, tourmaline, chlorite, zircon, pyrite, and monazite-(Ce).
(3) Spodumene pegmatites with quartz, feldspar, spodumene, muscovite, minor columbite group minerals (CGM), fluorapatite, montebrasite, triphylite, phosphoferrite, dufrénite, fairfieldite, chlorite, tourmaline, zircon, uraninite and sphalerite. The selected aplite-pegmatite body from Alijó belongs to this third group.
(4) Petalite pegmatites with quartz, feldspar, petalite, muscovite, minor cassiterite, CGM, fluorapatite, montebrasite, ferrisicklerite, eosphorite, pyrite, sphalerite, uraninite, monazite-(Ce), autunite and xenotime.
(5) Lepidolite pegmatites with albite, lepidolite, and muscovite, minor cassiterite, CGM, fluorapatite, zircon, and goyazite.

3. Materials and Methods

Representative samples from 8 different parts of the selected dyke were collected for the present study in a 32 m-thick section perpendicular to the strike of the dyke. A total of 23 host rock samples were collected following the procedures of the H2020 GREENPEG project, taking samples systematically in a section perpendicular to the strike of the dykes, up to 35 m far from the dyke (location in Supplementary Materials; more details for the sampling procedure are provided in [18]). Among those host rock samples, four were collected several hundred meters away from the pegmatite as 'control-samples', in order to represent non-metasomatized compositions for the metasedimentary rocks.

Thirty-five polished sections were made from representative samples of the mineralized pegmatite and its host rocks for petrographic study, which was carried out using a Leica DM LP model polarizing microscope, Leica Microsystems Pty Ltd, Lane Cove West, Australia fitted with a CCD camera at the Geology Department of the University of the Basque Country UPV/EHU.

3.1. Electron Microprobe (EMP) Analysis

Over 300 EMP analyses were obtained from polished thin sections using a Camebax SX-100 EMP at the Raimond Castaing Centre of the University of Paul Sabatier (Toulouse, France). The operating conditions were a voltage of 15 kV and a beam current of 10 nA (20 nA for phosphate minerals). The calibration standards used were synthetic SiO_2 (Si), synthetic $MnTiO_3$ (Ti, Mn) wollastonite (Ca) corundum (Al), hematite (Fe), albite (Na), orthoclase (K), fluorite (F), graftonite (P), periclase (Mg), synthetic chromite (Cr), synthetic glass of Rb_2O (Rb) and Cs_2O (Cs), synthetic $BaTiO_3$ (Ba), sphalerite (Zn), and tugtupite (Cl). Data were reduced using the procedure outlined in [46], and analytical errors estimated at $\pm$1%–2% for major elements and $\pm$10% for minor elements. Complete results can be found in the Supplementary Materials (SM).

3.2. Whole-Rock Geochemical Analysis

A total of 24 samples were sent to Activation Laboratories Ltd. (Ancaster, ON, Canada) for whole-rock major and trace element analyses. For the pegmatite, a representative sample was obtained by crushing 15 kg of different proportional parts of the pegmatite to a nominal grain size of 1 cm at the Geology Department of the UPV/EHU. In the case of the 23 host rock samples, they were cut to obtain representative sub-samples of 300 g. Each sample was then crushed up to 80% (passing 2 mm), split using a riffle splitter (250 g), and pulverized up to 95% (passing 105 μm) using a mild steel pulverizer.

Major and trace element concentrations were obtained via different analytical techniques. To obtain major element concentrations, as well as Zr, Sc, Hf, and Lu concentrations, a lithium metaborate/tetraborate fusion was performed to dissolve the entire sample. Major elements were measured using X-ray fluorescence (Actlabs, Ancaster, ON, Canada), whereas Sc and Zr were analyzed by inductively coupled plasma—optical emission spectrometry (ICP-OES, Actlabs, Ancaster, ON, Canada), and Hf and Lu by ICP—mass spectrometry (MS, Actlabs, Ancaster, ON, Canada). Concentrations of the remaining trace elements were also measured by ICP-MS, but with a previous sodium peroxide fusion. Fluorine concentrations were analyzed using an ion-selective electrode (ISE, Actlabs, Ancaster, ON, Canada) following a lithium borate fusion and dissolution in dilute nitric acid.

4. Results

4.1. Field Characterization of the Spodumene-Bearing Dyke from Alijó

The studied aplite-pegmatite body from Alijó is a subvertical discordant body with an approximate strike of N155° E and variable thickness (5 to 30 m). This heterogeneous dyke is composed mainly of quartz, plagioclase, K-feldspar, spodumene, and white mica, with accessory tourmaline, phosphate minerals, eucryptite, beryl, cookeite, and CGM. According to [47], this dyke may be classified as an LCT pegmatite of the complex type and the spodumene subtype. Aplitic and pegmatitic facies are distributed heterogeneously inside the dyke, typically with an abrupt transition between them. Quartz and plagioclase compose mainly the aplitic units, whereas K-feldspar occurs randomly distributed. On the contrary, plagioclase and K-feldspar (±quartz) are the main rock-forming minerals in the pegmatitic units. Locally, spodumene becomes the main mineral phase.

One of the most remarkable petrographic features of this dyke is the abundance of unidirectional solidification textures (UST). In the aplitic units, a subvertical layering is commonly observed parallel to the contact with the host rocks, with alternating (~1 cm thick) quartz + plagioclase-rich and plagioclase-rich layers (crystal sizes in the range of 1–3 mm) (Figure 2a). In the pegmatitic facies, plagioclase crystals often exhibit a combed texture, with the axis of the crystals growing perpendicularly to the contact with the host rocks (Figure 2b). Less frequently, spodumene crystals also appear as combed crystals.

Figure 2. (**a**) Layered structure in an aplitic sample alternating quartz + plagioclase and plagioclase-rich layers. (**b**) Hand sample with a pegmatitic texture, exhibiting combed feldspars marked with dashed lines. (**c**) Representative outcrop of the pelitic host metasediments. (**d**) Field photograph of an outcrop showing psammitic layers in a mainly pelitic lithology.

The Alijó spodumene-bearing aplite-pegmatite body intrudes discordantly into a metasedimentary sequence that includes phyllites and mica-schists interlayered with black schists, lydites, and some quartz-phyllites and calc-silicate rocks of upper Ordovician to lower Devonian age [30] (Figure 2c,d). Broadly, these metasediments can be classified as pelitic (mica-rich and quartz/feldspar-poor) and psammitic (mica-poor).

These metasediments were later affected by the Variscan Orogeny, with the respective metamorphic overprint.

4.2. Petrography and Mineral Chemistry of the Spodumene Pegmatite

The mineralogy of the studied pegmatitic body predominantly consists of quartz, plagioclase, K-feldspar, spodumene, and muscovite. Additionally, it contains accessory minerals such as tourmaline, phosphate minerals, eucryptite, beryl, cookeite, and members of the columbite group (Figure 3).

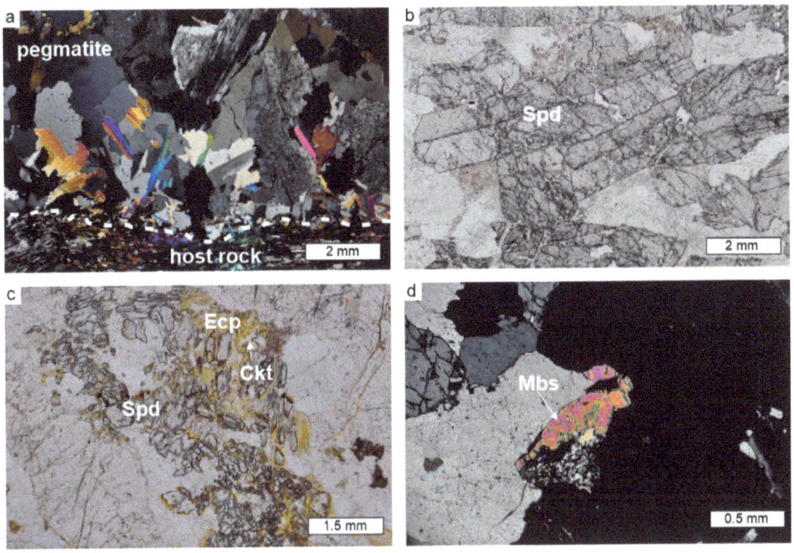

Figure 3. (**a**) Microphotograph of combed micas growing from the contact with the host rock marked by dashed line. (**b**) Euhedral spodumene crystals of up to 8 mm. (**c**) Eucryptite (Ecp) (yellowish) and cookeite (Ckt) (uncolored) replacing spodumene (Spd) crystals, with some relicts of Spd. (**d**) Anhedral crystal of montebrasite (Msb) with a slight twinning.

Feldspars constitute a volumetrically important part of the studied dyke, with albite (Ab_{98}–$Ab_{99.8}$) being the most common term (Supplementary Materials). Albite occurs as fine to coarse subhedral crystals, with the biggest ones often growing perpendicularly to the dyke contacts. The chessboard texture or myrmekite intergrowths have been described. Potassium feldspar (Or_{92}–Or_{98}) occurs as subhedral to anhedral crystals, varying in size from very fine to coarse (<5 cm). Some crystals display inclusions of quartz and/or micas, whereas others display Carlsbad or cross-hatched tartan pattern twinning. Phosphorous contents reach up to 0.43 wt.% and 0.36 wt.% (P_2O_5) in K-feldspar and plagioclase, respectively. A negative correlation between Si and P + Al suggests the influence of the berlinite substitution type for the incorporation of P in the structure of these alkali feldspars (Figure 4a). Rubidium concentrations reach up to 2469 ppm in K-feldspar and below the detection limit in plagioclase.

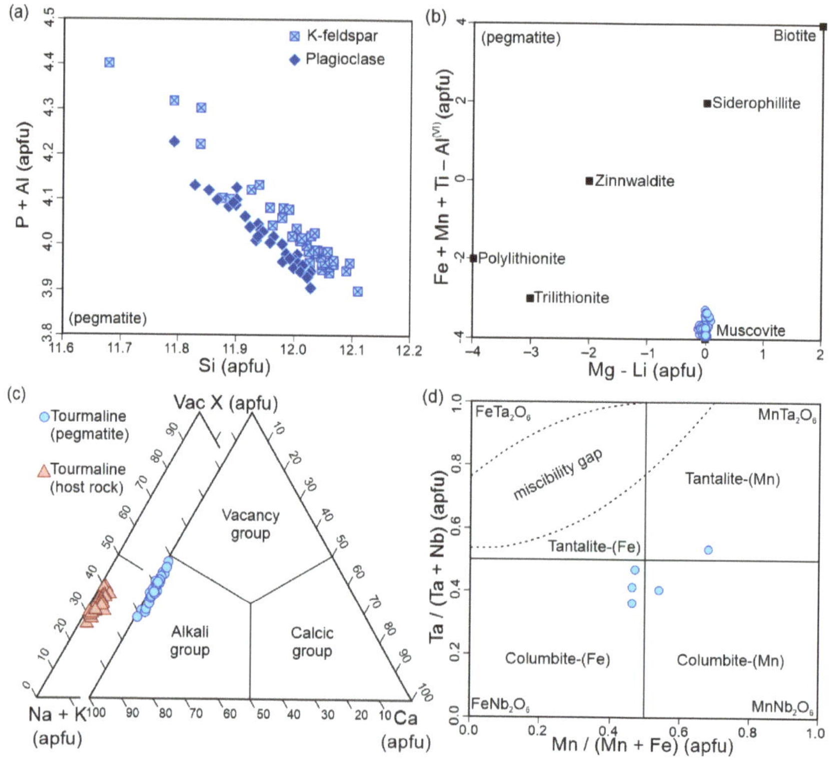

Figure 4. (**a**) Binary Si vs. P + Al diagram of K-feldspar and plagioclase from the Alijó pegmatite. (**b**) Binary Mg − Li vs. Fe + Mn + Ti + Al^{VI} diagram of analyzed micas (based on [48]). (**c**) Ternary Na + K vs. Ca vs. vacancy (X site) classification diagram of analyzed tourmaline crystals. (**d**) Quadrilateral Mn/(Mn + Fe) vs. Ta/(Ta + Nb) plot of analyzed CGM from the Alijó pegmatite. Analyses of CGM from other localities from BA are also shown in green for comparison (data from [44]).

White mica crystals are fine-grained (<1 cm), subhedral, and usually tabular (less frequently radial). Close to the contacts with the host rocks, micas with combed textures grow perpendicularly to them (Figure 3a). Occasionally, primary crystals show "patchy zoning". Analyzed micas fall in the muscovite field sensu stricto of the [48] classification diagram (Figure 4b; Supplementary Materials). Overall, Rb and Cs contents in micas reach up to 8961 ppm and 1792 ppm, respectively, whereas the Li content does not exceed 1000 ppm (Supplementary Materials).

Spodumene is the main Li-bearing rock-forming mineral phase in Alijó. The primary spodumene crystals are euhedral to subhedral with a medium to coarse grain size (<3 cm) (Figure 3b). As with other main mineral phases, these prismatic crystals may also grow perpendicularly to the contact with the host rock, showing a combed texture. Anhedral fine-grained secondary spodumene crystals are also common, containing small sub-rounded quartz crystals as a result of petalite replacement. Aggregates of spodumene crystals with an acicular habit, showing vermicular textures oriented around coarser crystals, have also been found. Eucryptite, as a low-PT Li-aluminosilicate, occurs as an accessory phase, as very fine-grained aggregates of anhedral crystals, together with albite and/or K-feldspar. These very fine-grained aggregates commonly replace pseudomorphically previous spodumene crystals, with small relicts of spodumene in the interior (Figure 3c).

Sometimes, eucryptite also occurs together with phyllosilicates such as cookeite, a frequent product of spodumene alteration (Figure 3c).

Tourmaline occurs as an accessory phase, mainly in the aplitic units of the dyke. It shows a fine to medium crystal size and appears as euhedral to subhedral crystals with a prismatic or acicular habit, apparently with no preferred orientation. According to the X-position occupancy [49,50], analyzed tourmalines are alkaline (Figure 4c). Chemical data showed 9.97–12.84 wt.% and <0.09 wt.% of FeO and MgO, respectively, allowing them to be classified as Fe-rich term schorl (Supplementary Materials). Instead of Mg, Li is present in tourmaline from the dyke (Li_2O 0.28–0.8 wt.%), belonging to the schorl-elbaite series and evolving via the $LiAl(Fe)_{-2}$ exchange vector.

Phosphate minerals appeared scattered, predominantly in the aplitic unit. In general, these accessory phases appear as subhedral to anhedral fine crystals and display brownish to yellowish color tones. Apatite is the most common phosphate, comprising chlorapatite (1.48–3.71 wt.% of Cl) with dark brown shades and a prismatic habit, and light brown to beige fluorapatite (2.3–3.46 wt.% of F) (Supplementary Materials). In general, MnO and FeO contents are high (1.41–6.75 wt.% and 1.25–4.44 wt.% respectively), with chlorapatites richer in Mn (~5 wt.%) and fluorapatites in Al (1.19–5.32 wt.%). Apatite crystals with low F and Cl could be classified as hydroxyapatites. Measured concentrations of Sr reach up to 1.12 wt.% (SrO). Regarding the amblygonite group minerals, they show a marked twinning (Figure 3d). Taking into account the low values of F (<1.4 wt%), these accessories correspond to almost pure montebrasite-term (Supplementary Materials).

Columbite group minerals from the studied dyke occur as an accessory phase. These fine-grained tabular crystals show homogeneous Mn/(Mn + Fe) and Ta/(Ta + Nb) ratios close to 0.5 (Supplementary Materials). Based on their composition, most of the analyzed crystals may be classified as columbite-(Fe) (Figure 4d).

4.3. Petrography and Mineral Chemistry of the Host Rock

The studied host rock samples were composed of mica-schists, metagreywackes, quartzites, and phyllites that form a metamorphosed psammopelitic sequence (Figure 2c,d). In the pelitic samples, a well-developed slaty cleavage was observed (Figure 5a). In these samples, muscovite, biotite, and quartz were the main mineral phases whereas oxides, apatite, and garnet corresponded to the most common accessory phases (Figure 5a). Contrarily, the psammitic samples were mainly composed of quartz, with significantly smaller amounts of muscovite and biotite, with oxides as accessory minerals that appeared essentially in the pelitic layers (Figure 5b). Some samples corresponded to intermediate terms, with nearly equal proportions of micas and quartz (Figure 5c). Scattered biotite and garnet porphyroblasts were common in many samples, with relatively greater crystal sizes (Bt up to 2 mm, Grt up to 1.5 mm) (Figure 5d).

Close to the contact with the pegmatite, well-developed tourmalinization was observed in the pelitic samples and/or layers, with up to 1.5 cm long euhedral/subhedral tourmaline crystals that become gradually smaller in the samples collected further from the contact (Figure 5e). There were notable compositional differences between the tourmaline from the dyke and that found in its host rocks. Analyzed tourmaline crystals from the host rocks showed relatively lower FeO (8.94–10.97 wt.%) and higher MgO (2.75–3.9 wt.%) contents, belonging to the schorl-dravite series (Supplementary Materials).

Also close to the pegmatite contact, irregular masses of greisen formed mainly by quartz and muscovite were locally observed. The crystal size in these bodies (up to 3 mm) was greater than that of the metasedimentary host rocks (Figure 5f).

4.4. Whole Rock Geochemistry

The studied aplite-pegmatite body showed relatively high SiO_2 (75.62 wt.%), Al_2O_3 (16.05 wt.%), Na_2O (3.18 wt.%), and P_2O_5 (0.23 wt.%) contents, resulting in a strongly peraluminous and perphosphorous nature (A/CNK = 2.1; normative apatite = 0.18; Table 1). As expected, MgO (0.03 wt.%), $Fe_2O_3^t$ (0.34 wt.%), TiO_2 (0.01 wt.%), and CaO (0.1 wt.%) concentrations are low, leading to a low maficity (B) parameter in the B-A diagram (Figure 6a). Moderate concentrations of K_2O (2.05 wt.%) lead to a K_2O/Na_2O ratio lower than unity (0.67). Regarding trace element composition, the studied dyke showed relatively low contents of F (100 ppm), but significantly high contents of Li (8480 ppm) (Figure 6b). The ratios K/Rb (27) and Nb/Ta (1.6) were in the range of Iberian-enriched aplite-pegmatite bodies (Figure 6c). In addition to Li, Nb, and Ta, concentrations of other trace elements such as Rb, Cs, Be, Sn, Ge, U, and Tl were significantly high as well, contrasting with commonly low contents of Ba, Sr, Zr, Y, Th, and REE (Figure 6e; Table 1).

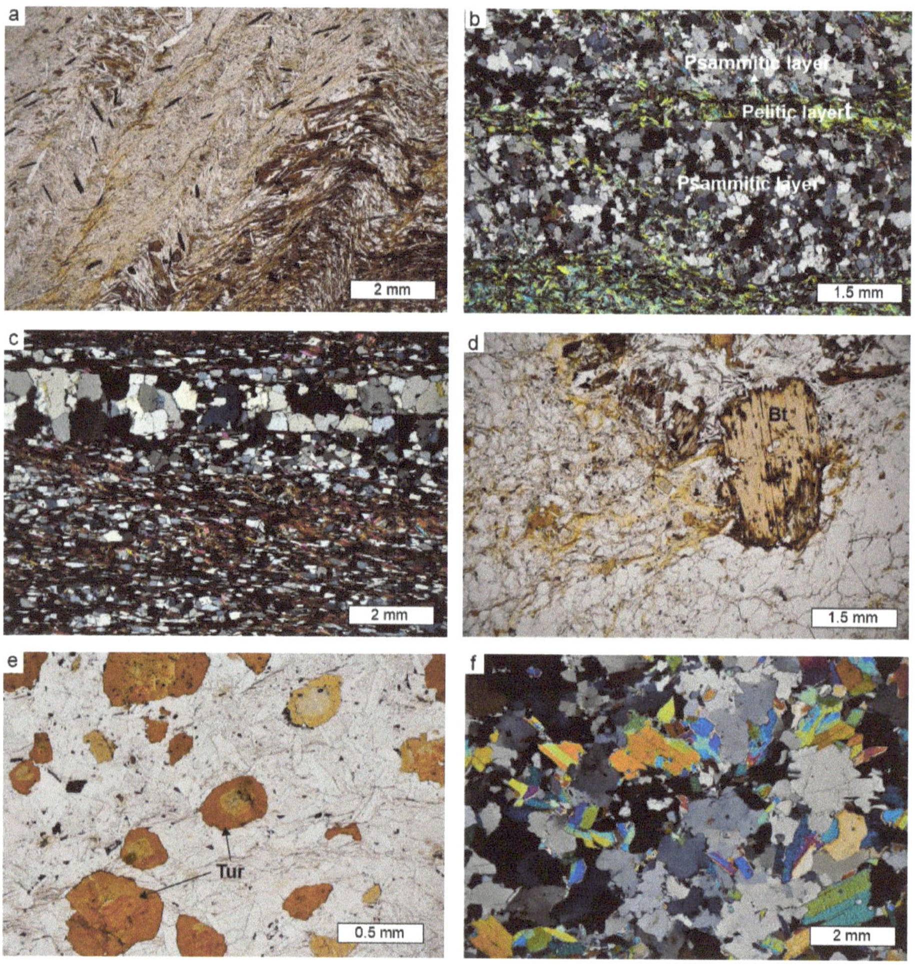

Figure 5. (**a**) Microphotograph of a pelitic (mica-rich) host rock sample, with a marked original slaty cleavage and a well-developed crenulation cleavage. (**b**) Alternating psammitic (quartz-rich) and pelitic layers in a psammitic host rock. (**c**) Microphotograph of an intermediate sample, with a greater mica proportion compared with the psammitic samples. (**d**) Porphyroblast of biotite (Bt) from the host rock. (**e**) Small tourmaline (Tur) crystals formed in the pelitic layers of the host rocks relatively close to the pegmatite. (**f**) Representative microphotograph of a greisen sample.

Table 1. Major- (wt.%) and trace-element (ppm) whole rock geochemical analyses of studied rocks from Alijó.

(a)

Type	**met.**	**met.**	**met.**	**met.**	**met.**	**met.**	**met.**	**met.**	**met.**	**met.**	**met.**	**met.**
Class.	lith.	wacke	sub-lith.	shale	lith.	lith.	lith.	shale	shale	shale	greisen	Shale
Sample	B-09	B-10	B-11	B-12	B-13	B-14	B-15	B-16	B-17	B-18	B-19	B-20
Dist. (m)	0 W	0.5 W	1 W	2 W	3 W	6 W	12 W	20 W	35 W	50 W	0 E	0.5 E
SiO_2	78.36	75.49	90.49	61.3	84.53	82.18	81.62	64.49	65.6	67.91	70.1	56.41
TiO_2	0.54	0.56	0.19	0.9	0.45	0.47	0.46	1.18	0.82	0.56	0.13	0.96
Al_2O_3	12.37	13.99	4.61	20.1	9.43	11.21	9.89	23.05	18.51	16.21	18.01	24.83
$Fe_2O_3^t$	3.08	2.95	2.69	8.02	1.56	1.71	3.31	3.65	5.92	7.57	2.06	6.54
MnO	0.024	0.026	0.026	0.085	0.026	0.015	0.037	0.055	0.044	0.052	0.04	0.026
MgO	0.62	0.54	0.26	1.69	0.18	0.18	0.42	0.37	0.77	0.49	0.2	0.58
CaO	<0.01	<0.01	0.02	<0.01	<0.01	<0.01	<0.01	<0.01	<0.01	<0.01	0.79	<0.01
Na_2O	0.22	0.2	0.06	0.23	0.14	0.16	0.17	0.28	0.28	0.23	0.37	0.28
K_2O	2.9	3.68	1.01	4.47	2.49	2.85	2.71	4.42	4.45	3.7	4.62	6.22
P_2O_5	0.04	0.04	0.02	0.04	0.04	0.03	0.05	0.09	0.1	0.1	0.93	0.11
LOI	1.82	2.26	0.71	3.79	1.49	1.79	1.69	2.69	4.09	3.65	2.73	4.7
Total	99.97	99.74	100.09	100.63	100.34	100.6	100.36	100.28	100.58	100.47	99.98	100.66
F	600	500	<100	200	<100	<100	<100	<100	<100	<100	600	300
As	22	66	12	30	15	12	8	104	46	19	14	32
B	1430	430	60	80	90	90	90	180	150	90	350	240
Ba	481	642	179	823	477	532	539	814	940	689	170	1170
Be	11	<3	<3	<3	<3	<3	<3	<3	<3	<3	147	6
Cr	90	110	110	150	80	100	90	130	130	120	70	150
Cs	26.8	22.8	4.3	16.8	8.8	8.2	11	9.1	17.1	7.9	66.5	30.2
Ga	15.9	19.8	6	31.4	15.9	13.3	15.9	29.9	25.4	24.5	30.3	32.6
Ge	2.8	3.1	1.3	3.7	2.3	2.1	2.7	2.9	2.7	2.6	6.4	3.3
Li	288	294	104	400	99	105	123	105	190	113	264	472
Mo	5	6	9	3	3	6	3	1	<1	3	2	2
Nb	12.5	11.8	5	24.6	10.8	9.8	10	20.5	17.7	13.7	58.3	19.3
Ni	40	40	40	50	20	30	20	30	50	40	140	20
Pb	27.9	19.6	17.2	25.2	18.3	21.4	19.9	32.2	26.5	26	19.4	34.8
Rb	215	216	52.1	229	104	104	102	147	178	146	834	334
Sn	38	22.8	3.5	7.6	3.7	2.7	3	8.3	5	4.3	185	24.4
Sr	58	53	23	69	52	54	56	82	62	60	197	85
Ta	1.3	0.9	0.3	1.8	0.8	1.1	0.9	2.1	1.5	1.2	51.4	1.8
Th	17.5	14.4	5.5	20.4	18.3	14.9	16.2	30.9	19.5	14.7	3.3	22
Tl	1.5	1.6	0.3	1.5	0.5	0.6	0.5	0.6	0.8	0.5	4.4	1.7
U	4	3.6	1.3	3.5	2.6	2.5	3.3	6.6	4.5	4.1	7.6	4.2
V	52	63	17	128	32	35	38	71	78	71	16	121
W	5.9	5.8	1.9	6.9	3.6	5.1	4.3	5.7	7.6	5.4	5.4	16.9
Y	32	24.7	9.3	35	29.4	24	28.9	41.5	30.4	25.7	6.3	38.2
Zn	60	30	30	80	<30	30	30	40	40	60	60	100
La	41.8	40.3	13.2	59.3	36.5	34.6	43.3	55.5	45	38.1	8	59.2
Ce	90.6	80.8	27.3	116	77.7	70.2	91	119	89.3	79.2	16.7	112
Pr	11.1	10.1	3.3	14.8	9.8	8.5	10.5	13.8	11.1	10.2	1.9	13.9
Nd	37	32.9	12.2	48.1	32.1	32	35.9	52.9	40.5	32.6	7.6	50.5
Sm	8.3	8	2.3	9.1	6	5.2	7	9.1	7.1	8.2	1.6	10
Eu	1.6	1.6	0.4	1.6	1.3	1.5	1.6	1.7	1.3	1.3	0.7	1.5
Gd	6.3	5.4	2.3	7.9	5.4	4.8	6	7.7	6	5.5	1.2	7.2
Tb	0.9	0.8	0.2	0.9	0.9	0.7	0.7	1.2	1	0.9	0.2	1.1
Dy	5.1	4.8	1.8	5.9	4.8	4.2	4.9	6.9	5.5	4.3	1	6.3
Ho	1	1.1	0.3	1.3	1.1	0.8	1.1	1.4	1.1	0.8	<0.2	1.4
Er	3.2	2	1.1	2.9	2.6	2.3	3.2	3.3	3.3	2.6	0.5	3.8
Tm	0.5	0.4	0.1	0.5	0.5	0.4	0.5	0.6	0.5	0.4	<0.1	0.5
Yb	4	3.6	1.4	4.2	3.8	2.7	4.3	5	4.3	2.9	1.1	3.9

(b)

Type	**met.**	**met.**	**met.**	**met.**	**met.**	**met.**	**met.**	**Ctrl.**	**Ctrl.**	**Ctrl.**	**Ctrl.**	**Peg.**
Class.	wacke	shale	lith.	lith.	wacke	shale	shale	lith.	shale	lith.	shale	Spd. Peg.
Sample	B-21	B-22	B-23	B-24	B-25	B-26	B-27	ALI-01	ALI-03	ALI-04	ALI-05	ALI-02
Dist. (m)	1 E	2 E	3 E	6 E	12 E	20 E	35 E	> 200	> 200	> 100	> 100	> 100
SiO_2	71.31	54.76	78.18	81.2	73.52	63.31	63.12	77.92	60.16	81.57	59.14	75.62
TiO_2	0.76	1.34	0.43	0.45	0.69	0.87	0.84	0.6	0.86	0.52	0.88	0.01
Al_2O_3	16.96	26.2	12.89	11.47	17.02	22.32	19.57	12.73	21.1	10.53	22.19	16.05
$Fe_2O_3^t$	3.5	6.67	2.19	1.63	3.14	5.89	7.41	2.9	6.82	2.36	6.74	0.34
MnO	0.024	0.056	0.016	0.015	0.026	0.051	0.048	0.054	0.065	0.039	0.074	0.029
MgO	0.3	0.85	0.36	0.24	0.44	0.87	0.62	0.44	0.91	0.41	1.48	0.03
CaO	<0.01	<0.01	<0.01	<0.01	0.01	0.02	<0.01	<0.01	0.13	0.11	0.06	0.1
Na_2O	0.21	0.24	0.16	0.15	0.18	0.24	0.38	0.14	0.26	0.15	0.28	3.18
K_2O	4.34	5.28	3.33	3.11	3.1	3.96	4.93	3.73	6.13	3.03	6.69	2.05
P_2O_5	0.09	0.15	0.04	0.02	0.04	0.1	0.1	0.04	0.27	0.12	0.1	0.23
LOI	2.93	4.32	2.12	1.8	2.24	2.93	3.69	1.83	3.36	1.52	3.07	0.84
Total	100.42	99.87	99.72	100.09	100.41	100.56	100.71	100.38	100.07	100.36	100.7	98.48

Table 1. *Cont.*

F	<100	<100	<100	<100	<100	<100	200	<100	300	<100	200	<100
As	16	102	15	19	22	78	17	18	175	24	40	<5
B	170	180	90	80	150	170	280	40	80	170	170	20
Ba	788	1020	614	560	568	755	1090	647	900	422	968	6
Be	<3	3	<3	<3	4	4	<3	<3	4	<3	5	141
Cr	150	160	80	90	100	160	130	80	160	100	140	70
Cs	17.1	17	8.8	5.8	13.5	32.6	19.6	7.8	25.2	7.2	22.5	65.5
Ga	22.7	33.4	16.6	17.5	21.7	28.8	27.9	15.4	34.9	15	30.8	17.5
Ge	3	3.8	2.4	2.3	3	3.3	2.7	1.6	3.2	2.1	2.5	6.8
Li	355	676	167	129	298	497	231	44	108	50	130	8480
Mo	4	<1	1	2	<1	3	<1	<1	<1	2	2	2
Nb	16.5	24.9	10.8	11.4	12.4	18.6	18.6	12.5	20.2	13.1	19.6	46.5
Ni	50	30	20	20	40	70	30	20	30	30	40	20
Pb	26.8	43.7	23.5	22	22.1	28.2	33	17.2	31.1	18.8	28.5	12
Rb	200	232	132	101	127	190	214	138	289	121	287	628
Sn	11.1	7	3.5	3.4	4.2	15.1	13.3	3.1	4.8	2.9	5	50.1
Sr	85	89	69	57	66	99	91	49	76	48	68	103
Ta	1.7	2.2	0.7	0.8	1.4	1.7	1.4	1.2	1.7	0.9	1.7	28.8
Th	20.5	27.7	9.4	12.9	22.5	19.8	21	24	22.5	21.3	19.7	0.5
Tl	1.2	1.1	0.7	0.4	0.8	1.1	1.1	0.9	1.3	0.6	1.6	3.7
U	4.1	6.3	2.3	2.6	4.2	4.4	4.6	5.1	4.9	4.5	4.1	12.7
V	63	113	38	35	46	82	94	37	104	36	101	<5
W	10.9	11	7	5.8	4.4	8.7	7.2	4.4	7.6	3	8.9	1.3
Y	30.8	42.7	26.2	22.7	26	39.2	33	19.7	52.1	30.8	30	0.2
Zn	40	90	30	<30	60	80	110	<30	70	40	60	40
La	48.6	78.2	28.7	31.1	39.1	52.1	52.6	30.7	93.5	21.5	31	<0.4
Ce	95.5	154	63	69.2	83	110	111	61.6	193	48.9	61.6	<0.8
Pr	12.1	18.9	7.2	8.2	10.6	14.2	13.7	8.8	24.8	6.1	7.9	<0.1
Nd	37.2	66	28.3	28.2	32.9	50.8	42	28.6	83.4	21.5	26.6	<0.4
Sm	7.6	13.3	4.1	5.7	6.5	9.2	6.7	4.8	15.9	5.1	5.8	<0.1
Eu	1.8	2.7	1.2	1.1	1.5	2	1.6	1	3.3	0.9	1.1	<0.1
Gd	6.2	10.6	4.7	5.2	5.5	7.5	7.6	3	15.2	5.6	4.5	0.2
Tb	1.1	1.5	0.7	0.7	0.8	1.2	1	0.6	2.2	0.9	0.8	<0.1
Dy	4.9	7.6	3.8	4.1	4.5	6.6	6.3	3.3	9.5	5.7	4.7	<0.3
Ho	1.2	1.5	0.9	0.9	1	1.2	1.3	0.8	1.9	1.3	0.9	<0.2
Er	3.7	4.6	2.8	2.6	3.2	3.5	3	2.3	5.6	3.5	3	<0.1
Tm	0.5	0.8	0.3	0.4	0.4	0.5	0.6	0.5	0.7	0.7	0.5	<0.1
Yb	4	5	2.6	2.8	3.8	4.1	3.8	3.7	5.2	4.1	3.3	0.1

met. = metasomatized host metasedimentary rock; Ctrl. = control sample (non-metasomatized composition); Peg. = pegmatite; lith. = litharenite; Spd. Peg. = spodumene-bearing pegmatite.

The studied set of metasedimentary samples could be classified geochemically as shales (eight pelitic samples), litharenites-arkoses (eight psammite samples), and two intermediate wackes (Figure 6e). The SiO_2 contents were lower in the pelitic samples (55–68 wt.%) compared with the psammitic samples (76–91 wt.%), which were inversely correlated with the Al_2O_3 contents (16–26 wt.% and 5–14 wt.%, respectively) (Table 1). The pelitic samples had the highest FeO^t (3.6–8.0 wt.%) and K_2O (3.7–6.2) concentrations (1.6–3.3 wt.% and 1.0–3.9 wt.% in psammitic samples). The contents of TiO_2, MgO, Na_2O, and P_2O_5 were also higher in the metapelites as expected, and all of them were extremely poor in CaO (most of them presented values below the detection limit). Trace element composition also differed depending on their pelitic or psammitic nature, with overall higher concentrations of Ba, Sr, Y, Th, and REE in the pelitic samples (Figure 6d,f).

In the case of other trace elements, significant variations were observed when metasomatized and non-metasomatized samples were compared. These differences were greater in the case of pelitic samples, as they held higher absolute concentrations compared with psammitic samples. Despite the similar low F contents of <600 ppm for the complete set of samples (<100 ppm for most of them), some tourmalinized samples displayed high B contents of 500–1430 ppm, and many of the metasomatized pelitic samples displayed Li concentrations greater than 300 ppm (up to 676 ppm) (Figure 7a). These Li concentrations decreased with the distance from the pegmatite contact, up to values that resembled those of the non-metasomatized samples (about 50 ppm in psammites and around 100–130 ppm in pelites). Maximum concentrations of other trace elements such as Rb (<350 ppm), Cs (<35 ppm), and Sn (<40 ppm) were also measured in host metasediments close to the pegmatite contact (Figure 7b–d).

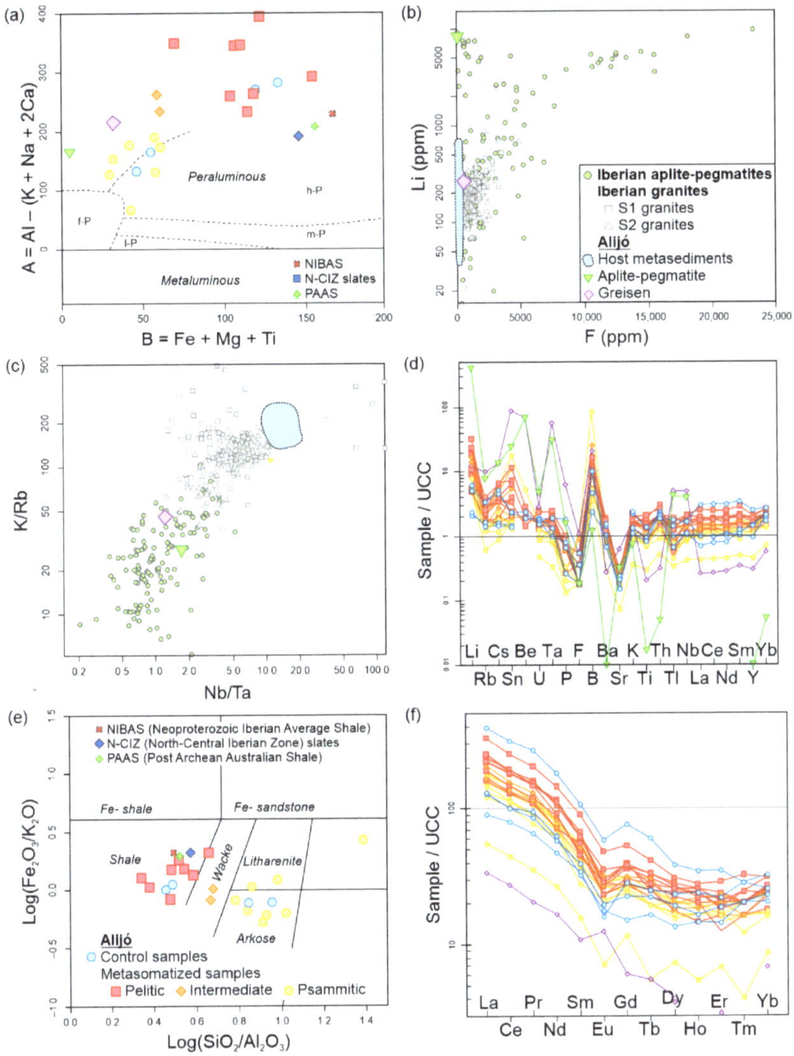

Figure 6. (**a**) Multicationic [B] vs. [A] diagram from [51] modified by [52], showing compositions of the selected aplite-pegmatite, greisen, and host rocks from Alijó. (**b**) Bivariate plot displaying F vs. Li (logarithmic) of the same set of samples (Iberian aplite-pegmatites, S1 granites, and S2 granites also shown for comparison; data taken from [13]. (**c**) Nb/Ta vs. K/Rb diagram (logarithmic) of the same three types of studied rocks. (**d**) Spider plots of the studied spodumene-pegmatite, host metasediments, and greisen normalized to the upper continental crust (UCC; [53]). (**e**) Classification diagram of siliciclastic sediments [54], showing compositions of the studied metasedimentary host rocks. Compositions of the Neoproterozoic Iberian Average Shale (NIBAS; [55]), average of North-Central Iberian Zone (N-CIZ) slates [56], and Post Archean Australian Shale (PAAS; [57]) are also displayed for comparison. (**f**) Chondrite-normalized [58] REE diagrams of the host metasediments.

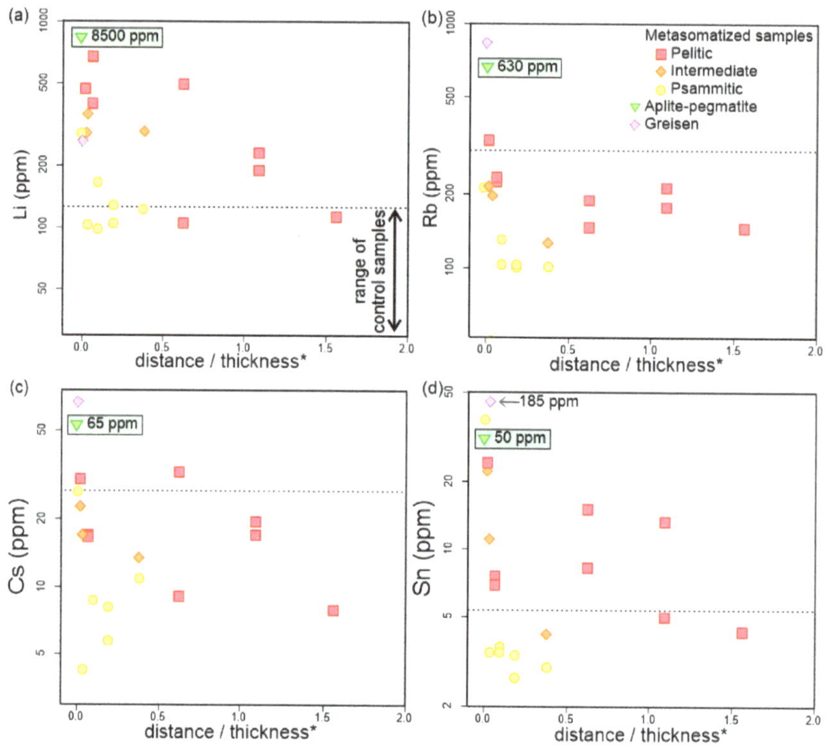

Figure 7. Binary diagrams showing (**a**) lithium, (**b**) rubidium, (**c**) cesium, and (**d**) tin concentrations in the studied host metasediments displayed relative to the distance from the dyke (*), i.e., the distance from the dyke divided by the thickness of the dyke (range of non-metasomatized control samples also displayed in grey for comparison).

Regarding the greisen sample collected immediately after contact with the pegmatite, it showed remarkably high concentrations of P_2O_5 (0.93 wt.%), F (600 ppm), B (350 ppm), Li (264 ppm), Be (147 ppm), Rb (834 ppm), Cs (66.5 ppm), Nb (58.3 ppm), Ta (51.4 ppm), Sn (185 ppm), and Tl (4.4 ppm) (Figure 7; Table 1).

5. Discussion

5.1. Internal Evolution of the Alijó Pegmatite

Several studies support that the structural state and texture of feldspars, as well as their chemical composition, may help to clarify the crystallization conditions of pegmatites [59–62]. The coexistence of albite and K-feldspar as individual phases suggests the subsolvus character of the studied body [63,64], as observed in the aplite-pegmatite body in Alijó. The crystallization of the two feldspars may be due to the increase in temperature of the solvus, associated with abundant Ca [65], or to the relatively high H_2O activity in the system, which would decrease the temperature of the solidus [63,66]. Considering the low Ca contents observed in the different mineral phases of the studied pegmatite, which are consistent with the extremely low Ca concentrations in the whole rock (CaO = 0.1 wt.%), the second option seems to be the most plausible. The formation of greisen bodies and the tourmalinization observed close to the contact also attest to the relatively high H_2O contents in the pegmatitic system. On the other hand, the presence of cross-hatched twinning in microcline, as a result of the monoclinic–triclinic phase transition, indicates that the crystallization temperature of the dyke might have been above 450–500 °C [67,68].

Moreover, Li-aluminosilicates may help in interpreting the crystallization conditions [20,69]. The presence of both primary and secondary spodumene indicates that, in addition to crystallizing directly from the pegmatitic melt, some of the spodumene crystals in the dyke have been formed during subsolidus processes. These secondary crystals were most likely formed as the result of the replacement of primary petalite by a spodumene + quartz intergrowth [SQI texture] [70–73]. As shown by the P-T phase diagram for Li-aluminosilicates (Figure 8), the textural and paragenetic relationships suggest a rapid T decrease to allow the coexistence of primary and secondary spodumene with late eucryptite. Considering that the maximum conditions of regional metamorphism for the host rocks are ~3 kbar [30] and that the minimum crystallization temperature for K-feldspar to show lattice twins is 450–500 °C, the primary crystallization of the Alijó pegmatite is proposed to have occurred within the range of 450–500 °C and 2–3 kbar (blue rectangle with dashed-line in Figure 8). This is consistent with the simultaneous crystallization of spodumene and 'non-preserved' petalite. The secondary mineral paragenesis of the studied dyke also indicates that the magmatic–hydrothermal transition in the evolution of the studied spodumene-bearing pegmatite continued up to eucryptite stability conditions (Figure 8). The replacement of spodumene crystals by eucryptite and cookeite is evidence of low-temperature metasomatism in a subsolidus stage. However, recent studies show that the stability of Li-aluminosilicates is more complex than previously thought, suggesting that P and T may not be the only factors controlling the crystallization of these minerals [72,73].

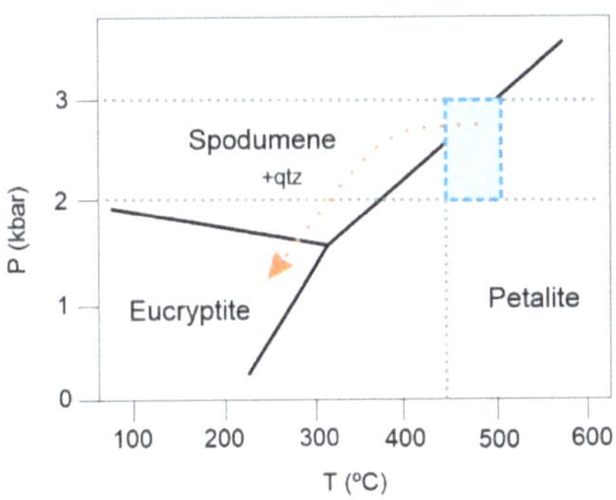

Figure 8. Temperature vs. pressure phase diagram for Li-aluminosilicates (modified from [69]).

Apart from textures, mineral chemistry provides valuable insights into the crystallization processes that occurred within the Alijó dyke. The ratio K/Rb is considered a good indicator of the degree of fractionation in granitic-pegmatitic magmas, decreasing as the fractionation proceeds [74–76]. The K/Rb ratio in micas of spodumene pegmatites from BA is about 29 [45]. This value is relatively low, but it cannot be considered as extremely low as seen in highly fractionated pegmatites (K/Rb = 9 for lepidolite-bearing pegmatites, [45]). In other pegmatitic fields from the CIZ, the observed values are consistent. In Tres Arroyos, micas from the lepidolite-bearing pegmatites show K/Rb ratios of 20-4 [60]. Micas of the Li-mica-rich pegmatites from Fregeneda–Almendra show ratios of about 8–12, whereas spodumene pegmatites are in the range of 33–39 [77]. Considering these values, the ratio of 29 for micas from BA suggests a higher degree of fractionation compared with other spodumene-bearing pegmatites from the CIZ, but lower when compared with Li-mica and/or lepidolite-rich pegmatites from the same area. Whole-rock geochemistry is in line with this, with a K/Rb ratio of 27 calculated for the studied dyke, which is relatively lower than that obtained by [19] and that of spodumene-bearing pegmatites from Fregeneda–Almendra [K/Rb = 39–46], and higher than those of the Li-mica-rich pegmatites from Fregeneda–Almendra (K/Rb = 16–17) [78] and lepidolite-bearing dykes from Tres Arroyos (K/Rb = 7) [79].

Phosphate minerals also play a significant role as petrogenetic indicators in these granitic-pegmatitic environments [80–83]. Considering that Cl is a highly volatile element, and so, would easily separate from the melt once the system was opened [84], the crystallization of chlorapatite must have occurred before the opening of the system. The analyzed CGM crystals show slightly higher Ta/[Ta + Nb] and Mn/[Mn + Fe] ratios than the ones proposed for the spodumene subtype pegmatites from the Barroso-Alvão region [44]. However, all of them are in agreement with other CGM from spodumene-bearing pegmatites from the CIZ, such as the Fregeneda–Almendra pegmatite Field [77]. The small variation in the fractionation of Nb–Ta could be attributed to the presence of other Fe competitors such as schorl [85].

According to experimental data based on the correlation $C_F^{Montebrasite} = 3.65C_F^{melt} + 0.07$ obtained by [86], the F concentration in the melt during crystallization would be below 0.27%. Such a low F contents in the melt, also evidenced by the absence of topaz and the presence only of montebrasite, would determine the principal Li-bearing minerals, crystallizing Li-aluminosilicate instead of lepidolite in the Alijó dyke.

5.2. External Metasomatism Caused by the Alijó Pegmatite

The magmatic–hydrothermal transition in pegmatite evolution is a complex, multi-stage process where silicate melt–hydrosaline melt–aqueous fluid immiscibility may occur as the pegmatitic melt cools (e.g., [87–90]). This transition in magmatic systems of granitic composition usually occurs when crystallization is already in an advanced stage, but in the case of volatile-rich shallow magmatic systems, early exsolution of fluid phases is also likely [91]. The nature of the exsolved fluid strongly depends on the nature of the pegmatitic melt, conditioning the availability of the elements that can partition into the fluid phase [92,93]. In the case of evolved terms of LCT pegmatites, the studied cases have shown that the pegmatite-derived metasomatizing fluids may cause enrichment in elements such as Li, Rb, Cs, F, B, Sn, Be, Tl, and W in the host rocks [15,17,18,94,95]. Despite varying in abundance compared to host rocks of other types of LCT-type enriched pegmatites, the host rocks that accommodate similar spodumene-rich pegmatites in this case study show enrichment in these same elements, especially in B, F, Li, Rb, Cs, and Sn (e.g., [17,18]).

The scarce presence of tourmaline within the studied spodumene-bearing dyke, together with the tourmalinization developed in the host metasedimentary rocks, particularly in close proximity to the pegmatite, suggest a relatively early opening of the system. This tourmalinization indicates B metasomatism in the host metasedimentary rocks, which caused the enrichment of up to 1430 ppm in the whole rock. Fluorine contents are also markedly higher in the host rocks close to the pegmatite contact (up to 600 ppm) and the greisen (up to 600 ppm) when compared with that in the spodumene-bearing pegmatite (<100 ppm) (Table 1). Although concentrations are not strictly correlated, Li enrichment seems to have occurred together with that of B and F. Lithium enrichment decreases gradually as the distance from the pegmatite increases, resembling a geochemical halo that differs significantly between pelitic (mica-rich) and psammitic (mica-poor) metasomatized samples (Figures 7a and 9a). Tin concentration is also remarkable outside the dyke, with a maximum value of 185 ppm observed in the greisen sample. Similarly to Li, Sn contents decrease with the distance in the metasomatized host metasediments, from a maximum value of 38 ppm in samples collected close to the dyke (Figure 7d). Rubidium and Cs enrichment in the metasedimentary host rocks is surprisingly minimal and is only recognizable in some of the pelitic samples. On the contrary, in the greisen sample, Rb and Cs concentrations reach high values of 834 and 66 ppm, respectively (Table 1). Despite the relatively low Li contents in the greisen (264 ppm) compared with metasomatized pelitic samples (400–700 ppm), the whole-rock geochemistry of the greisen properly reflects the overall enrichment prompted by the pegmatite-derived metasomatizing fluids. The studied greisen sample shows notably high contents of B, F, Li, Rb, Cs, Sn, Be, Nb, Ta, and Tl, with a total sum of B, F, Li, Rb, Cs, Sn of up to 2300 ppm (Figure 9b). The sum of these key trace elements shows differences between mica-rich and mica-poor samples as well, with greater values in mica-rich samples (Figure 9b).

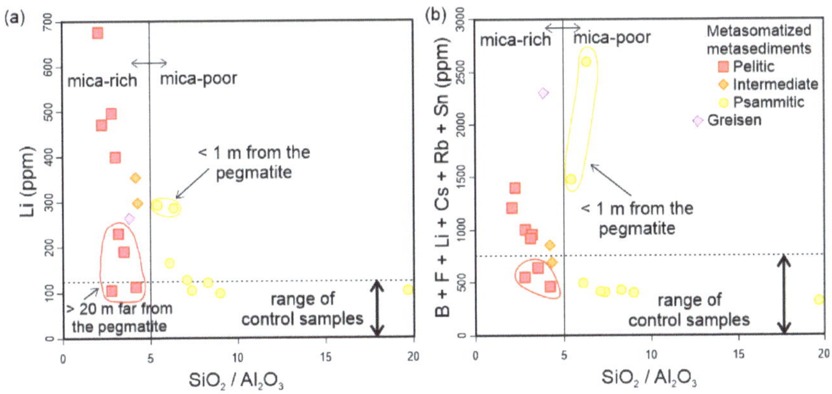

Figure 9. Binary SiO_2/Al_2O_3 vs. Li (**a**) and SiO_2/Al_2O_3 vs. B + F + Li + Cs + Rb + Sn (**b**) diagrams that show geochemical differences between mica-rich (pelitic) and mica-poor (psammitic) metasedimentary host rocks.

Despite the absence of clear concentration decreasing trends that may define geochemical halos around the studied spodumene-bearing pegmatite (the only exception could be Li), the metasomatic signature related to the pegmatite in the metasedimentary host rocks is consistent with other studied cases in central Iberia (e.g., [18]). The geochemistry of the pelitic samples collected close to the studied pegmatite, together with the geochemical composition of the greisen, indicate that at least B, F, Li, Rb, Cs, Sn, Be, Nb, Ta, and Tl were mobilized from the pegmatitic melt to the host rocks by the exsolved fluids. The early exsolution of a fluid phase from the pegmatitic melt and its subsequent escape into the nearby host rocks were probably facilitated by shallow emplacement conditions. The loss of flux elements such as H_2O, Li, F, and B may have induced a strong undercooling of the system, resulting in the formation of the UST observed in the Alijó pegmatite. This lowering of the solidus, together with a sudden decrease in pressure and temperature due to the shallow emplacement of the pegmatitic melt into open fractures, may have led to the formation of such textures, as suggested in experimental works [96–99]. In addition, the high concentrations of B and F in the exsolved fluids probably favored the formation of metal–ligand complexes, facilitating their transport from the pegmatite to the surrounding host rocks during fluid exsolution [100–103].

Based on the obtained geochemical data, certain inferences can be made regarding lithium exploration for the BA region. Considering the significant differences observed between pelitic and psammitic host metasedimentary rocks in terms of elemental enrichment, only the pelitic samples should be considered for lithogeochemical exploration studies. Apart from the visible tourmalinization in the field, which may only indicate close proximity to a dyke, the whole rock geochemical gains in pelitic host rocks may trace mineralized pegmatites at greater distances. Lithium contents greater than 300 ppm in this type of sample may indicate the presence of a Li-rich pegmatite but should be used in combination with other elements due to the lack of clear geochemical halos. We suggest the sum of B, F, Li, Rb, Cs, and Sn as a possible parameter to consider for Li exploration in BA, with an approximate minimum value of 1000 ppm (with all due caution) that may relate to a mineralized pegmatite in this region.

6. Conclusions

The spodumene-bearing pegmatite dyke from Alijó, located in the Portuguese Barroso–Alvão region, is an unzoned pegmatite enriched in elements such as Li, Cs, Rb, Be, Sn, Nb, Ta, Ge, U, and Tl, among others. Mineralogical data suggest a moderate to high degree of differentiation of the pegmatitic melt, which was emplaced under pressure conditions of approximately 2–3 kbar. Primary crystallization likely occurred at temperatures between 450–500 °C, with significant undercooling caused by (1) a sudden decrease in P-T conditions due to the shallow emplacement of the pegmatitic melt into open fractures, (2) the early exsolution of a fluid phase carrying flux elements such as F, B, and Li, or (3) the combination of both of them.

The escape of the exsolved fluid phase generated greisen bodies and induced external metasomatism in the surrounding metasedimentary host rocks. This study shows that the degree of metasomatism varies not only with the composition of the intruded pegmatite, as observed in other pegmatitic fields from CIZ [18], but also with the nature of the host rocks, which are generally classified as either pelitic (mica-rich) or psammitic (mica-poor). Whole-rock geochemistry reveals that pelitic samples exhibit a significantly higher enrichment of fluid-mobilized elements (B, F, Li, Sn, Be, ±Rb, and ±Cs) compared to psammitic elements collected at the same distances from the dyke. This suggests that, for lithogeochemical studies aimed at Li exploration, pelitic samples should be prioritized over psammitic ones. With all caution, we propose that whole rock Li concentrations greater than 300 ppm, together with a minimum value of 1000 ppm for the sum of B, F, Li, Rb, Cs, and Sn in pelitic metasediments from Barroso–Alvão, may indicate the presence of a mineralized pegmatite dyke in the area.

Supplementary Materials: The following supporting information can be downloaded at https://www.mdpi.com/article/10.3390/min14070701/s1.

Author Contributions: Conceptualization, I.G.-O., J.E.-M. and E.R.-R.; methodology, J.E.-M.; formal analysis, N.S.-L. and J.E.-M.; writing—original draft preparation, I.G.-O., N.S.-L. and J.E.-M.; writing—review and editing, I.G.-O., E.R.-R., A.L. and T.M.; visualization, I.G.-O. and J.E.-M.; supervision, E.R.-R.; project administration, E.R.-R.; funding acquisition, E.R.-R. All authors have read and agreed to the published version of the manuscript.

Funding: Financial support was provided by the European Commission's Horizon 2020 Innovation Programme [grant agreement No 869274, project GREENPEG: New Exploration Tools for European Pegmatite Green-Tech Resources]; and University of the Basque Country UPV/EHU [grant GIU21/008]. N.S.L. is beneficiary of the Predoctoral Programme for the Formation of Non-Doctoral Research Staff of the Basque Government's Department of Education.

Data Availability Statement: Data are contained within the article and Supplementary Materials.

Acknowledgments: The authors are grateful to three anonymous reviewers for their suggestions. The authors would like to acknowledge also Eric Xu for the editorial handling. The authors thank Grupo Lagoa for access to the Alijó pegmatite during this study.

Conflicts of Interest: The authors declare no conflicts of interest.

References

1. Jaskula, B.W. *Minerals Yearbook: Lithium*; U.S. Geological Survey: Reston, VA, USA, 2017.
2. Gourcerol, B.; Gloaguen, E.; Melleton, J.; Tuduri, J.; Galieguea, X. Re-assessing the European lithium resource potential—A review of hard-rock resources and metallogeny. *Ore Geol. Rev.* **2019**, *109*, 494–519. [CrossRef]
3. Bibienne, T.; Magnan, J.-F.; Rupp, A.; Laroche, N. From Mine to Mind and Mobiles: Society's Increasing Dependence on Lithium. *Elements* **2020**, *16*, 265–270. [CrossRef]
4. Moss, R.L.; Tzimas, E.; Kara, H.; Willis, P.; Kooroshy, J. *Critical Metals in Strategic Energy Technologies*; Institute for Energy and Transport European Commission: Luxembourg, 2011.
5. Linnen, R.L.; Van Litchervelde, M.; Černý, P. Granitic pegmatites as sources of strategic metals. *Elements* **2012**, *8*, 275–280. [CrossRef]
6. Simons, B.; Andersen, J.C.Ø.; Shail, R.K.; Jenner, F.E. Fractionation of Li, Be, Ga, Nb, Ta, In, Sn, Sb, W and Bi in the peraluminous Early Permian Variscan granites of the Cornubian Batholith: Precursor processes to magmatic-hydrothermal mineralization. *Lithos* **2017**, *278–281*, 491–512. [CrossRef]

7. European Comission. *Critical Raw Materials Resilience: Charting a Path towards Greater Security and Sustainability*; European Commission: Brussels, Belgium, 2020.
8. Müller, A.; Reimer, W.; Wall, F.; Williamson, B.; Menuge, J.; Brönner, M.; Haase, C.; Brauch, K.; Pohl, C.; Lima, A.; et al. GREENPEG—Exploration for pegmatite minerals to feed the energy transition: First steps towards the Green Stone Age. *Geol. Soc. Lond. Spec. Publ.* **2022**, *526*, 27. [CrossRef]
9. Roda-Robles, E.; Pesquera, A.; Gil-Crespo, P.P.; Vieira, R.; Lima, A.; Garate-Olave, I.; Martins, T.; Torres-Ruiz, J. Geology and mineralogy of Li mineralization in the Central Iberian Zone (Spain and Portugal). *Mineral. Mag.* **2016**, *80*, 103–126. [CrossRef]
10. Simmons, W.B.; Webber, K.L. Pegmatite genesis: State of the art. *Eur. J. Mineral.* **2008**, *20*, 421–438. [CrossRef]
11. London, D. Ore-forming processes within granitic pegmatites. *Ore Geol. Rev.* **2018**, *101*, 349–383. [CrossRef]
12. Blundy, J.; Wood, B. Partitioning of trace elements between crystals and melts. *Earth Planet. Sci. Lett.* **2003**, *210*, 383–397. [CrossRef]
13. Roda-Robles, E.; Villaseca, C.; Pesquera, A.; Gil-Crespo, P.P.; Vieira, R.; Lima, A.; Garate-Olave, I. Petrogenetic relationships between variscan granitoids and Li–(F–P)-rich aplite-pegmatites in the Central Iberian Zone: Geological and geochemical constraints and implications for other regions from the European variscides. *Ore Geol. Rev.* **2018**, *95*, 408–430. [CrossRef]
14. Breiter, K.; Ďurišová, J.; Korbelová, Z.; Lima, A.; Vašinová Galiová, M.; Hložková, M.; Dosbaba, M. Rock textures and mineralzoning—A clue to understanding rare-metal granite evolution: Argemela stock, Central-Eastern Portugal. *Lithos* **2022**, *410–411*, 106562. [CrossRef]
15. Linnen, R.; McNeil, A.; Flemming, R. Some thoughts on metasomatism in pegmatites. *Can. Mineral.* **2019**, *57*, 765–766. [CrossRef]
16. Roza-Llera, A.; Fuertes-Fuente, M.; Cepedal, A.; Martin-Izard, A. Barren and Li–Sn–Ta Mineralized Pegmatites from NW Spain (Central Galicia): A Comparative Study of Their Mineralogy, Geochemistry, and Wallrock Metasomatism. *Minerals* **2019**, *9*, 739. [CrossRef]
17. Barros, R.; Kaeter, D.; Menuge, J.F.; Škoda, R. Controls on chemical evolution and rare element enrichment in crystallising albite-spodumene pegmatite and wallrocks: Constraints from mineral chemistry. *Lithos* **2020**, *352–353*, 105289. [CrossRef]
18. Errandonea-Martin, J.; Garate-Olave, I.; Roda-Robles, E.; Cardoso-Fernandes, J.; Lima, A.; Ribeiro, M.A.; Teodoro, A.C. Metasomatic effect of Li-bearing aplite pegmatites on psammitic and pelitic metasediments: Geochemical constraints on critical raw material exploration at the Fregeneda–Almendra Pegmatite Field (Spain and Portugal). *Ore Geol. Rev.* **2022**, *150*, 105155. [CrossRef]
19. Lima, A. Estrutura, Mineralogia e Génese dos Filões Aplitopegmatíticos com Espodumena da Região do Barroso-Alvão (Norte de Portugal). Ph.D. Thesis, University Porto, Porto, Portugal; INPL, Nancy, France, 2000.
20. Charoy, B.; Noronha, F.; Lima, A.M.C. Spodumene—Petalite—Eucryptite: Mutual relationships and alteration style in Li-rich aplite–Pegmatite dykes from northern Portugal. *Can. Mineral.* **2001**, *39*, 729–746. [CrossRef]
21. Martins, T. Multidisciplinary Study of Pegmatites and Associated Li and Sn–Nb–Ta Mineralisation from the Barroso-Alvão Region. Ph.D. Thesis, University Porto, Porto, Portugal, 2009.
22. Farias, P.; Gallastegui, G.; González Lodeiro, F.; Marquínez, J.; Martín-Parra, L.M.; Martínez Catalán, J.R.; Pablo Maciá, J.G.; Rodríguez-Fernández, L.R. Aportaciones al conocimiento de la litoestratigrafía y estructura de Galicia Central. In *Memorias del Museo y Laboratorio Mineralógico y Geológico, de la Facultad de Ciencias*; Universidad de Oporto: Porto, Portugal, 1987; Volume 1, pp. 411–431.
23. Julivert, M.; Marcos, A.; Truyols, J. L'évolution paléogéographique du NW de l'Espagne pendant l'Ordovicien–Silurien. *Bull. Soc. Géol. Mineral. Bretagne* **1972**, *4*, 1–7.
24. Ribeiro, A.; Pereira, E.; Dias, R.; Gil Ibarguchi, J.I.; Arenas, R. Allochthonous Sequences. In *Pre-Mesozoic Geology of Iberia. IGCP-Project 233*; Dallmeyer, R.D., Garcia, E.M., Eds.; Springer: Berlin/Heidelberg, Germany, 1990. [CrossRef]
25. Valverde-Vaquero, P.; Marcos, A.; Farias, P. U–Pb dating of Ordovician felsic volcanics in the Schistose domain of the Galicia-Trás-os-Montes Zone near Cabo Ortegal (NW Spain). *Geol. Acta* **2005**, *3*, 27–37. [CrossRef]
26. Martínez Catalán, J.R.; Rubio Pascual, F.J.; Díez Montes, A.; Díez Fernández, R.; Gómez Barreiro, J.; Dias da Silva, I.; González Clavijo, E.; Ayarza, P.; Alcock, J.E. The late Variscan HT/LP metamorphic event in NW and Central Iberia: Relationships to crustal thickening, extension, orocline development and crustal evolution. In *The Variscan Orogeny: Extent, Timescale and the Formation of the European Crust*; Schulmann, K., Martínez Catalán, J.R., Lardeaux, J.M., Janousek, V., Oggiano, G., Eds.; Geological Society: London, UK, 2014; pp. 225–247. [CrossRef]
27. Ribeiro, M.A.; Martins, H.C.; Almeida, A.; Noronha, F. *Notícia Explicativa da Carta Geológica de Portugal na Escala 1:50,000—Folha 06C—Cabeceiras de Basto*; Serviços Geológicos de Portugal: Amadora, Portugal, 2000.
28. Noronha, F.; Ramos, J.M.F.; Rebelo, J.; Ribeiro, A.; Ribeiro, M.L. Essai de corrélation des phases de d'eformation hercyniennes dans le NW de la péninsule Ibérique. *Leidse Geol. Meded.* **1981**, *52*, 87–91.
29. Dias, R.; Ribeiro, A. The Ibero-Armorican Arc: A collisional effect against an irregular continent? *Tectonophysics* **1995**, *246*, 113–128. [CrossRef]
30. Ribeiro, M.A.; Ramos, R.; Noronha, F. Pegmatite-aplite veins of Barroso-Alvão Field. Lithostratigraphy and metamorphism of host rocks. In *Granitic Pegmatites: The State of the Art, Field Trip Guidebook*; Lima, A., Roda-Robles, E., Eds.; Universidade do Porto: Porto, Portugal, 2007.
31. Noronha, F.; Ribeiro, M.L. *Notícia Explicativa da Carta Geológica de Portugal na Escala 1:50,000—Folha 06A—Montalegre*; Serviços Geológicos de Portugal: Amadora, Portugal, 1983.
32. Teixeira, C. *Notícia Explicativa da Carta Geológica de Portugal na Escala 1:50,000—Folha 06B—Chaves*; Serviços Geológicos de Portugal: Amadora, Portugal, 1974.

33. Sant'Ovaia, H.; Ribeiro, M.A.; Martins, H.C.B.; Noronha, F. *Notícia Explicativa da Carta Geológica de Portugal na Escala 1:50,000—Folha 06D—Vila Pouca de Aguiar*; Serviços Geológicos de Portugal: Amadora, Portugal, 2011.
34. Ferreira, N.; Iglesias, M.; Noronha, F.; Pereira, E.; Ribeiro, A.; Ribeiro, M.L. Granitoides da Zona Centro Ibérica e o seu enquadramento geodinâmico. In *Geología de los Granitóides y Rocas Asociadas del Macizo Hesperico. Homenage a L.C. García de Figuerola*; Bea, F., Carnicero, A., Gonzalo, J.C., López-Plaza, M., Rodríguez Alonso, M.D., Eds.; Libro Editorial Rueda: Madrid, Spain, 1987; pp. 37–51.
35. Dias, G.; Leterrier, J.; Mendes, A.; Simões, P.P.; Bertrand, J.M. U–Pb zircon and monazite geochronology of post-collisional Hercynian granitoids from the Central Iberian Zone (Northern Portugal). *Lithos* **1998**, *45*, 349–369. [CrossRef]
36. Sant'Ovaia, H.; Bouchez, J.L.; Noronha, F.; Leblanc, D.; Vigneresse, J.L. Composite-laccolith emplacement of the post-tectonic Vila Pouca de Aguiar granite pluton (northern Portugal): A combined AMS and gravity study. In *The Fourth Hutton Symposium on the Origin of Granites and Related Rocks*; Barbarin, B., Stephens, W.E., Bonin, B., Bouchez, J.-L., Clarke, D.B., Cuney, M., Martin, H., Eds.; Geological Society of America: Boulder, CO, USA, 2000; Volume 350. [CrossRef]
37. Almeida, A.; Martins, H.C.; Noronha, F. Hercynian acid magmatism and related mineralizations in northern Portugal. *Gondwana Res.* **2002**, *5*, 423–434. [CrossRef]
38. Cruz, C.; Sant'Ovaia, H.; Barolomeu Raposo, M.I.; Lourenço, J.M.; Almeida, F.; Noronha, F. Unraveling the emplacement history of a Portuguese post-tectonic Variscan pluton using magnetic fabrics and gravimetry. *J. Struct. Geotech.* **2021**, *153*, 104470. [CrossRef]
39. Almeida, A.; Leterrier, J.; Noronha, F.; Bertrand, J.M. U-Pb zircon and monazite geochronology of the Hercynian two mica granite composite pluton of Cabeceiras de Basto (Northern Portugal). *C. R. Acad. Sci.* **1998**, *326*, 779–785. [CrossRef]
40. Mendes, A.; Dias, G. Mantle-like Sr-Nd isotope composition of Fe-K subalkaline granites: The Peneda-Gerês Variscan massif (NW Iberian Peninsula). *Terra Nova* **2004**, *16*, 109–115. [CrossRef]
41. Teixeira, R.J.S.; Neiva, A.M.R.; Gomes, M.E.P.; Corfu, F.; Cuesta, A.; Croudace, I.W. The role of fractional crystallization in the genesis of early syn-D3, tin-mineralized Variscan two-mica granites from the Carrazeda de Ansiães area, northern Portugal. *Lithos* **2012**, *153*, 177–191. [CrossRef]
42. Dias, G.; Simões, P.P.; Ferreira, N.; Leterrier, J. Mantle and Crustal Sources in the Genesis of Late-Hercynian Granitoids (NW Portugal): Geochemical and Sr-Nd Isotopic Constraints. *Gondwana Res.* **2002**, *5*, 287–305. [CrossRef]
43. Villaseca, C.; Bellido, F.; Pérez-Soba, C.; Billström, K. Multiple crustal sources for post-tectonic I-type granites in the Hercynian Iberian Belt. *Miner. Petrol.* **2009**, *96*, 197–211. [CrossRef]
44. Martins, T.; Lima, A.; Simmons, B.; Falster, A.U.; Noronha, F. Geochemical fractionation of Nb-Ta oxides in Li-bearing pegmatites from the Barroso-Alvão Pegmatite Field, northern Portugal. *Can. Mineral.* **2011**, *49*, 777–791. [CrossRef]
45. Martins, T.; Roda-Robles, E.; Lima, A.; Parseval, P. Geochemistry and evolution of micas in the Barroso-Alvão pegmatite field, northern Portugal. *Can. Mineral.* **2012**, *50*, 1117–1129. [CrossRef]
46. Pouchou, J.L.; Pichoir, F. "PAP" φ(ρZ) procedure for improved quantitative microanalysis. In *Microbean Analysis*; Armstrong, J.T., Ed.; San Francisco Press: San Francisco, CA, USA, 1985; pp. 104–106.
47. Cerný, P.; Ercit, T.S. The classification of granitic pegmatites revisited. *Can. Mineral.* **2005**, *43*, 2005–2026. [CrossRef]
48. Tischendorf, G.; Gottesmann, B.; Förster, H.-J.; Trumbull, R.B. On Li-bearing micas; estimating Li from electron microprobe analyses and an improved diagram for graphical representation. *Mineral. Mag.* **1997**, *61*, 809–834. [CrossRef]
49. Henry, D.J.; Novak, M.; Hawthorne, F.C.; Ertl, A.; Dutrow, B.L.; Uher, P.; Pezzotta, F. Nomenclature of the tourmaline-supergroup minerals. *Am. Mineral.* **2011**, *9*, 895–913. [CrossRef]
50. Bosi, F. Tourmaline crystal chemistry. *Am. Mineral.* **2018**, *2018103*, 298–306. [CrossRef]
51. Debon, F.; Le Fort, P. A chemical–mineralogical classification of common plutonic rocks and associations. *Trans. R. Soc. Edinburgh Earth Sci.* **1983**, *73*, 135–149. [CrossRef]
52. Villaseca, C.; Barbero, L.; Herreros, V. A re-examination of the typology of peraluminous granite types in intracontinental orogenic belts. *Trans. R. Soc. Edinburgh Earth Sci.* **1998**, *89*, 113–119. [CrossRef]
53. Rudnick, R.L.; Gao, S. Composition of the continental crust. In *Treatise on Geochemistry*, 2nd ed.; Holland, H.D., Turekian, K.K., Eds.; Elsevier Science: Oxford, UK, 2014; Volume 4, pp. 1–51. [CrossRef]
54. Herron, M.M. Geochemical classification of terrigenous sands and shales from core or log data. *J. Sediment. Petrol.* **1988**, *58*, 820–829. [CrossRef]
55. Ugidos, J.M.; Sánchez-Santos, J.M.; Barba, P.; Valladares, M.I. Upper Neoproterozoic series in the Central Iberian, Cantabrian and West Asturian Leonese Zones (Spain): Geochemical data and statistical results as evidence for a shared homogenised source area. *Precambrian Res.* **2010**, *178*, 51–58. [CrossRef]
56. Villaseca, C.; Merino, E.; Oyarzun, R.; Orejana, D.; Pérez-Soba, C.; Chicharro, E. Contrasting chemical and isotopic signatures from Neoproterozoic metasedimentary rocks in the Central Iberian Zone (Spain) of pre-Variscan Europe: Implications for terrane analysis and early Ordovician magmatic belts. *Precambrian Res.* **2014**, *245*, 131–145. [CrossRef]
57. Taylor, S.R.; McLennan, S.M. *The Continental Crust: Its Composition and Evolution*; Blackwell Scientific Publications: Oxford, UK, 1985.
58. McDonough, W.F.; Sun, S.-S. The composition of the Earth. *Chem. Geol.* **1995**, *120*, 223–253. [CrossRef]
59. Neiva, A.M.R. Distribution of trace elements in feldspars granitic aplites and pegmatites from Alijó-Sanfins, northern Portugal. *Mineral. Mag.* **1995**, *59*, 35–45. [CrossRef]

60. Garate-Olave, I.; Roda-Robles, E.; Gil-Crespo, P.P.; Pesquera, A. Mica and feldspar as indicators of the evolution of a highly evolved granite-pegmatite system in the Tres Arroyos area (Central Iberian Zone, Spain). *J. Iberian Geol.* **2018**, *44*, 375–403. [CrossRef]
61. London, D.; Hunt, L.E.; Schwing, C.R.; Gutter, B.M. Feldspar thermometry in pegmatites: Truth and consequences. *Contrib. Mineral. Petr.* **2020**, *175*, 8. [CrossRef]
62. Liu, Y.; Qin, K.; Zhao, J.; Zhou, O.; Shi, R. Feldspar traces mineralization processes in the Qongjiagang giant lithium ore district, Himalaya, Tibet. *Ore Geol. Rev.* **2023**, *157*, 105451. [CrossRef]
63. Tuttle, O.F.; Bowen, N.L. Origin of Granite in the Light of Experimental Studies in the System $NaAlSi_3O_8$–$KAlSi_3O_8$–SiO_2–H_2O. *Geol. Soc. Am. Mem.* **1958**, *74*, 1–146. [CrossRef]
64. Parson, I. Feldspars defined and described: A pair of posters published by the Mineralogical Society. Sources and supporting information. *Mineral. Mag.* **2010**, *74*, 529–551. [CrossRef]
65. Carmichael, I.S.E.; Turner, F.J.; Verhoogen, J. *Igneous Petrology*; McGraw—Hill: New York, NY, USA, 1974; p. 739.
66. Tuttle, O.F. Origin of the contrasting mineralogy of extrusive and plutonic salic rocks. *J. Geol.* **1952**, *60*, 107–124. [CrossRef]
67. Smith, J.V. Chemical and textural properties. In *Feldspar Minerals*; Smith, J.V., Ed.; Springer: Berlin/Heidelberg, Germany, 1974.
68. Ribbe, P.H. *Feldspar Mineralogy*, 2nd ed.; De Gruyter: Berlin, Germany, 1983; p. 362.
69. London, D. Experimental phase equilibria in the system $LiAlSiO_4$-SiO_2-H_2O: A petrogenetic grid for lithium-rich pegmatites. *Am. Mineral.* **1984**, *69*, 995–1004.
70. Černý, P.; Ferguson, R.B. The Tanco pegmatite at Bernic Lake, Manitoba. IV. Petalite and spodumene relations. *Can. Mineral.* **1972**, *11*, 660–678.
71. Rossovskyi, L.N.; Matrosov, I.I. Pseudomorphs of quartz and spodumene after petalite and their importance to the pegmatite-forming process. *Dokl. Akad. Nauk SSSR* **1974**, *216*, 1135–1137.
72. Dias, F.; Ribeiro, R.; Gonçalves, F.; Lima, A.; Roda-Robles, E.; Martins, T. Calibrating a Handheld LIBS for Li Exploration in the Barroso–Alvão Aplite-Pegmatite Field, Northern Portugal: Textural Precautions and Procedures When Analyzing Spodumene and Petalite. *Minerals* **2023**, *13*, 470. [CrossRef]
73. Dias, F.; Lima, A.; Roda-Robles, E. Mutual relationships between spodumene and petalite from the Iberian Massif pegmatites: More than PT changes? *Can. Mineral.* **2019**, *57*, 731–732. [CrossRef]
74. Roda, E.; Pesquera, A.; Gil-Crespo, P.P.; Torres-Ruiz, J.; Fontan, F. Origin and internal evolution of the Li-F-Be-B-P-bearing Pinilla de Fermoselle pegmatite (Central Iberian Zone, Zamora, Spain). *Am. Mineral.* **2005**, *90*, 1887–1899. [CrossRef]
75. Marchal, K.L.; Simmons, W.B.; Falster, A.U.; Webber, K.L.; Roda-Robles, E. Geochemistry, mineralogy, and evolution of Li-Al micas and feldspars from the Mount Mica pegmatite, Maine, USA. *Can. Mineral.* **2014**, *52*, 221–233. [CrossRef]
76. Cao, C.; Shen, P.; Bai, Y.; Luo, Y.; Feng, H.; Li, C.; Pan, H. Chemical evolution of micas and Nb-Ta oxides from the Koktokay pegmatites, Altay, NW China: Insights into rare-metal mineralization and genetic relationships. *Ore Geol. Rev.* **2022**, *146*, 104933. [CrossRef]
77. Vieira, R.; Roda-Robles, E.; Pesquera, A.; Lima, A. Chemical variation and significance of micas from the Fregeneda-Almendra pegmatitic field (Central-Iberian Zone, Spain and Portugal). *Am. Mineral.* **2011**, *96*, 637–645. [CrossRef]
78. Roda-Robles, E.; Vieira, R.; Lima, A.; Errandonea-Martin, J.; Pesquera, A.; Cardoso-Fernandes, J.; Garate-Olave, I. Li-rich pegmatites and related peraluminous granites of the Fregeneda-Almendra field (Spain-Portugal): A case study of magmatic signature for Li enrichment. *Lithos* **2023**, *452–453*, 107195. [CrossRef]
79. Garate-Olave, I.; Roda-Robles, E.; Gil-Crespo, P.P.; Pesquera, A.; Errandonea-Martin, J. The Tres Arroyos Granitic Aplite-Pegmatite Field (Central Iberian Zone, Spain): Petrogenetic Constraints from Evolution of Nb-Ta-Sn Oxides, Whole-Rock Geochemistry and U-Pb Geochronology. *Minerals* **2020**, *10*, 1008. [CrossRef]
80. Keller, P.; Von Knorring, O. Pegmatites at the Okatjimukuju farm, Karibib, Namibia Part I: Phosphate mineral associations of the Clementine II pegmatite. *Eur. J. Mineral.* **1989**, *1*, 567–593. [CrossRef]
81. Baijot, M.; Hatert, F.; Philippo, S. Mineralogy and geochemistry of phosphates and silicates in the Sapucaia pegmatite, Minas Gerais, Brazil: Genetic implications. *Can. Mineral.* **2012**, *50*, 1531–1554. [CrossRef]
82. Vignola, P.; Zucali, M.; Rotiroti, N.; Marotta, G.; Risplendente, A.; Pavese, A.; Boscardin, M.; Mattioli, V.; Bertoldi, G. The chrysoberyl- and phosphate-bearing albite pegmatite of Malga Garbella Val Di Rabbi, Trento province, Italy. *Can. Mineral.* **2018**, *56*, 411–424. [CrossRef]
83. Roda-Robles, E.; Gil-Crespo, P.P.; Pesquera, A.; Lima, A.; Garate-Olave, I.; Merino-Martínez, E.; Cardoso-Fernandes, J.; Errandonea-Martin, J. Compositional Variations in Apatite and Petrogenetic Significance: Examples from Peraluminous Granites and Related Pegmatites and Hydrothermal Veins from the Central Iberian Zone (Spain and Portugal). *Minerals* **2022**, *12*, 1401. [CrossRef]
84. Piccoli, P.; Candela, P. Apatite in Igneous Systems. *Rev. Mineral. Geochem.* **2002**, *48*, 255–292. [CrossRef]
85. Van Lichtervelde, M.; Linnen, R.L.; Salvi, S.; Beziat, D. The role of metagabbro rafts on tantalum mineralization in the Tanco pegmatite, Manitoba. *Can. Mineral.* **2006**, *44*, 625–644. [CrossRef]
86. London, D.; Morgan Vi, G.B.; Wolf, M.B. Amblygonite-montebrasite solid solutions as monitors of fluorine in evolved granitic and pegmatitic melts. *Am. Mineral.* **2001**, *86*, 225–233. [CrossRef]
87. Kaeter, D.; Barros, R.; Menuge, J.F.; Chew, D.M. The magmatic hydrothermal transition in rare-element pegmatites from southeast Ireland: LA-ICP-MS chemical mapping of muscovite and columbite-tantalite. *Geochim. Cosmochim. Acta* **2018**, *240*, 96–130. [CrossRef]
88. Ballouard, C.; Elburg, M.A.; Tappe, S.; Reinke, C.; Ueckermann, H.; Doggart, S. Magmatic-hydrothermal evolution of rare metal pegmatites from the Mesoproterozoic Orange River pegmatite belt (Namaqualand, South Africa). *Ore Geol. Rev.* **2020**, *116*, 103252. [CrossRef]

89. Hulsbosch, N.; Muchez, P. Tracing fluid saturation during pegmatite differentiation by studying the fluid inclusion evolution and multiphase cassiterite mineralisation of the Gatumba pegmatite dyke system (NW Rwanda). *Lithos* **2020**, *354–355*, 105285. [CrossRef]
90. Shaw, R.A.; Goodenough, K.M.; Deady, E.; Nex, P.; Ruzvidzo, B.; Rushton, J.C.; Mounteney, I. The Magmatic–Hydrothermal Transition in Lithium Pegmatites: Petrographic and Geochemical Characteristics of Pegmatites from the Kamativi Area, Zimbabwe. *Can. Mineral.* **2022**, *60*, 957–987. [CrossRef]
91. Candela, P.A. A Review of Shallow, Ore-related Granites: Textures, Volatiles, and Ore Metals. *J. Petrol.* **1997**, *38*, 1619–1633. [CrossRef]
92. Thomas, R.; Davidson, P. Water in granite and pegmatite-forming melts. *Ore Geol. Rev.* **2012**, *46*, 32–46. [CrossRef]
93. Thomas, R.; Davidson, P. Revisiting complete miscibility between silicate melts and hydrous fluids, and the extreme enrichment of some elements in the supercritical state—Consequences for the formation of pegmatites and ore deposits. *Ore Geol. Rev.* **2016**, *72*, 1088–1101. [CrossRef]
94. Selway, J.B.; Breaks, F.W.; Tindle, A.G. A Review of Rare-Element (Li-Cs-Ta) Pegmatite Exploration Techniques for the Superior Province, Canada, and Large Worldwide Tantalum Deposits. *Explor. Min. Geol.* **2005**, *14*, 1–30. [CrossRef]
95. Martins, T.; Linnen, R.L.; Fedikow, M.A.F.; Singh, J. Whole-rock and mineral geochemistry as exploration tools for rare-element pegmatite in Manitoba: Examples from the Cat Lake–Winnipeg River and Wekusko Lake pegmatite fields (parts of NTS 52L6, 63J13). In *Report of Activities 2017*; Manitoba Growth, Enterprise and Trade, Manitoba Geological Survey: Winnipeg, MB, Canada, 2017; pp. 42–51.
96. Webber, K.L.; Falster, A.U.; Simmons, W.B.; Foord, E.E. The role of diffusion-controlled oscillatory nucleation in the formation of line rock in pegmatite-aplite dikes. *J. Petrol.* **1997**, *38*, 1777–1791. [CrossRef]
97. Webber, K.L.; Simmons, W.B.; Falster, A.U.; Foord, E.E. Cooling rates and crystallization dynamics of shallow level pegmatite-aplite dikes, San Diego County, California. *Am. Mineral.* **1999**, *84*, 708–717. [CrossRef]
98. Sirbescu, M.-L.C.; Hartwick, E.E.; Student, J.J. Rapid crystallization of the Animikie Red Ace Pegmatite, Florence county, northeastern Wisconsin: Inclusion microthermometry and conductive-cooling modeling. *Contrib. Mineral. Petrol.* **2008**, *156*, 289–305. [CrossRef]
99. Nabelek, P.I.; Whittington, A.G.; Sirbescu, M.-L.C. The role of H_2O in rapid emplacement and crystallization of granite pegmatites: Resolving the paradox of large crystals in highly undercooled melts. *Contrib. Mineral. Petrol.* **2010**, *160*, 313–325. [CrossRef]
100. Pollard, P.J.; Pichavant, M.; Charoy, B. Contrasting evolution of fluorine- and boron-rich tin systems. *Mineral. Depos.* **1987**, *22*, 315–321. [CrossRef]
101. Thomas, R.; Webster, J.D.; Heinrich, W. Melt inclusions in pegmatite quartz: Complete miscibility between silicate melts and hydrous fluids at low pressure. *Contrib. Mineral. Petrol.* **2000**, *139*, 394–401. [CrossRef]
102. Thomas, R.; Förster, H.-J.; Heinrich, W. The behaviour of boron in a peraluminous granite–pegmatite system and associated hydrothermal solutions: A melt and fluid inclusion study. *Contrib. Mineral. Petrol.* **2003**, *144*, 457–472. [CrossRef]
103. Migdisov, A.A.; Williams-Jones, A.E. Hydrothermal transport and deposition of the rare earth elements by fluorine-bearing aqueous liquids. *Mineral. Depos.* **2014**, *49*, 987–997. [CrossRef]

minerals

MDPI

Article

Geochemistry, Mineralogy, and Geochronology of the NYF Pegmatites, Jiaolesayi, Northern Qaidam Basin, China

Long Zhang [1,2], Xianzhi Pei [1,*], Yongbao Gao [2], Zuochen Li [1], Ming Liu [2], Yongkang Jing [2], Yuanwei Wang [2], Kang Chen [2,3], Nan Deng [2], Yi Zhang [2] and Junwei Wu [2]

1 School of Earth Science and Resources, Chang'an University, Xi'an 710054, China; zhanglong242016@163.com (L.Z.); lizuochen2003@163.com (Z.L.)

2 Innovation Center for Gold Ore Exploration, Xi'an Center of Mineral Resources Survey, China Geological Survey, Xi'an 710100, China; gaoyongbao2006@126.com (Y.G.); mingl2005@163.com (M.L.); jingyk0112@163.com (Y.J.); wyuanwei2023@163.com (Y.W.); kangchen320@163.com (K.C.); dengn0706@outlook.com (N.D.); 15501175101@163.com (Y.Z.); micwujunwei@163.com (J.W.)

3 State Key Laboratory of Continental Dynamics, Department of Geology, Northwest University, Xi'an 710069, China

* Correspondence: peixzh@chd.edu.cn

Abstract: A significant amount of pegmatite has been discovered on the northwest margin of the Qaidam basin. Among this, the Jiaolesayi pegmatite, located in the northwestern margin of the Quanji Massif (Oulongbuluke micro-continent), shows rare element mineralization potential. Detailed field investigations, along with mineralogical, geochemical, and zircon U-Pb geochronological studies, were carried out on the pegmatite. The results show that the Jiaolesayi pegmatite is syenite, without obvious compositional zoning in the outcrop. It exhibits a peraluminous, high-K calc-alkaline nature with strong depletions in Eu, Sr, Ba, Ti, and P, and high contents of Nb, Ta, Y, Ti, U, Th, and heavy rare earth elements (HREEs), which are primarily concentrated in allanite-(Ce), euxenite-(Y), limonite, thorite, and zircon. The geochemical and mineralogical features of the syenite pegmatite indicate it belongs to the euxenite-type in the rare element class (REE) of the NYF family, with the characteristic accessory mineral being euxenite-(Y). Its 10,000 Ga/Al ratios (2.46 to 2.96), Zr + Nb + Ce + Y contents (998 to 6202 ppm), Y/Nb ratios (0.62 to 0.75), and Yb/Ta ratios (0.80 to 1.49) show an affinity with A_1-type granite. Zircons from the syenite sample yielded a weighted mean $^{206}Pb/^{238}U$ age of 413.6 ± 1.4 Ma, while the elevated U and Th concentrations in the zircons and Th/U ratios (0.04 to 0.16) suggest the possible influence of hydrothermal processes in the late-stage fractional crystallization. In the context of the regional tectonic evolution, the syenite pegmatite may have formed from a basic alkaline magma derived from an OIB-like melt with minor crustal contamination, under the post-collisional extension setting.

Keywords: NYF pegmatite; rare element; euxenite-(Y); northern Qaidam basin

1. Introduction

Competition among countries for strategic resources such as lithium (Li), niobium (Nb), tantalum (Ta), and rare earth elements (REEs) is intensifying due to their critical role in emerging technologies and industries [1]. About 1%–2% of all pegmatites contain these rare elements, making these pegmatite deposits have significant economic value [2]. Pegmatites are primarily classified into two chemical families based on their distinctive rare element enrichment: the LCT (Lithium–Cesium–Tantalum) family and the NYF (Niobium–Yttrium–Fluorine) family [3,4]. Recent studies [5–7] have also proposed a mixed family (LCT + NYF) for pegmatites exhibiting characteristics of both.

Despite advancements in classification, LCT pegmatites dominate in both quantity and subtype diversity among rare element pegmatites. They also represent the most significant type of rare earth deposits and have become a primary focus of research [8]. LCT pegmatites are enriched in Li, Cs, and Ta, exhibit peraluminous features, and show compositional affinity with S-type granites [9]. They are believed to form during the late-stage crystallization differentiation of granitic melts, primarily in late orogenic and post-orogenic settings [4]. Globally renowned deposits such as Bikita [10] in Zimbabwe; Tanco [11] in Canada; Greenbushes [12] in Australia; and Dahongliutan [13] in China all belong to the LCT family. In contrast, NYF pegmatite signatures are relatively obscure. Pegmatites that fit this type enrich Nb, Ta, Y, Zr, U, Th, HREEs, and F. They often contain exotic minerals such as gadolinite, euxenite-(Y), polycrase-(Y), aeschynite-(Y), samarskite, thorite, allanite-(Ce), or fergusonite [14–18], and rarely exhibit regional zoning patterns [8,19,20]. NYF pegmatites typically show subaluminous or peraluminous characteristics, are closely associated with A-type granites, and form under non-orogenic settings [4]. NYF pegmatites also hold significant value for rare element resources, for instance, the Strange Lake deposit in Canada hosts over one million tons of high-grade REEs, ZrO_2, Nb_2O_5, and BeO [21].

Since 2020, researchers have discovered over two hundred pegmatites with rare element mineralization potential in the western Quanji Massif, the northwest margin of the Qaidam basin (Figure 1C–E). These pegmatites were identified in the Niubiziliang, Dachaigou, and Jiaolesayi areas using remote sensing interpretation and identification technology [13,22]. Many pegmatites exhibit Nb, Ta, Rb, and HREE mineralization. The largest continuous vein verified in the field extends approximately 7 km. Previous researchers [23] reported a small Nb-Ta deposit in the Jiaolesayi area, but detailed studies were not conducted. Recently, the authors found several rare element-bearing pegmatites while examining the Nb-Ta deposit. Besides known Nb-Ta enrichment, the newly discovered pegmatites also significantly enrich Zr, Y, U, Th, and HREEs, showing the characteristics of NYF pegmatite. Therefore, through petrological, mineralogical, and geochemical analyses, combined with electron probe microanalysis (EPMA) and zircon U-Pb dating, this study aims to determine the geological features, mineral assemblages, possible petrogenesis, and geochronology of newly discovered rare element-bearing pegmatites in Jiaolesayi.

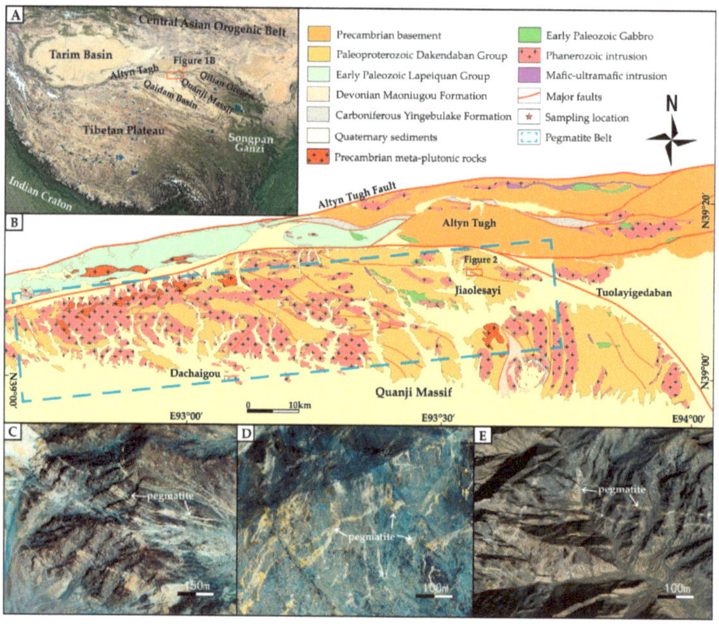

Figure 1. (A) Location of the study area (based on Google Earth image). (B) Geological map of the northwest margin of Quanji Massif. Figure 2 is shown as red rectangle. (C–E) Remote sensing images of pegmatites in Niubiziliang, Dachaigou, and Jiaolesayi (unpublished images from Xi'an Center of China Geological Survey, 2020).

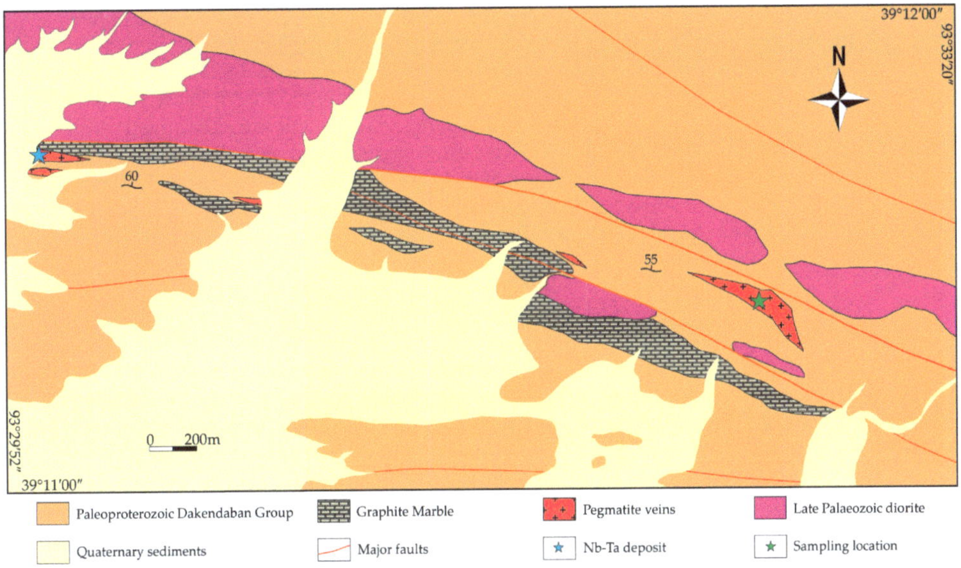

Figure 2. Simplified geological map of Jiaolesayi.

2. Geological Background

2.1. Regional Geology

The Jiaolesayi area is on the north margin of the Quanji Massif (Figure 1B), close to the Altun Tagh Fault which marks the northern margin of the Qinghai–Tibet Plateau [24]. The Quanji Massif, also known as the Oulongbuluke micro-continent, is one of several micro-massifs situated between the North China Craton (NCC) and Tarim Craton (TC) [25–27]. It extends WNW-ESE for nearly 500 km and measures about 50–100 km wide. Bounded by the Qilian orogenic belt to the northeast and the early Paleozoic north Qaidam high- to ultrahigh-pressure (HP-UHP) belt to the southwest [28,29] (Figure 1A), the Quanji Massif features a Paleoproterozoic medium to high-grade metamorphosed crystalline basement overlain by unmetamorphosed Neoproterozoic to Mesozoic strata [25,28,30–32]. This typical cratonic double-layered structure along with similarities to the Tarim Craton in terms of stratigraphic sequences, magmatic evolution, and metamorphic histories [33], suggests that this narrow ancient cratonic fragment was once part of the TC [26,34].

The metamorphic basement of the Quanji Massif is mainly composed of the Delingha complex and the Dakendaban Group paragneisses. The Delingha complex is primarily a suite of metamorphosed granitoids with ca. 2.41 Ga [35] amphibolite enclaves, grouped into Quanjishan, Delingha, Mohe, and Hudesheng granitoid plutons [36,37]. The Dakendaban Group comprises amphibolite-facies volcano-sedimentary rocks for its lower subgroup and metasedimentary rocks for its upper subgroup. The primary rock types include graphite–marble, sillimanite–garnet–biotite–quartz schist, graphite–biotite–quartz schist, amphibolite–plagioclase schist, garnet–biotite–plagioclase gneiss, and some garnet–amphibole pyroxene granulite lenses [38]. These aluminum-rich metamorphic rocks are identified as khondalite-series and have suffered multiple regional and dynamic metamorphic events [39,40]. The unconformably overlaying sedimentary cover, the Quanji Group, mainly consists of quartz sandstone with minor tuff beds [29]. Its maximum deposition age is determined at ca.1.73 Ga [41]. The long-term left-lateral strike-slip activity [42] of the Altun Tagh Fault has induced significant structural deformations in the adjacent Quanji Massif, resulting in intense deformation of the rocks. Combined with multiple metamorphic events, the original geological features within the rocks have been completely altered.

The orientation of structural fabric within the Dakendaban Group is notably shifted from predominantly N-S to NEE-SWW as it approaches the Altun Tagh Fault.

The Silurian diorite is the only known massive intrusion in Jiaolesayi, as Mesozoic–Cenozoic sediments covered almost half of the area (Figure 2). Additionally, there are several small granites, granodiorite, alkali feldspar granite, granitic pegmatites, and quartz veins, including the previously reported Nb-Ta-bearing pegmatite, intruded in the Dakendaban Group and the Silurian diorite. Most of their ages remain uncertain due to the limited investigation, while the early studies suggested these veins formed during the Yanshanian Movement [23], without detailed geochronology data. However, this hypothesis has yet to be verified.

2.2. Geological Characteristics of Jiaolesayi Pegmatite and Sample Description

Jiaolesayi pegmatites are generally exposed as lenses intruded into the garnet–biotite–plagioclase gneiss and marble of the Dakendaban Group, with clear boundaries and no signs of migmatization or anatexis. Four primary outcrops are located within a 2 km range from west to east, with widths ranging from 10 to 80 m (Figure 2). The previously reported Nb-Ta deposit is at the western end (Figure 3A,B), while the newly found pegmatite is at the eastern end. Pervasive joints and brittle fractures are well-developed in the pegmatite due to the activities of the Altun Tagh Fault (Figure 3C). Since the newly discovered pegmatite is close to the regional NEE-NSS trending faults, its northern margin exhibits brecciation and has developed some quartz veins that indicate hydrothermal activities caused by later tectonic activities (Figure 3E). Magnetites are widespread in the pegmatite, with a maximum grain size exceeding 3 cm (Figure 3D), and exhibit strong magnetic properties. No obvious mineral zoning features were observed in the pegmatite outcrops, distinguishing them from the significant zoning patterns of LCT-type pegmatite.

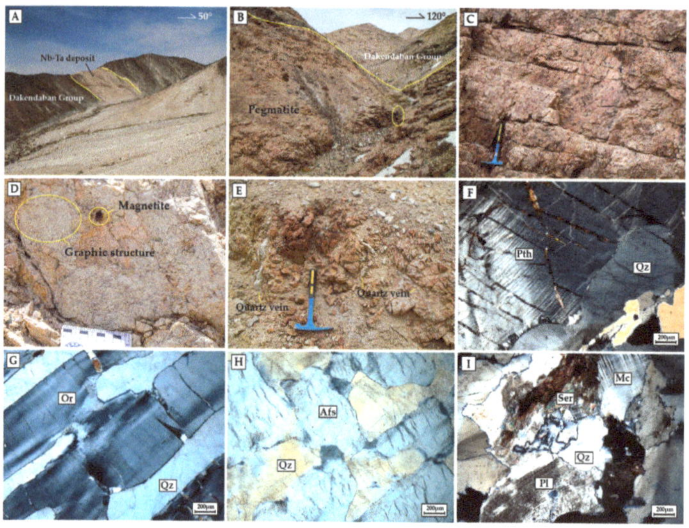

Figure 3. Photographs and photomicrographs of pegmatite outcrops and rock samples from Jiaolesayi. (**A**) Previous reported Nb-Ta deposit, represented by the blue star in Figure 2. (**B**) Newly discovered pegmatite, represented by the green star in Figure 2, a person in the yellow circle as a scale. (**C**) Conjugated joints and brittle fractures developed in the flesh-red pegmatite. (**D**) Graphic structure and magnetite aggregates in the pegmatite. (**E**) Quartz veins intruded into the brecciated pegmatite. (**F**) Perthitic texture of perthite under a polarizing microscope. (**G**,**H**) Photomicrographs of graphic structure under a polarizing microscope. (**I**) Pegmatite with sericitization under a polarizing microscope. Abbreviations: Afs—alkali feldspar; Mc—mica; Or—orthoclase; Pl—plagioclase; Pth—perthite; Qz—quartz; Ser—sericite.

To investigate the rare element mineralization of the pegmatite, eight fresh pegmatite samples (D3301-1 to D3301-8) were collected from the newly discovered outcrop (Figure 2). All samples underwent bulk rock major and trace element analyses. D3301-1 was selected for zircon U-Pb dating. D3301-1 and D3301-4 were chosen for petrographic observation and EPMA analyses. These samples are generally flesh-red or brick-red in color, displaying a massive structure (Figure 3C) and typical graphic texture (Figure 3D). They mainly consist of fine- to coarse-grained, subhedral-to-euhedral alkali feldspar, orthoclase, quartz, mica, biotite, and plagioclase. The maximum grain size of orthoclase is about 8.3 cm, and quartz exceeds 10 cm. In the photomicrographs, quartz, alkali feldspar, and orthoclase exhibit typical graphic structures (Figure 3G,H), and sericite indicates the sericitization of plagioclase (Figure 3I). Accessory minerals including magnetite, allanite, zircon, limonite, pyrite, and thorite are observed under a reflecting microscope.

3. Analytical Methods

3.1. Major and Trace Element Analyses

Major element compositions were determined using a Zetium X-ray fluorescence spectrometer (XRF) manufactured by Malvern Panalytical Ltd., Almelo, The Netherlands, at the Yanduzhongshi Geological Analysis Laboratories Ltd., Beijing, China. The analytical errors for major elements were better than 1%. Trace elements of these samples were analyzed using an Aurora M90 inductively coupled mass spectrometer (ICP-MS) manufactured by Analytik Jena Ltd., Jena, Germany, with an analytical uncertainty better than 5%. The obtained trace element values in the GSR-2 standard are all consistent with their recommended values.

3.2. LA-ICP-MS Zircon U-Pb Dating

Zircon U-Pb dating and element analysis were conducted using a Coherent GeoLasPro 193 nm laser ablation system, Saxonburg, PA, USA, coupled to an Agilent 7700× ICP-MS at the Key Laboratory for the Study of Focused Magmatism and Giant Ore Deposits, Xi'an Center of Geological Survey (CGS), China. Helium was the carrier gas, and argon was used as the make-up gas to adjust sensitivity. The laser beam is focused at 36 μm. Each analysis included a background acquisition of approximately 10 s (gas blank), 40 s of sample signal acquisition, and 10 s for rinsing. U-Pb dating employed 91,500 and GJ-1 zircon standards, with common lead correction based on measured ^{204}Pb [43]. Trace element compositions of zircons were calibrated using artificial synthetic silicate glass NIST610 of the American National Standard Substance Bureau as an external standard, with Si as the internal standard. Data analysis and element concentration calculations were performed using Glitter (ver. 4.4) [44]. The recommended element content values for NIST glass are sourced from the GeoReM database (http://georem.mpch-mainz.gwdg.de accessed on 28 March 2022). Detailed analytical procedures were described by Li [45].

3.3. EPMA Analysis

Mineral compositions were determined using a JEOL JXA-8230 Electron Probe Micro Analyzer (EPMA), Tokyo, Japan, equipped with four wavelength-dispersive spectrometers at the Yanduzhongshi Geological Analysis Laboratories Ltd., Beijing, China. Before the analyses, the samples were coated with a ca. 20 nm thin conductive carbon film. The analyses include an accelerating voltage of 15 kV, a beam current of 20 nA, and a 5 μm spot size. Natural minerals and synthetic oxides were used as standards. Data were corrected online using a modified ZAF (atomic number, absorption, fluorescence) correction procedure. Detailed EPMA methods are based on Yang [46].

4. Results

4.1. Major and Trace Elements

Whole-rock major and trace element compositions of the Jiaolesayi pegmatite samples are listed in Supplementary Table S1.

The Jiaolesayi pegmatite samples (D3301-1 to D3301-8) have medium SiO_2 (60.61~63.02 wt.%) and Al_2O_3 (20.53~21.14 wt.%) contents, both in a narrow range, and high total alkali abundances ($Na_2O + K_2O$ = 9.82~11.26 wt.%), with low MgO (0.07~0.18 wt.%), FeO^T (1.04~2.68 wt.%), MnO (0.04~0.05 wt.%), TiO_2 (0.09~0.35 wt.%), and P_2O_5 (<0.01~0.10 wt.%) contents. All samples plot in the syenite field of the TAS diagram (Figure 4A) and show weakly peraluminous characteristics in the A/NK vs. A/CNK diagram (A/CNK values of 1.00~1.05) (Figure 4B). They are located in the high-K calc-alkaline field in the K_2O vs. SiO_2 diagram (Figure 4C).

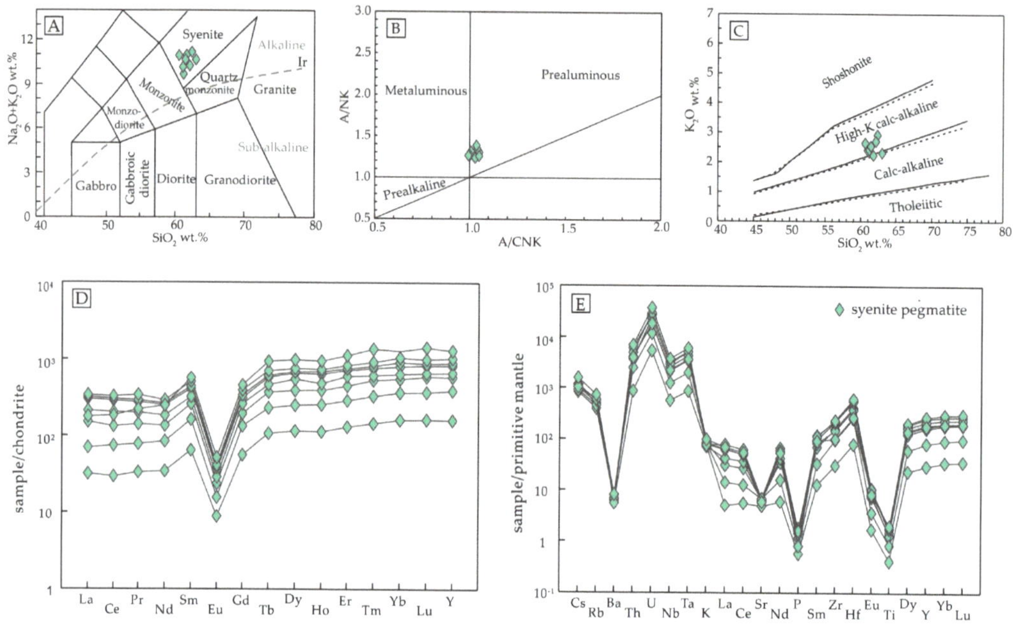

Figure 4. Classification diagrams for the Jiaolesayi pegmatite. (**A**) Total alkalis ($Na_2O + K_2O$) vs. SiO_2 (TAS) diagram after Middlemost [47]; (**B**) (A/NK) vs. (A/CNK) diagram, solid lines are after Peccerillo [48]; (**C**) K_2O vs. SiO_2 diagram after Rickwood [49], and (**D**) Chondrite-normalized REE patterns. (**E**) Primitive mantle-normalized trace element spider diagrams for the Jiaolesayi pegmatite samples. Normalizing data for the chondrite and primitive mantle are from Sun and McDonough [50]. Symbols: green diamond—syenite pegmatite samples.

The chondrite-normalized rare earth element (REE) patterns (Figure 4D) of the syenite pegmatite samples exhibit a medium left-sloping REE distribution, showing strong enrichment of HREEs, with $(La/Yb)_N$ ratios ranging from 0.14 to 0.37, and pronounced negative Eu anomalies (Eu/Eu* = 0.09–0.15). The ΣREE (total REE concentrations) of all samples range from 162 to 1286 ppm. On the primitive mantle-normalized trace element spider diagrams (Figure 4E), all samples exhibit similar wave-like patterns, characterized by pronounced negative anomalies of Ba, Sr, P, and Ti, and strong enrichment in large ion lithophile elements (LILEs, e.g., Cs and Rb) and high field strength elements (HFSEs, e.g., Nb, Ta, Zr, Hf, U, and Th).

4.2. Zircon U-Pb Geochronology

Zircons from the syenite pegmatite sample D3301-1 are subhedral to euhedral, transparent, and dark brown (Figure 5A). Most of them are incomplete crystal fragments. The complete ones have lengths ranging from 100 to 300 μm and widths from 50 to 150 μm, with a length/width ratio between 3:1 and 1.5:1. Core mantling structures and weak oscillatory zoning in the rims are visible. Their cathodoluminescence (CL) images are opaque and spongy (Figure 5A), which is common when zircons contain elevated rare elements [51]. The uranium (U) concentrations of the zircons range from 5795 to 12,393 ppm, thorium (Th) contents vary from 268 to 1169 ppm, and the Th/U ratios range from 0.04 to 0.16, indicating the possible influence of metamorphism or hydrothermal processes [52]. Therefore, the obtained data have been carefully checked, and 24 out of 30 concordant data points were chosen, resulting in a weighted mean $^{206}Pb/^{238}U$ age of 413.6 ± 1.4 Ma with a mean square weighted deviation (MSWD) of 1.2 (Figure 5B). Zircon data are listed in Supplementary Table S2.

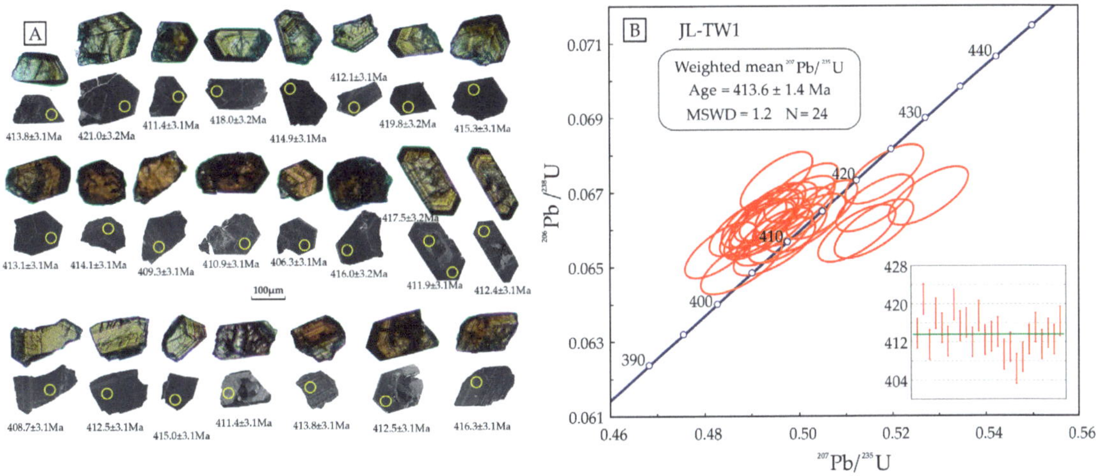

Figure 5. (**A**) Photomicrographs (cross-polarized light) and cathodoluminescence images of all tested zircons from Jiaolesayi pegmatite. (**B**) U-Pb Concordia diagram of sample D3301-1.

4.3. EPMA Analysis

Under reflected light microscopy, some anhedral to euhedral, transparent dark brown, columnar, or needle-like minerals were observed in samples D3301-1 and D3301-4. They are embedded in the blocky K-feldspar and quartz with a grain size of 100 to 500 μm (Figure 6A–C,G). EPMA later identified these minerals as euxenite-(Y). Their chemical compositions mainly consist of Nb_2O_5 (28.88 to 32.64 wt.%), TiO_2 (21.97 to 24.11 wt.%), Y_2O_3 (15.65 to 21.20 wt.%), UO_2 (5.52 to 11.88 wt.%), ΣREE (7.16 to 8.70 wt.%), ThO_2 (3.06 to 4.22 wt.%), and Ta_2O_5 (1.05 to 2.57 wt.%). Other typical minerals in NYF pegmatites, including zircon, thorite, limonite, and allanite-(Ce), were also observed and identified (Figure 6C–G). The thorite is translucent yellow-brown, subhedral, with a grain size over 200 μm and ThO_2 content up to 71.96 wt.%. The allanite-(Ce) is significantly enriched in REE (ΣREE content 20.54 to 24.95 wt.%) with Ce_2O_3 up to 12.00 wt.%. The limonite may be the oxidized product of pyrite from the late mineralization period. These HREE-bearing minerals could explain the significant enrichment of HREE in the chondrite-normalized REE patterns.

In the BSE images (Figure 6K,L), euhedral euxenite-(Y) grains appear almost homogenous without any obvious zoning. Some cracks and fractures are well-developed, extending into the host minerals (Figure 6H,I). High concentrations of UO_2 and ThO_2 in the euxenite-(Y) could produce radiogenic helium, which may form nano-bubbles within the crystal. These bubbles can contain pressures up to 500 MPa, causing fractures in the euxenite-(Y) and the surrounding minerals [53]. Detailed EPMA data are listed in Supplementary Table S3.

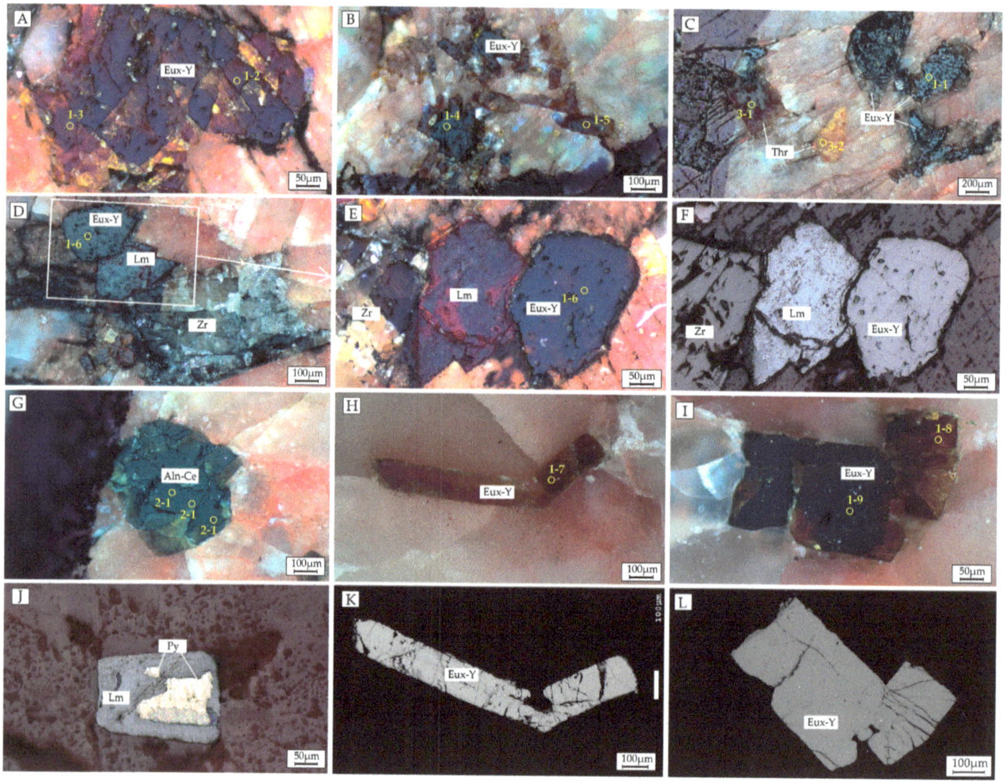

Figure 6. Photomicrographs under a reflecting microscope. (**A**) Typical subhedral euxenite-(Y) grain. (**B**) Short columnar euxenite-(Y) aggregates. (**C**) Thorite and anhedral euxenite-(Y). (**D**) Zircon, euxenite-(Y), and later-formed limonite in between. (**E**) A partial enlargement of Figure (**D**). (**F**) Image (**E**) in plane-polarized light. (**G**) An allanite-(Ce) grain. (**H**,**I**) Euhedral euxenite-(Y) grains and the EPMA test locations. (**J**) Limonite pseudomorph with residual pyrite in the core. (**K**,**L**) BSE images of euhedral euxenite-(Y) grains. Abbreviations: Aln-(Ce)—allanite-(Ce); Eux—euxenite-(Y); Lm—limonite; Py—pyrite; Thr—thorite; Zr—zircon.

5. Discussion

5.1. Pegmatite Classification and Petrogenetic Type

Pegmatites are categorized based on their distinctive characteristics, such as the host rock's metamorphic environment, mineralogy, element composition, and internal structure [14]. Since the subdivision of pegmatite classes was first proposed in 1979 [54], the classification of granitic pegmatites has become increasingly complex and detailed. The acknowledged LCT and NYF families have now been extended into the class–subclass-type–subtype hierarchy to describe their specific features and diversities [14].

In the rare element class of the NYF family, euxenite-type pegmatites typically contain aeschynite-group minerals (AGMs) and euxenite-group minerals (EGMs). These Y, REE, Nb, Ta, and Ti-bearing minerals share a common formula of AB_2O_6 and exhibit similar mineralogical characteristics, making them indistinguishable in hand specimens or under the microscope. Therefore, canonical discrimination analysis is used to distinguish between the AGMs and EGMs [55,56].

In the syenite pegmatite samples, those columnar, or needle-like minerals, with chemical compositions acquired by the EPMA (N = 9), were plotted in the EGM field (Figure 7A) and identified as euxenite-(Y) in the triangular discriminant graph (Figure 7B). According to the classification of Černý and Ercit [4], the geochemical and mineralogical features of syenite pegmatite indicate its affinity to the euxenite-type in the rare element class (REE) of the NYF family, with the characteristic accessory mineral being euxenite-(Y).

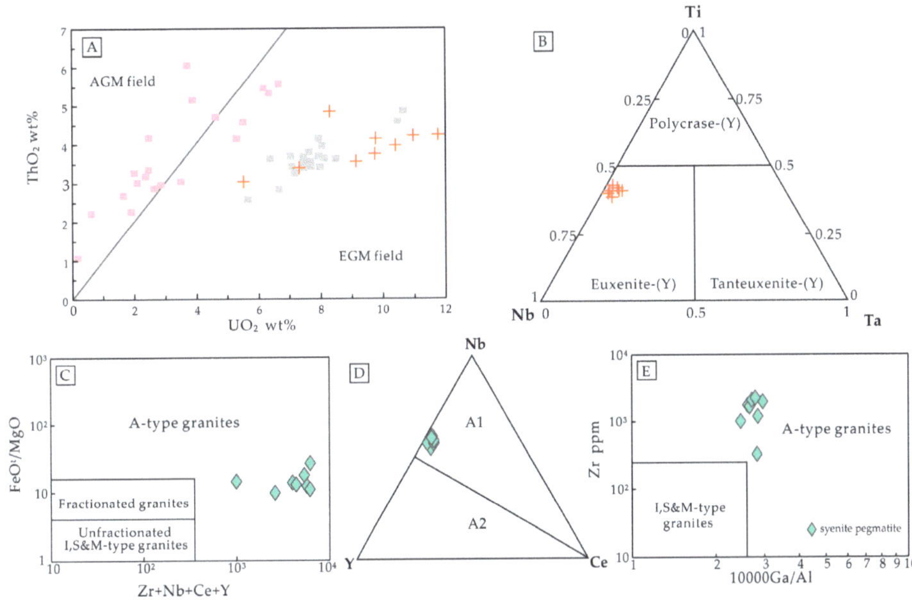

Figure 7. (**A**) Canonical discrimination analysis of AGM and EGM and (**B**) triangular discriminant graph after Škoda [55]; (**C**) FeO*/MgO vs. Zr + Nb + Ce + Y; (**D**) ternary plot of Nb-Y-Ce after Eby [57] and (**E**) Zr vs. 10,000 Ga/Al, A-type granite discrimination diagrams after Whalen [58]. Symbols: pink squares—AGM, grey squares—EGM from Škoda [55], orange crosses—EPMA data in this study (Supplementary Table S3).

Most NYF pegmatites are associated with A-type granite, sharing similar features with their granite origin. While the definition and petrogenesis of A-type granite, first proposed in 1979 [59], remain debated [60], there is consensus on certain characteristics based on previous studies. These characteristics include high Fe*, K_2O, Ga/Al ratios, enrichment in fluorine, high field strength elements (HFSEs), REEs, and low concentrations of Al_2O_3, Sr, Ba, Ti, P, with pronounced negative Eu anomalies [61–64].

However, whether syenite belongs to the A-type granite remains controversial due to its relatively low SO_2 content. The Jiaolesayi syenite pegmatite's SO_2 content ranges from 60.61 to 63.2 wt.%, within the SO_2 content range (60.4 to 79.8%) of the initial definition of the A-type granite [57]. Moreover, the syenite pegmatites are significantly enriched in HSFEs and REEs, exhibiting low tetrad effect in REE patterns (Figure 4D), and displaying some non-CHARAC (CHArge- and RAdius-Controlled) trace element ratios such as Zr/Hf (11.83 to 14.60), Nb/Ta (11.24 to 19.17), K/Rb (42.51 to 57.50), and La/Nb (0.01 to 0.04) [65]. Their 10000 Ga/Al ratios range from 2.46 to 2.96, and Zr + Nb + Ce + Y contents range from 998 to 6202 ppm, typical of A-type granite [57,62,64,66] (Figure 7C,E). Combined with their peraluminous, high-K calc-alkaline nature (Figure 4C), and the high content of Nb-Ta, low Y/Nb (0.62 to 0.75) and Yb/Ta (0.80 to 1.49) ratios, the Jiaolesayi pegmatites can be classified as A_1-type granite in the ternary plot of Nb-Y-Ce after Eby [56] (Figure 7D). In some instances, Ga enrichments could be observed in the highly fractionated S- and I-type granites [67]. However, the Jiaolesayi syenite pegmatites have a shallow content of P_2O_5 (0.01~0.10 wt.%), which is significantly different from the fractionated S-type granite [68,69].

Their enrichment of HFSEs distinguishes them from the HFSE-depleted I-type granite [57]. Therefore, we consider the Jiaolesayi syenite pegmatites indicative of A_1-type affinity [60].

5.2. Petrogenesis and Tectonic Setting

Granitic pegmatites are primarily formed by fractional crystallization of granitic magmas or by anatexis during high-grade metamorphism [70]. Although the Jiaolesayi pegmatites occur within the amphibolite- to granulite-facies metamorphosed Dakendaban Group, no migmatization or anatexis veins are observed along the contact boundaries, distinguishing them from pegmatites of anatexis origin [71]. Additionally, the syenite pegmatites have very low MgO (<0.18%) and P_2O_5 (<0.10%) contents, along with strong negative anomalies in Eu, Sr, P, and Ti, indicating intensive crystal fractionation [66,72]. Therefore, the Jiaolesayi syenite pegmatites are likely formed by fractional crystallization of their parental magma.

Generally, syenite/trachytic rocks are considered to form through three possible processes: fractional crystallization of the mantle-derived magma [73,74], anatexis of the thickened lower crust [75], and mixing of mafic and felsic magmas [76]. To reveal the geochemical signatures of the parental magma, incompatible element ratios such as Nb/Yb, Th/Yb, Y/Nb, and Yb/Ta are used as geochemical indicators due to their resistance to fractionation during partial melting and fractional crystallization [77].

As shown in Figure 8, the syenite pegmatite samples fall into the MORB-OIB array in the Nb/Yb vs. Th/Yb plot and the OIB-A_1-type granite field, exhibiting OIB-like geochemical properties. This suggests a possible mantle-derived parental magma. The enrichment of incompatible elements (e.g., Nb, Ta) in the syenite pegmatite, with high Nb/La ratios (27 to 67), also supports a mantle-deep source [72,78]. Furthermore, melts derived from lower crust typically have high Sr/Y ratios (>40) [79], while the Jiaolesayi syenite pegmatites show low Sr/Y ratios (0.07 to 0.65), suggesting insignificant crustal material involvement. The elevated Nb and Ta contents also indicate minor crustal contamination, as Nb and Ta would significantly decrease during the crustal contamination process [80].

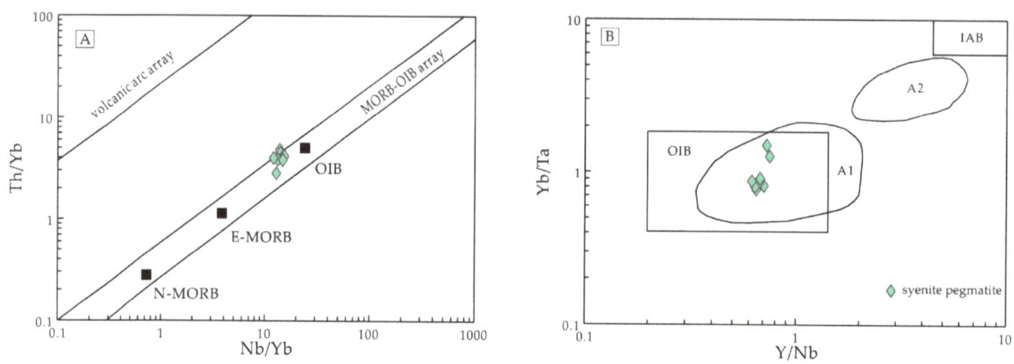

Figure 8. (**A**) Plots of Nb/Yb vs. Th/Yb, after Wang [78], and (**B**) plots of Y/Nb vs. Yb/Ta, after Eby [56]. Black squares represent the three end-members in the MORB-OIB array. Abbreviations: OIB, oceanic island basalt; IAB, island arc basalt; N-MORB, normal middle oceanic ridge basalt; E-MORB, enriched middle oceanic ridge basalt; A1, A_1 type granite; A2, A_2 type granite.

Combined with the restricted SiO_2 content, high alkali, and weakly peraluminous nature, the syenite pegmatite is likely formed from a basic alkaline magma derived from OIB-like melt, with minor crustal contamination. However, constraints on the magma source provided by incompatible elements are limited. It is necessary to combine these with isotopic characteristics of elements such as Pb, Nd, and Hf to more accurately reveal the magma source and the processes of magma evolution.

Previous studies suggest that the North Qaidam UHP continental subduction belt formed at ca. 440 to 420 Ma, during the transition from subduction to continental collision between the Qaidam and Central Qilian Block [81,82]. The Quanji Massif, located on the northern margin of the Qaidam Block, experienced post-collisional extension at ca. 415 Ma. This is indicated by the formation of the Subei pluton, which formed through low-pressure processes accompanied by a significant influx of mantle-derived OIB-like magma during the crustal thinning [83]. Therefore, under the same tectonic regime, the parental magma of the Jiaolesayi syenite pegmatite could have been derived from the same source, explaining the OIB-like geochemical signature in the syenite pegmatite, and experienced the protracted fractional crystallization with minor crustal contamination.

5.3. Implications for REE Mineralization in the Northwest Margin of the Qaidam Basin

NYF-type pegmatites have significant metallogenic potential to form rare earth/rare element deposits. The Jiaolesayi syenite pegmatite is the first reported NYF-type pegmatite in the western Quanji Massif in the northern margin of the Qaidam basin. The characteristic accessory minerals in the pegmatite, such as euxenite-(Y) and thorite, can be used to extract Nb, Ta, U, and other rare earth elements, providing great economic value. During 410–420 Ma, the Quanji Massif was under an extensional tectonic setting, providing a favorable geological environment for the formation of NYF-type pegmatites. The discovery of the Jiaolesayi syenite pegmatite has indicative significance for the exploration of rare earth/rare element mineralization in the northern margin of the Qaidam basin.

6. Conclusions

The Jiaolesayi syenite pegmatite contains high contents of Nb, Ta, Y, Ti, U, Th, and HREE, forming typical rare element-bearing minerals such as allanite-(Ce), euxenite-(Y), limonite, thorite, and zircon. The geochemical and mineralogical features of the syenite indicate it belongs to the euxenite-type in the rare element class (REE) of the NYF family, with the characteristic accessory mineral being euxenite-(Y) and showing an affinity with A_1-type granite. The syenite pegmatite may have formed from a basic alkaline magma derived from an OIB-like melt, with minor crustal contamination, under the post-collisional extension setting at ca. 413.6 ± 1.4 Ma.

Supplementary Materials: The following supporting information can be downloaded at: https://www.mdpi.com/article/10.3390/min14080805/s1, Table S1: Bulk rock major and trace element geochemical data for Jiaolesayi pegmatite; Table S2: The LA-ICP-MS analytical results of U-Pb isotopes for zircons of Jiaolesayi pegmatite; Table S3: EPMA compositions of minerals from the pegmatite samples in Jiaolesayi.

Author Contributions: Conceptualization, L.Z. and X.P.; methodology, Y.G. and M.L.; software, Y.W.; validation, N.D., Y.Z. and Y.J.; formal analysis, K.C.; investigation, L.Z., Y.J., N.D., Y.Z. and J.W.; data curation, Y.J.; writing—original draft preparation, L.Z.; writing—review and editing, Z.L.; visualization, L.Z.; funding acquisition, L.Z. All authors have read and agreed to the published version of the manuscript.

Funding: This research was financially supported by the Natural Science Basis Research Program of Shaanxi (Grant No. 2023-JC-QN-0342), the China Geology Survey project (Grant No. DD20211551 and DD20243309), and the Youth Innovation Team of Shaanxi Universities (Grant No. 2022-36).

Data Availability Statement: The authors confirm that the data supporting the findings of this study are available within the article.

Acknowledgments: The authors thank Zonghui Li and Huanhuan Wu of the Xi'an Center of Mineral Resources Survey, CGS, for their guidance, and appreciate all the editors and reviewers for their constructive suggestions.

Conflicts of Interest: The authors declare no conflict of interest.

References

1. Linnen, R.L.; Van Lichtervelde, M.; Cerny, P. Granitic Pegmatites as Sources of Strategic Metals. *Elements* **2012**, *8*, 275–280. [CrossRef]
2. London, D. A petrologic assessment of internal zonation in granitic pegmatites. *Lithos* **2014**, *184–187*, 74–104. [CrossRef]
3. Černý, P. Rare-element granitic pegmatites. Part I: Anatomy and internal evolution of pegmatite deposits. *Geosci. Can.* **1991**, *18*, 49–67.
4. Černý, P.; Ercit, T.S. The classification of granitic pegmatites revisited. *Can. Mineral.* **2005**, *43*, 2005–2026. [CrossRef]
5. Müller, A.; Romer, R.L.; Pedersen, R.-B. The Sveconorwegian pegmatite province–thousands of pegmatites without parental granites. *Can. Mineral.* **2017**, *55*, 283–315. [CrossRef]
6. Turlin, F.; André-Mayer, A.-S.; Moukhsil, A.; Vanderhaeghe, O.; Gervais, F.; Solgadi, F.; Groulier, P.-A.; Poujol, M. Unusual LREE-rich, peraluminous, monazite-or allanite-bearing pegmatitic granite in the central Grenville Province, Québec. *Ore Geol. Rev.* **2017**, *89*, 627–667. [CrossRef]
7. Bonzi, W.M.-E.; Van Lichtervelde, M.; Vanderhaeghe, O.; André-Mayer, A.-S.; Salvi, S.; Wenmenga, U. Insights from mineral trace chemistry on the origin of NYF and mixed LCT+ NYF pegmatites and their mineralization at Mangodara, SW Burkina Faso. *Miner. Depos.* **2023**, *58*, 75–104. [CrossRef]
8. Ercit, T.S. REE-enriched granitic pegmatites. In *Rare-Element Geochemistry and Mineral Deposits*; Linnen, R.L., Samson, I.M., Eds.; GAC Short Course Notes; Geological Association of Canada: St. John's, NL, Canada, 2005; Volume 17, pp. 175–199.
9. Chappell, B.W.; White, A.J. Two contrasting granite types: 25 years later. *Aust. J. Earth Sci.* **2001**, *48*, 489–499. [CrossRef]
10. Dittrich, T.; Seifert, T.; Schulz, B.; Hagemann, S.; Gerdes, A.; Pfänder, J. *Archean Rare-Metal Pegmatites in Zimbabwe and Western Australia: Geology and Metallogeny of Pollucite Mineralisations*; Springer: Berlin/Heidelberg, Germany, 2019; Volume 134.
11. Stilling, A.; Ćerný, P.; Vanstone, P.J. The Tanco pegmatite at Bernic Lake, Manitoba. XVI. Zonal and bulk compositions and their petrogenetic significance. *Can. Mineral.* **2006**, *44*, 599–623. [CrossRef]
12. Partington, G.; McNaughton, N.; Williams, I. A review of the geology, mineralization, and geochronology of the Greenbushes pegmatite, Western Australia. *Econ. Geol.* **1995**, *90*, 616–635. [CrossRef]
13. Gao, Y.; Bagas, L.; Li, K.; Jin, M.; Liu, Y.; Teng, J. Newly Discovered Triassic Lithium Deposits in the Dahongliutan Area, NorthWest China: A Case Study for the Detection of Lithium-Bearing Pegmatite Deposits in Rugged Terrains Using Remote-Sensing Data and Images. *Front. Earth Sci.* **2020**, *8*, 591966. [CrossRef]
14. Černý, P.; London, D.; Novak, M. Granitic Pegmatites as Reflections of Their Sources. *Elements* **2012**, *8*, 289–294. [CrossRef]
15. Zhang, H.-X.; Jiang, S.-Y.; Yuan, F.; Liu, S.-Q. LA-(MC)-ICP-MS U-Th-Pb dating and Nd isotopes of allanite in NYF pegmatite from lesser Qingling orogenic belt, central China. *Ore Geol. Rev.* **2022**, *145*, 104893. [CrossRef]
16. Simmons, W.B.; Webber, K.; Falster, A. NYF pegmatites of the South Platte district, Colorado. *Can. Mineral.* **1999**, *37*, 836–838.
17. Novák, M.; Filip, J. Unusual (Na,Mg)-enriched beryl and its breakdown products (beryl ii, bazzite, bavenite) from euxenite-type nyf pegmatite related to the orogenic ultrapotassic trebíc pluton, Czech republic. *Can. Mineral.* **2010**, *48*, 615–628. [CrossRef]
18. Colombo, F.; Simmons, W.; Falster, A.U.; Lira, R. Occurrence, crystal chemistry and alteration of thorite from the NYF-type miarolitic pegmatites of the El Portezuelo Granite, Papachacra (Catamarca, NW Argentina). *Asoc. Geol. Argent. Ser. D Publ. Espec.* **2011**, *14*, 65–67.
19. Simmons, W.B.; Lee, M.T.; Brewster, R.H. Geochemistry and evolution of the South Platte granite-pegmatite system, Jefferson County, Colorado. *Geochim. Cosmochim. Acta* **1987**, *51*, 455–471. [CrossRef]
20. Smeds, S.A. Zoning and fractionation trends of a peraluminous NYF granitic pegmatite field at Falun, south-central Sweden. *GFF* **1994**, *116*, 175–184. [CrossRef]
21. Vasyukova, O.; Williams-Jones, A.E. Fluoride–silicate melt immiscibility and its role in REE ore formation: Evidence from the Strange Lake rare metal deposit, Québec-Labrador, Canada. *Geochim. Cosmochim. Acta* **2014**, *139*, 110–130. [CrossRef]
22. Jin, M.S.; Gao, Y.B.; Li, C.; Song, Z.B.; Yan, Z.Q. Remote Sensing Prospecting Method for Pegmatite Type Rare Metal Deposit—Taking Dahongliutan Area in Western Kunlun for Example. *Northwest. Geol.* **2019**, *52*, 222–231.
23. Yang, S.Q. Ore Body features and prospecting indicators of Jiaolesayi Nb-Ta deposit in Gansu Province. *Gansu Sci. Technol.* **2009**, *25*, 29–30.
24. Liu, L.; Wang, C.; Cao, Y.-T.; Chen, D.-L.; Kang, L.; Yang, W.-Q.; Zhu, X.-H. Geochronology of multi-stage metamorphic events: Constraints on episodic zircon growth from the UHP eclogite in the South Altyn, NW China. *Lithos* **2012**, *136*, 10–26. [CrossRef]
25. Lu, S.; Li, H.; Zhang, C.; Niu, G. Geological and geochronological evidence for the Precambrian evolution of the Tarim Craton and surrounding continental fragments. *Precambr. Res.* **2008**, *160*, 94–107. [CrossRef]
26. Gong, S.; Chen, N.; Wang, Q.; Kusky, T.; Wang, L.; Zhang, L.; Ba, J.; Liao, F. Early Paleoproterozoic magmatism in the Quanji Massif, northeastern margin of the Qinghai–Tibet Plateau and its tectonic significance: LA-ICPMS U–Pb zircon geochronology and geochemistry. *Gondwana Res.* **2012**, *21*, 152–166. [CrossRef]
27. Liao, F.; Zhang, L.; Chen, N.; Sun, M.; Santosh, M.; Wang, Q.; Mustafa, H.A. Geochronology and geochemistry of meta-mafic dykes in the Quanji Massif, NW China: Paleoproterozoic evolution of the Tarim Craton and implications for the assembly of the Columbia supercontinent. *Precambr. Res.* **2014**, *249*, 33–56. [CrossRef]
28. Chen, N.; Zhang, L.; Sun, M.; Wang, Q.; Kusky, T. U–Pb and Hf isotopic compositions of detrital zircons from the paragneisses of the Quanji Massif, NW China: Implications for its early tectonic evolutionary history. *J. Asian Earth Sci.* **2012**, *54*, 110–130. [CrossRef]

29. Gong, S.; He, C.; Wang, X.-C.; Chen, N.; Kusky, T. No plate tectonic shutdown in the early Paleoproterozoic: Constraints from the ca. 2.4 Ga granitoids in the Quanji Massif, NW China. *J. Asian Earth Sci.* **2019**, *172*, 221–242. [CrossRef]
30. Yu, S.; Zhang, J.; Li, S.; Sun, D.; Li, Y.; Liu, X.; Guo, L.; Suo, Y.; Peng, Y.; Zhao, X. Paleoproterozoic granulite-facies metamorphism and anatexis in the Oulongbuluke Block, NW China: Respond to assembly of the Columbia supercontinent. *Precambr. Res.* **2017**, *291*, 42–62. [CrossRef]
31. Yu, S.; Li, S.; Zhang, J.; Liu, Y.; Peng, Y.; Sun, D.; Li, Y. Grenvillian orogeny in the Oulongbuluke block, NW China: Constraints from an ~1.1 Ga Andean-type arc magmatism and metamorphism. *Precambr. Res.* **2019**, *320*, 424–437. [CrossRef]
32. Wang, C.; Yu, S.; Sun, D.; Lv, P.; Feng, Z.; Wang, G.; Gou, J. Mesoproterozoic tectonic-thermal events in the Oulongbuluke Block, NW China: Constraints on the transition from supercontinent Columbia to Rodinia. *Precambr. Res.* **2021**, *352*, 106010. [CrossRef]
33. Lu, S.; Zhao, G.; Wang, H.; Hao, G. Precambrian metamorphic basement and sedimentary cover of the North China Craton: A review. *Precambr. Res.* **2008**, *160*, 77–93. [CrossRef]
34. Zhang, L.; Wang, Q.; Chen, N.; Sun, M.; Santosh, M.; Ba, J. Geochemistry and detrital zircon U–Pb and Hf isotopes of the paragneiss suite from the Quanji massif, SE Tarim Craton: Implications for Paleoproterozoic tectonics in NW China. *J. Asian Earth Sci.* **2014**, *95*, 33–50. [CrossRef]
35. Lu, S.N.; Yang, C.L.; Li, H.K.; Chen, Z.K. North China Continent and Columbia Supercontinent. *Earth Sci. Front.* **2002**, *9*, 225–233, (In Chinese with English abstract).
36. Gong, S.; Chen, N.; Geng, H.; Sun, M.; Zhang, L.; Wang, Q. Zircon Hf isotopes and geochemistry of the early Paleoproterozoic high-Sr low-Y quartz-diorite in the Quanji Massif, NW China: Crustal growth and tectonic implications. *J. Earth Sci.* **2014**, *25*, 74–86. [CrossRef]
37. He, C.; Gong, S.; Wang, L.; Chen, N.; Santosh, M.; Wang, Q. Protracted post-collisional magmatism during plate subduction shutdown in early Paleoproterozoic: Insights from post-collisional granitoid suite in NW China. *Gondwana Res.* **2018**, *55*, 92–111. [CrossRef]
38. Yusheng, W.; Jianxin, Z.; Jingsui, Y.; Zhiqin, X. Geochemistry of high-grade metamorphic rocks of the North Qaidam mountains and their geological significance. *J. Asian Earth Sci.* **2006**, *28*, 174–184. [CrossRef]
39. Li, M.; Wang, C.; Li, R.; Meert, J.G.; Peng, Y.; Zhang, J. Identifying late Neoproterozoic—Early Paleozoic sediments in the South Qilian Belt, China: A peri-Gondwana connection in the northern Tibetan Plateau. *Gondwana Res.* **2019**, *76*, 173–184. [CrossRef]
40. Wang, Q.; Chen, N.; Li, X.; Hao, S.; Chen, H. LA-ICPMS zircon U-Pb geochronological constraints on the tectonothermal evolution of the Early Paleoproterozoic Dakendaban Group in the Quanji Block, NW China. *Chin. Sci. Bull.* **2008**, *53*, 2849–2858. [CrossRef]
41. Haijun, Z.; Xunlian, W.; Xun, W.; Hongrui, Z. U-Pb zircon ages of tuff beds from the Hongzaoshan Formation of the Quanji Group in the north margin of the Qaidam Basin, NW China, and their geological significances. *Earth Sci. Front.* **2016**, *23*, 202.
42. Yue, Y.; Ritts, B.D.; Graham, S.A. Initiation and long-term slip history of the Altyn Tagh fault. *Int. Geol. Rev.* **2001**, *43*, 1087–1093. [CrossRef]
43. Andersen, T. Correction of common lead in U–Pb analyses that do not report 204Pb. *Chem. Geol.* **2002**, *192*, 59–79. [CrossRef]
44. Griffin, W. GLITTER: Data reduction software for laser ablation ICP-MS. In *Laser Ablation ICP-MS in the Earth Sciences: Current Practices and Outstanding Issues*; Mineralogical Association of Canada: Quebec, QC, Canada, 2008; pp. 308–311.
45. Li, Y.; Song, S.; Yang, X.; Zhao, Z.-F.; Dong, J.; Gao, X.; Wingate, M.T.; Wang, C.; Li, M.; Jin, M. Age and composition of Neoproterozoic diabase dykes in North Altyn Tagh, northwest China: Implications for Rodinia break-up. *Int. Geol. Rev.* **2023**, *65*, 1000–1016. [CrossRef]
46. Yang, S.Y. Electron Probe Microanalysis in Geosciences: Analytical Procedures and Recent Advances. *Atom. Spectrosc.* **2022**, *43*, 186–200. [CrossRef]
47. Middlemost, E.A. Naming materials in the magma/igneous rock system. *Earth-Sci. Rev.* **1994**, *37*, 215–224. [CrossRef]
48. Peccerillo, A.; Taylor, S. Geochemistry of Eocene calc-alkaline volcanic rocks from the Kastamonu area, northern Turkey. *Contrib. Mineral. Petr.* **1976**, *58*, 63–81. [CrossRef]
49. Rickwood, P.C. Boundary lines within petrologic diagrams which use oxides of major and minor elements. *Lithos* **1989**, *22*, 247–263. [CrossRef]
50. Sun, S.S.; McDonough, W.F. Chemical and isotopic systematics of oceanic basalts: Implications for mantle composition and processes. *Geol. Soc. Lond. Spec. Publ.* **1989**, *42*, 313–345. [CrossRef]
51. Zhou, Q.; Qin, K.; Tang, D.; Tian, Y.; Cao, M.; Wang, C. Formation Age and Evolution Time Span of the Koktokay No. 3 Pegmatite, Altai, NW China: Evidence from U–Pb Zircon and 40Ar–39Ar Muscovite Ages. *Resour. Geol.* **2015**, *65*, 210–231. [CrossRef]
52. Hoskin, P.; Black, L. Metamorphic zircon formation by solid-state recrystallization of protolith igneous zircon. *J. Metamorph. Geol.* **2000**, *18*, 423–439. [CrossRef]
53. Seydoux-Guillaume, A.-M.; David, M.-L.; Alix, K.; Datas, L.; Bingen, B. Trapping of helium in nano-bubbles in euxenite: Positive identification and implications. *Earth Planet. Sci. Lett.* **2016**, *448*, 133–139. [CrossRef]
54. Ginsburg, A.; Timofeyev, I.; Feldman, L. *Principles of Geology of the Granitic Pegmatites*; Nedra: Moscow, Russia, 1979; Volume 296.
55. Škoda, R.; Novák, M. Y, REE, Nb, Ta, Ti-oxide (AB2O6) minerals from REL–REE euxenite-subtype pegmatites of the Třebíč Pluton, Czech Republic; substitutions and fractionation trends. *Lithos* **2007**, *95*, 43–57. [CrossRef]
56. Szuszkiewicz, A.; Pieczka, A.; Szełęg, E.; Turniak, K.; Ilnicki, S.; Nejbert, K. The euxenite-group minerals and products of their alteration in the hybrid Julianna granitic pegmatite, Piława Górna, Sudetes, southwestern Poland. *Can. Mineral.* **2016**, *54*, 879–898. [CrossRef]

57. Eby, G.N. Chemical subdivision of the A-type granitoids: Petrogenetic and tectonic implications. *Geology* **1992**, *20*, 641–644. [CrossRef]
58. Whalen, J.B.; Currie, K.L.; Chappell, B.W. A-type granites: Geochemical characteristics, discrimination and petrogenesis. *Contrib. Mineral. Petr.* **1987**, *95*, 407–419. [CrossRef]
59. Loiselle, M. Characteristics and origin of anorogenic granites. *Geol. Soc. Am.* **1979**, *11*, 468.
60. Bonin, B. A-type granites and related rocks: Evolution of a concept, problems and prospects. *Lithos* **2007**, *97*, 1–29. [CrossRef]
61. King, P.L.; White, A.; Chappell, B.; Allen, C. Characterization and origin of aluminous A-type granites from the Lachlan Fold Belt, southeastern Australia. *J. Petrol.* **1997**, *38*, 371–391. [CrossRef]
62. Collins, W.J.; Beams, S.D.; White, A.; Chappell, B. Nature and origin of A-type granites with particular reference to southeastern Australia. *Contrib. Mineral. Petr.* **1982**, *80*, 189–200. [CrossRef]
63. Creaser, R.A.; Price, R.C.; Wormald, R.J. A-type granites revisited: Assessment of a residual-source model. *Geology* **1991**, *19*, 163–166. [CrossRef]
64. Eby, G.N. The A-type granitoids: A review of their occurrence and chemical characteristics and speculations on their petrogenesis. *Lithos* **1990**, *26*, 115–134. [CrossRef]
65. Bau, M. Controls on the fractionation of isovalent trace elements in magmatic and aqueous systems: Evidence from Y/Ho, Zr/Hf, and lanthanide tetrad effect. *Contrib. Mineral. Petr* **1996**, *123*, 323–333. [CrossRef]
66. Li, X.H.; Li, Z.X.; Li, W.X.; Liu, Y.; Yuan, C.; Wei, G.; Qi, C. U–Pb zircon, geochemical and Sr–Nd–Hf isotopic constraints on age and origin of Jurassic I-and A-type granites from central Guangdong, SE China: A major igneous event in response to foundering of a subducted flat-slab? *Lithos* **2007**, *96*, 186–204. [CrossRef]
67. Myint, A.Z.; Cai, Y.; Robb, L.; Yonezu, K. Petrogenesis of Sn–W mineralized granites in the Dawei and Mawchi, regions of Myanmar. *Geol. J.* **2024**, *59*, 1186–1198. [CrossRef]
68. Champion, D.C.; Bultitude, R.J. The geochemical and SrNd isotopic characteristics of Paleozoic fractionated S-types granites of north Queensland: Implications for S-type granite petrogenesis. *Lithos* **2013**, *162*, 37–56. [CrossRef]
69. Meng, X.; Zhang, D.; Zhang, R.; Wang, J.; Zhang, F.; Yu, Z.; Noel, W.C.; Zhou, T. Geochronology and geochemical characteristics of ore-forming granite in Maopengdian Sn deposit, northern Jiangxi Province. *Ore Geol. Rev.* **2022**, *149*, 105098. [CrossRef]
70. Zhou, Q.F.; Qin, K.Z.; Zhu, L.Q.; Zhao, J.X. Overview of magmatic differentiation and anatexis: Insights into pegmatite genesis. *Earth Sci. Front.* **2023**, *30*, 26–39.
71. Simmons, W.; Falster, A.; Webber, K.; Roda-Robles, E.; Boudreaux, A.P.; Grassi, L.R.; Freeman, G. Bulk composition of Mt. Mica pegmatite, Maine, USA: Implications for the origin of an LCT type pegmatite by anatexis. *Can. Mineral.* **2016**, *54*, 1053–1070. [CrossRef]
72. Li, X.-H.; Chen, Z.; Liu, D.; Li, W.-X. Jurassic gabbro-granite-syenite suites from Southern Jiangxi Province, SE China: Age, origin, and tectonic significance. *Int. Geol. Rev.* **2003**, *45*, 898–921. [CrossRef]
73. Ackerman, L.; Ulrych, J.; Řanda, Z.; Erban, V.; Hegner, E.; Magna, T.; Balogh, K.; Frána, J.; Lang, M.; Novák, J.K. Geochemical characteristics and petrogenesis of phonolites and trachytic rocks from the České Středohoří Volcanic Complex, the Ohře Rift, Bohemian Massif. *Lithos* **2015**, *224*, 256–271. [CrossRef]
74. Kumar, K.V.; Frost, C.D.; Frost, B.R.; Chamberlain, K.R. The Chimakurti, Errakonda, and Uppalapadu plutons, Eastern Ghats Belt, India: An unusual association of tholeiitic and alkaline magmatism. *Lithos* **2007**, *97*, 30–57.
75. Chen, J.-l.; Xu, J.-f.; Kang, Z.-q.; Jie, L. Origin of Cenozoic alkaline potassic volcanic rocks at KonglongXiang, Lhasa terrane, Tibetan Plateau: Products of partial melting of a mafic lower-crustal source? *Chem. Geol.* **2010**, *273*, 286–299. [CrossRef]
76. Lucassen, F.; Pudlo, D.; Franz, G.; Romer, R.L.; Dulski, P. Cenozoic intra-plate magmatism in the Darfur volcanic province: Mantle source, phonolite-trachyte genesis and relation to other volcanic provinces in NE Africa. *Int. J. Earth Sci.* **2013**, *102*, 183–205. [CrossRef]
77. Weaver, B.L. The origin of ocean island basalt end-member compositions: Trace element and isotopic constraints. *Earth Planet. Sci. Lett.* **1991**, *104*, 381–397. [CrossRef]
78. Wang, K.; Wang, L.-X.; Ma, C.-Q.; Zhu, Y.-X.; She, Z.-B.; Deng, X.; Chen, Q. Mineralogy and geochemistry of the Zhuxi Nb-rich trachytic rocks, South Qinling (China): Insights into the niobium mineralization during magmatic-hydrothermal processes. *Ore Geol. Rev.* **2021**, *138*, 104346. [CrossRef]
79. Ding, L.-X.; Ma, C.-Q.; Li, J.-W.; Robinson, P.T.; Deng, X.-D.; Zhang, C.; Xu, W.-C. Timing and genesis of the adakitic and shoshonitic intrusions in the Laoniushan complex, southern margin of the North China Craton: Implications for post-collisional magmatism associated with the Qinling Orogen. *Lithos* **2011**, *126*, 212–232. [CrossRef]
80. Wu, H.; Huang, H.; Zhang, Z.; Wang, T.; Guo, L.; Gao, Y.; Zhang, Z. Highly differentiated trachytic magma linked with rare metal mineralization: A case study from the Shuanghekou Nb deposit, South Qinling. *Lithos* **2023**, *438*, 106990. [CrossRef]
81. Song, S.; Niu, Y.; Su, L.; Zhang, C.; Zhang, L. Continental orogenesis from ocean subduction, continent collision/subduction, to orogen collapse, and orogen recycling: The example of the North Qaidam UHPM belt, NW China. *Earth-Sci. Rev.* **2014**, *129*, 59–84. [CrossRef]

82. Zhang, J.; Yu, S.; Mattinson, C.G. Early Paleozoic polyphase metamorphism in northern Tibet, China. *Gondwana Res.* **2017**, *41*, 267–289. [CrossRef]
83. Wang, C.; Li, R.S.; Smithies, R.H.; Li, M.; Peng, Y.; Chen, F.N.; He, S.P. Early Paleozoic felsic magmatic evolution of the western Central Qilian belt, Northwestern China, and constraints on convergent margin processes. *Gondwana Res.* **2017**, *41*, 301–324. [CrossRef]

minerals

MDPI

Article

Genesis of the Mengshan Granitoid Complex in an Early Mesozoic Intracontinental Subduction Tectonic Setting in South China: Evidence from Zircon U-Pb-Hf Isotopes and Geochemical Composition

Jin Wei [1,2,3,†], Yongpeng Ouyang [2,3,4,5,†], Jing Zou [2,3], Runling Zeng [2,3], Xinming Zhang [4], Tao Zhang [4], Shenao Sui [4], Da Zhang [4], Xiaolong He [1,*] and Yaoyao Zhang [6,*]

1 School of Environment and Resources, Xiangtan University, Xiangtan 411105, China; 18870160816@163.com
2 The Tenth Geological Brigade of Jiangxi Geological Bureau, Yingtan 335001, China; yongpeng0524@163.com (Y.O.); 15070014585@163.com (J.Z.); zengrl_jx@163.com (R.Z.)
3 Yingtan Key Laboratory of Exploration and Research of Scarce and Advantage Minerals, Yingtan 335001, China
4 School of Earth Sciences and Resources, China University of Geosciences, Beijing 100083, China; 13701031529@163.com (X.Z.); zhangtao073611@163.com (T.Z.); s1812203686@163.com (S.S.); zhangda@cugb.edu.cn (D.Z.)
5 Jiangxi Province Key Laboratory of Exploration and Development of Critical Mineral Resources, Nanchang 330009, China
6 Chinese Academy of Geological Sciences, Beijing 100037, China
* Correspondence: geohxl@xtu.edu.cn (X.H.); zhangyy@cags.ac.cn (Y.Z.)
† These authors contributed equally to this work and should be considered co-first authors.

Abstract: The Mengshan granitoid complex is located in the central part of Jiangnan Orogen and belongs to the western part of Jiangxi Province, where several phases of granitic magmatism record the crustal evolution of the late-Indosinian. However, its petrogenesis remains uncertain, largely due to controversies over its origin, evolutionary process and tectonic setting during intrusion. In this study, the lithological features and contact relationships observed in the systematic field geological investigations indicate that the late magmatic phases of the Mengshan granitoid complex are mainly composed of medium–fine-grained biotite monzogranite and fine-grained monzogranite, both of which developed primary fine-grained muscovite. Petrological, petrographic, geochemical and zircon U-Pb-Hf isotopic studies were further carried out on these rocks. Zircon U-Pb isotopic analyses suggest that the medium–fine-grained biotite monzogranite mainly formed at 220.7 ± 1.0 Ma to 218.0 ± 0.8 Ma, and that the fine-grained monzogranite formed at 211.5 ± 2.9 Ma to 212.9 ± 1.0 Ma. Whole-rock geochemical analysis results suggest that these rocks are rich in Cs, Rb, and U, and deficient in Ba, Sr, and Ti, and that they have properties characteristic of rocks with high silica, low P content and high K calc-alkali. Mineralogical and geochemical analysis results suggest that they are S-type granites. The $\varepsilon_{Hf}(t)$ values of the early-stage medium–fine-grained biotite monzogranite and late-stage fine-grained monzogranite range from –4.7 to 0.3 and from –3.2 to 0.7, respectively. Geochemical and isotopic data suggest that these granitoids were derived from the partial melting of Proterozoic continent basement rocks, and that minor mantle materials were involved during their generation. The presence of the early Mesozoic Mengshan granitoid complex reflects a reduplicated far-field converge effect of the collision of the North China and South China blocks and the subduction of the Palaeo-Pacific plate into the South China block. The thickening of the Earth's crust facilitated crustal delamination, underplating of mantle-derived magma, and crustal heating, triggering intense partial melting of the lithosphere and magma enrichment.

Keywords: S-type granite; zircon U-Pb-Hf isotopes; intracontinental subduction; Mengshan granitoid complex; South China

1. Introduction

Granite is widely distributed in continental areas and is mainly produced via the partial melting of the earth's crust [1–4]. This process can be induced via the anatectic melting of crustal protolith, heat from the convective asthenospheric mantle, thickened crust delamination, and resultant lithospheric mantle upwelling [5–8]. Investigating the details of the petrogenesis of granites is of great significance to understanding the crustal evolution process and its connection with the related contemporaneous polymetallic mineralization [9–15]. The various geochemical signatures reflect the derivation of granite from different sources and differing degrees of melting and differentiation, and have shaped the evolution of orogeny [16–19]. According to the current plate tectonic paradigm, orogenesis occurs either during accretion at a subducting continental margin or due to continent–continent collision [20–23]. However, the process of collision orogenesis includes several unresolved questions, such as the duration, magmatism architecture, and relationship with non-metal and polymetallic mineralization.

The Mengshan granitoid complex is located in the western part of Jiangxi Province. This area is situated near the intersection of the Yangtze and Cathaysia blocks in the interior of the South China Plate and has undergone complex tectonic evolution since the Neoproterozoic [24,25]. This area is also the middle section of the Jiangnan Orogenic Belt and the Pingxiang–Leping Depression Belt, adjacent to the Jiuling Thrust Fault Belt in the north and the Wugongshan Granite Dome in the south [26–29]. It is a large-scale thrust structure that was developed on the Yangtze Block's edge in the Late Paleozoic's Middle Triassic (Figure 1). The recently discovered Shizhushan wollastonite deposit in the Mengshan district contains about 54 Mt, making it the world's largest [30]. Several W-Au-Ag-Pb deposits are also located around the Mengshan granitoid complex. The results of many previous isotopic dating studies indicate that the Mengshan granitoid complex was formed mainly at ca. 230–220 Ma [28,31,32]. Some researchers further propose that some of the surrounding mineralization has also resulted from a combination of magmatism over a long period and multiple phases of evolution [26]. However, the petrogenesis of the Mengshan granitoid complex remains unclear, with much of the debate centered on the time of its formation, the period of magmatic intrusion, and whether it is an S-type or I-type granite [28,31,32]. In addition, field investigations have shown that some granites have intruded into the aforementioned granitoids formed at ca. 230–220 Ma. However, relevant geochronological and geochemical work has not yet been carried out on them, limiting the petrogenesis studies of the Mengshan granitoid complex. Their petrogenesis and relationship with mineralization also remain uncertain, largely due to controversy over their origin, evolutionary process and tectonic setting during intrusion.

In this study, we aim to investigate the mechanisms controlling the formation of the Mengshan granitoid complex through a careful study of the mineralogy, whole-rock geochemistry, zircon U-Pb geochronology, and zircon Lu-Hf isotopes of the late magmatic phases of the Mengshan granitoid complex. Our results and previous data provide essential insights into South China's petrogenesis and tectonic setting during the early Mesozoic that may aid in developing conceptual models and exploration strategies.

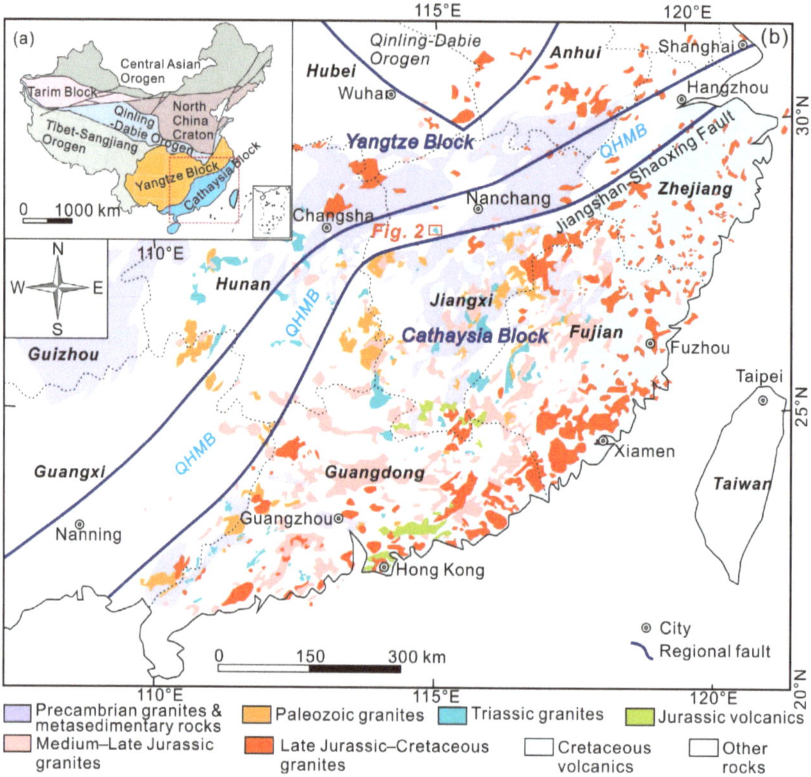

Figure 1. (**a**) Simplified tectonic map of China showing major terranes around the South China Block; (**b**) simplified geological map showing the distribution of Mesozoic granites and volcanic rocks (modified from [14,33]).

2. Geological Setting

The Mengshan granitoid complex is in the western part of Jiangxi Province. Regarding geotectonic position, the Mengshan granitoid complex is located in the interior of the South China Plate, at the junction of the Cathaysia Block and the Yangtze Block. This area is also the middle section of the Jiangnan Orogenic Belt and the Pingle Depression, adjacent to the Jiuling Thrust Fault Belt in the north and the Wugongshan Granite Dome in the south [26–29]. The Pingle Depression is bounded by the Pingxiang–Guangfeng–Yifeng–Jingdezhen Fault. It is a large-scale thrust structure that was developed on the Yangtze Block's edge in the Late Paleozoic's Middle Triassic [30]. In general, the distribution of the Mengshan granitoid complex and related rare metal deposits in the Pingle Depression is controlled by the northeast-trending structure.

The mineralization conditions in the Mengshan area are favorable, and the mineralization potential is considerable. The abundance values of mineralization elements such as tungsten, copper, lead, zinc, and molybdenum in the strata are higher than those in the crust of eastern China [31,34,35]. The area where the Mengshan granitoid complex is located exposes strata from the late Paleozoic to the Cenozoic, mainly including Carboniferous, Permian, Triassic, Jurassic, Cretaceous, and Quaternary (Figure 2). The Carboniferous strata are mainly siltstone, silty mudstone, dolomite, and limestone. The Permian strata include carbonate rocks such as limestone and dolomite, and siliceous rock. The Triassic strata are mainly sedimentary rocks such as conglomerate, sandstone, mudstone, and siltstone. The Cretaceous strata include mudstone, gravel-bearing siltstone, and conglomerate. The

Quaternary strata are distributed in modern river depressions and comprise sand, gravel, and sub-clay layers.

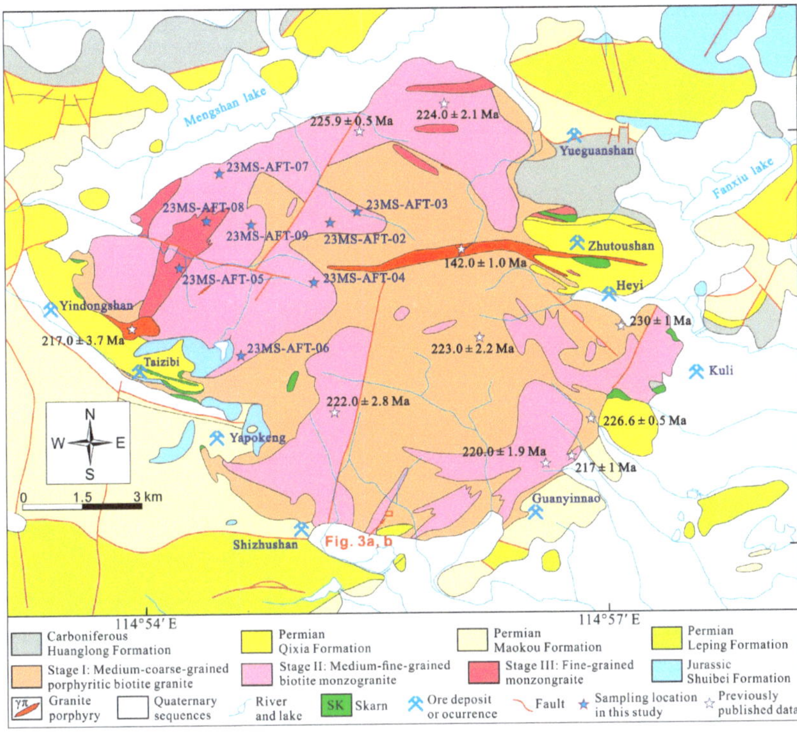

Figure 2. Simplified regional geological map showing the sampling locations in this study and the relative position of the three stages of granitic magmatism of the Mengshan granitoid complex (modified from [27,35]).

The Mengshan granitoid complex is elliptical and has an exposed area of 39 square kilometers. Surrounding the complex is a metamorphic halo formed via obvious hydrothermal replacement and thermal metamorphism, in which skarn-type deposits are distributed [26,31,36]. Previous studies have postulated that two phases of magmatic intrusion formed it, the first phase being Triassic granite (~217 to ~236 Ma) and the second phase being Jurassic–Cretaceous dykes (~135 Ma to ~142 Ma) [27]. The first phase of magmatic intrusion can be divided into three stages. The first stage is white medium–coarse-grained porphyritic biotite granite, with an age of 236–223 Ma [27,32]; the second stage is flesh-red biotite granite, with an age of 222–220 Ma [31,32,37]; the third stage is gray-white fine-grained granite, with an age of 218–217 Ma [31,32]. In the present study, systematic field geological investigations and the observed cross-cutting relationships have shown that the first phase of magmatism of the Mengshan granitoid complex can be divided into three stages, the first stage being medium–coarse-grained biotite granite, the second stage being medium–fine-grained biotite monzogranite, and the third stage being fine-grained monzogranite (Figure 3). Medium–coarse-grained biotite granite was observed to have been introduced via an intrusion of the second and third stages of granitoids (Figure 2), for example, by medium–fine-grained biotite monzogranite (Figure 3a). In addition, the contact zone between the Mengshan granitoid complex and the Permian carbonate sequences is generally strongly transformed into skarn, and is where the largest wollastonite deposit (Shizhushan deposit) in Asia has been discovered [30].

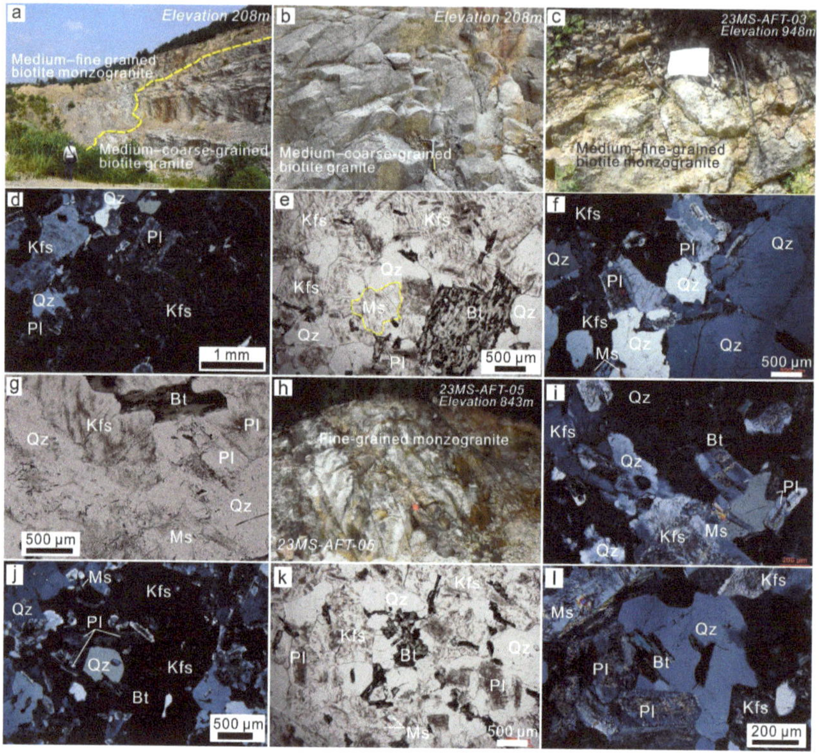

Figure 3. Graph and micrograph of the Mengshan granitoid complex. (**a**) Grayish medium–coarse-grained biotite granite in stage I introduced via an intrusion by the grayish-white medium–fine-grained biotite monzogranite in stage II; (**b**) photograph of medium–coarse-grained biotite granite (stage I); (**c**–**g**) photographs and micrographs of the medium–fine-grained biotite monzogranite in stage II (sampling number 23MS-AFT-03); (**h**–**l**) photographs and micrographs of the fine-grained monzogranite in stage III (23MS-AFT-05). Bt—biotite; Kfs—K-feldspar; Ms—muscovite; Pl—plagioclase; Qz—quartz.

Affected by the thrust-nappe structure of the South China Orogenic Belt in the Early Paleozoic and the thrust-slip structure since the Indosinian period, the structural deformation in the area mainly includes NE-trending, near-EW-trending, NE–NEE-trending, NNE-trending, and thrust-nappe (-slip) structures [38–43]. The NE-trending structure mainly manifests as a series of tight linear folds and derived faults. The near-EW structure is manifested as strong folds and normal faults in the south and strong folds and thrust faults in the north. The thrust-nappe structure is an external system overthrown on the local strata to form the klippe structure. Folds dominate the NE–NEE-trending structure. The north–northeast-trending structure is dominated by faults, mainly distributed in the western and central parts of the area.

3. Sample Collection and Analytical Methods

3.1. Sample Characteristics

The first stage of the medium–coarse-grained biotite granite is grayish in color, with a granite texture and massive structure (Figure 3a,b). It has the widest distribution in the district, and is generally observed to have been introduced via an intrusion of the later two stages of granitoids (Figure 3a). It is also the most studied granitoid in the region [30–32], so this phase was not investigated in this study. Representative samples of the medium–fine-grained biotite monzogranite and fine-grained monzogranite were selected for analysis. The sampling locations are shown in Figure 2.

Medium–fine-grained biotite monzogranite is grayish white, with a granite texture and massive structure (Figure 3a,c). The rock comprises plagioclase, potassium feldspar, quartz, biotite, and muscovite, with a 7–10 mm mineral size (Figure 3d–g). The plagioclase is hypidiomorphic and plate-shaped, undergoing sericitization, carbonation, and chloritization, makes up about 30% of the rock (Figure 3d). The potassium feldspar is hypidiomorphic and plate-shaped, with sericitization and carbonation, and makes up about 30% of the rock (Figure 3d–g). The quartz is xenomorphic granular, making up about 20% of the rock (Figure 3d–g). The biotite is flaky and has mostly transformed into chlorite (Figure 3e,g), making up 10%–15% of the rock. The muscovite is poikilitic in the feldspar, with obvious pleochroism, and makes up <5% of the rock (Figure 3e–g).

The fine-grained monzogranite has a light grayish-white granite texture and massive structure (Figure 3h). The rock comprises plagioclase, potassium feldspar, quartz, and muscovite, with a 3–5 mm mineral size (Figure 3i–l). The plagioclase is hypidiomorphic and plate-shaped, with sericitization and carbonate, and makes up 30–35% of the rock (Figure 3i–l). The potassium feldspar is also hypidiomorphic and plate-shaped, with sericitization; it makes up about 30%–35% of the rock (Figure 3i–l). The quartz is xenomorphic and granular, making up 25%–30% of the rock (Figure 3i–l). The muscovite is poikilitic in the feldspar, has obvious pleochroism, and makes up 1%–3% of the rock (Figure 3i–k).

3.2. Analytical Methods

The laboratory of Hebei Provincial Regional Survey Institute was used to conduct the geochemical analysis. The rocks were crushed into centimeter-sized blocks, and fresh samples without alterations or interspersed veins were selected, rinsed with purified water, dried, and crushed into a 200 mesh for testing. For the significant element test, the powder sample was weighed and mixed with $Li_2B_4O_7$ (1:8) flux and heated to 1150 °C using a melting machine to melt it into a uniform glass sheet in a gold–platinum crucible. Then, it was tested using an Axios ^{max} X-ray fluorescence spectrometer. The 200-mesh powder sample was weighed and placed in a polytetrafluoroethylene dissolving tank for the trace element test, and HF+HNO_3 was added. The high-pressure digestion tank was kept at 190 °C in a drying oven for 72 h, and then taken out; acid was driven out, and the solution was fixed to a dilute solution for testing. The test was completed using an ICAPQ inductively coupled plasma mass spectrometer. The measured data showed an error of less than 5% according to the monitoring standard GSR-2, and the analysis error of some volatile and extremely-low-content elements was less than 10%. The experimental temperature was 20 °C–25 °C, and the relative humidity was 10%–30%.

Zircon target preparation and cathodoluminescence (CL) were completed in the Hebei Provincial Institute of Regional Survey laboratory, and Beijing Zhongke Mining Research Testing Technology Co., Ltd. (Beijing, China) completed age testing. The age test analysis instruments used were the laser ablation multi-receiver inductively coupled plasma mass spectrometer (LA-ICP-MS; Agilent 7500, Agilent, Tokyo, Japan). The zircon dating method used was LA-ICP-MS, with a beam spot diameter of 30 μm, a laser energy density of 13–14 J·cm^{-2}, and a frequency of 8–10 Hz. Plešovice (age of 337 ± 0.37 Ma) was used as an external standard for matrix correction, and the common lead correction method was performed using the ComPbCorr#3.17 correction program. The data were processed using the ICP-Ms DataCal 10.2 software, and the concordance diagram and other figures were drawn with a confidence level of 95%. Detailed instrument operating conditions and data processing methods are shown in [44,45].

The zircon Hf isotope test was conducted at Beijing Zhongke Mining Research Testing Technology Co., Ltd. The analytical instrument was a New Wave-213 nm ArF-excimer laser-ablation system linked to a Neptune multiple- collector inductively coupled plasma mass spectrometer (NEPTUNE plus, Thermo Fisher Scientific, Waltham, MA, USA). The ablation protocol employed a spot diameter of 55 μm at a 10 Hz repetition rate and a fluence of 7–8 J/cm^2 for zircon, and He was used as the carrier gas for the ablation material. Zircon standard 91,500 ($^{176}Hf/^{177}Hf$ = 0.282308 ± 12 (2σ)) was used as an external standard for matrix correction [45]. In calculating the mantle model age of Hf, the present value of $^{176}Hf/^{177}Hf$ of the depleted mantle was 0.28325, and that of $^{176}Lu/^{177}Hf$ was 0.0384 [46]. When calculating the crustal model age, the average crustal value of $^{176}Lu/^{177}Hf$ = 0.015 was used [47].

4. Results

4.1. Geochemistry

The make-up of SiO_2 in the medium–fine-grained biotite monzogranite amounts to 72.51%–75.68%, suggesting it is an acidic rock. The make-up of Al_2O_3amounts to 14.37%–15.50%, that of K_2O amounts to 4.45%–5.19%, that of Na_2O amounts to 0.91%–2.96%, and the total alkali contents (K_2O+Na_2O) make up 5.36%–8.15% of the monzogranite. The amount of TiO_2 is 0.16%–0.39%, that of TFe_2O_3 is 0.91%–1.82%, that of MnO is 0.01%–0.05%, that of CaO is 0.19%–0.32%, that of MgO content is 0.49%–0.87%, and that of P_2O_5 is 0.04–0.07% (Table S1). After deducting the volatile matter and the loss on ignition, the samples were recalculated to 100%. The TAS diagram (Figure 4a) plots the samples in the granite range with sub-alkaline characteristics. The samples are plotted in the monzogranite area in the QAP diagram (Figure 4b). In the K_2O–SiO_2 diagram (Figure 4c), the samples are mainly plotted in the high-potassium calc-alkaline area, and some are plotted near the high-potassium calc-alkaline area and the boundary of potassium shoshonite, with characteristics of high potassium content. The rock differentiation index (DI) values are 87.07–89.77, indicating a high degree of differentiation. The aluminum saturation index A/CNK ratios are 1.38–2.19, indicating it belongs to the category of peraluminous rock (Figure 4d).

The amounts of SiO_2 in fine-grained monzogranite are 74.96%–83.73%, suggesting it is an acidic rock. The Al_2O_3 amounts have a variation range of 9.52%–13.25%. The K_2O amounts are 3.00%–5.13%, and the Na_2O amounts are 0.32%–2.82%, with an extensive variation range. The total alkali content (K_2O+Na_2O) amounts to 3.32%–7.98%, the TiO_2 content amounts to 0.10%–0.20%, the TFe_2O_3 content amounts to 1.17%–1.49%, the MnO content amounts to 0.01%–0.03%, the CaO content amounts to 0.19%–0.78%, the MgO content amounts to 0.43%–0.47%, and the P_2O_5 content amounts to 0.05–0.07% (Table S1). After deducting volatiles and loss on ignition, the samples were plotted and recalculated to 100%. The TAS diagram (Figure 4a) plots the samples in the granite range. Some points near the boundary between granite and quartzite show sub-alkaline characteristics. The samples are plotted in the monzogranite area in the QAP diagram (Figure 4b). The K_2O–SiO_2 diagram (Figure 4c) shows that the samples are plotted in the high-potassium calc-alkaline area and near the boundary of potassium shosphorite, showing characteristics of high potassium content. The differentiation index (DI) values are 90.14–90.91, indicating a high degree of differentiation. The aluminum saturation index A/CNK ratios are 1.15–2.38, also indicating that it belongs to the category of peraluminous rock (Figure 4d).

The rare earth content of the medium–fine-grained biotite monzogranite samples is low, with ΣREE ranging from 97.72 × 10^{-6} to 290.89 × 10^{-6}, the light rare earth LREE content ranges from 84.86 × 10^{-6} to 271.98 × 10^{-6}, and the heavy rare earth HREE content ranges from 12.86 × 10^{-6} to 25.02 × 10^{-6} (Table S1). The light/heavy rare earth ratio is 3.19–14.38, $(La/Yb)_N$ is 2.15–19.69, and the degree of differentiation between light and heavy rare earth is low; the value of δEu is 0.17–0.36, with a strong negative anomaly, and that of δCe is 0.74–3.36, with a weak anomaly. The rare earth distribution curve is seagull-shaped, with a relative abundance of light rare earths and relative scarcity of heavy rare earths (Figure 5a). The rare earth content in fine-grained monzogranite samples is low, with ΣREE ranging from 24.04 × 10^{-6} to 309.70 × 10^{-6}, the light rare earth LREE content ranges from 10.03 × 10^{-6} to 264.61 × 10^{-6}, and the heavy rare earth HREE content ranges from 14.01 × 10^{-6} to 45.09 × 10^{-6} (Table S1). The light/heavy rare earth ratio is 0.72–5.87, that of $(La/Yb)_N$ is 0.37–6.96, and the degree of differentiation between light and heavy rare earth is low; the value of δEu is 0.13–0.22, with a strong negative anomaly, and the value of δCe is 0.67–0.91, with a negative anomaly. The rare earth distribution curve is seagull-shaped, with a relative abundance of light rare earths and a relative scarcity of heavy rare earths (Figure 5a). In the N-MORB-normalized spider diagram (Figure 5b), the Mengshan granites exhibit positive Rb, K, Th, and U anomalies and negative Ba, Sr, P, and Ti anomalies.

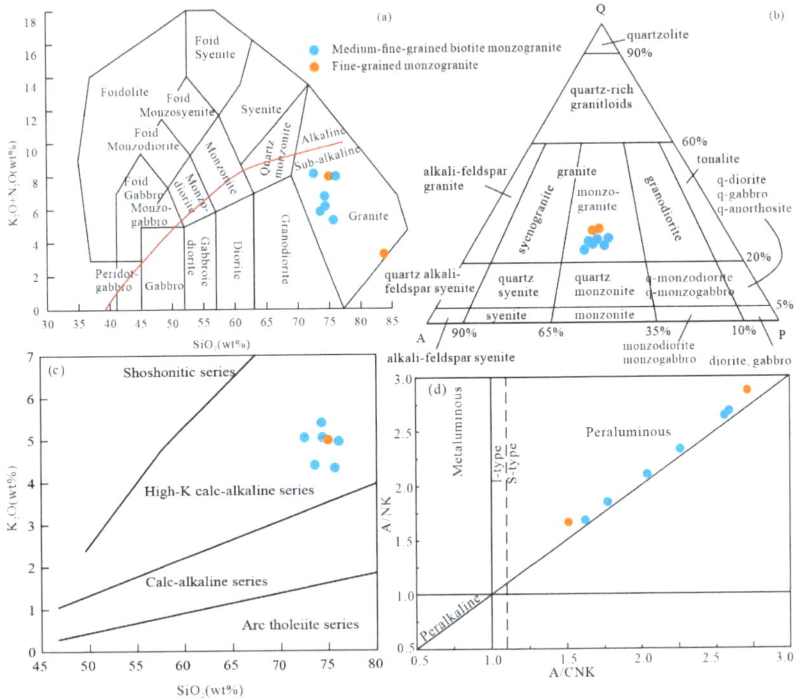

Figure 4. Classification and discriminant diagrams of magmatic rocks. (**a**) K_2O+Na_2O vs. SiO_2 diagram [48]; (**b**) QAP (quartz–akali-feldspar–plagioclase) diagram; (**c**) K_2O vs. SiO_2 diagram [48]; (**d**) A/NK vs. A/CNK diagram [48].

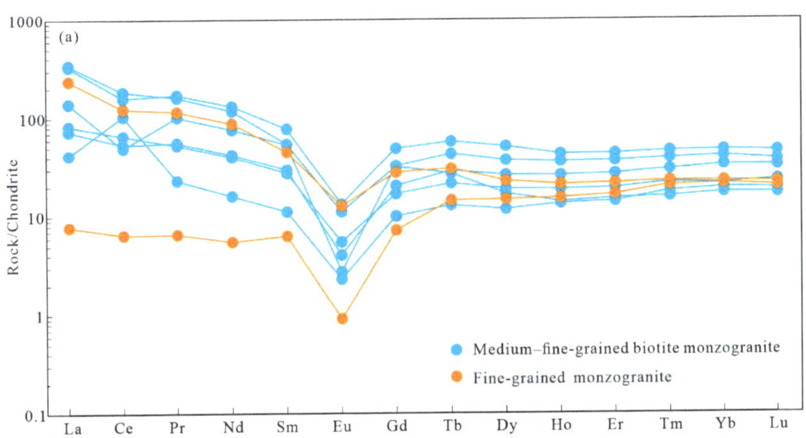

Figure 5. *Cont.*

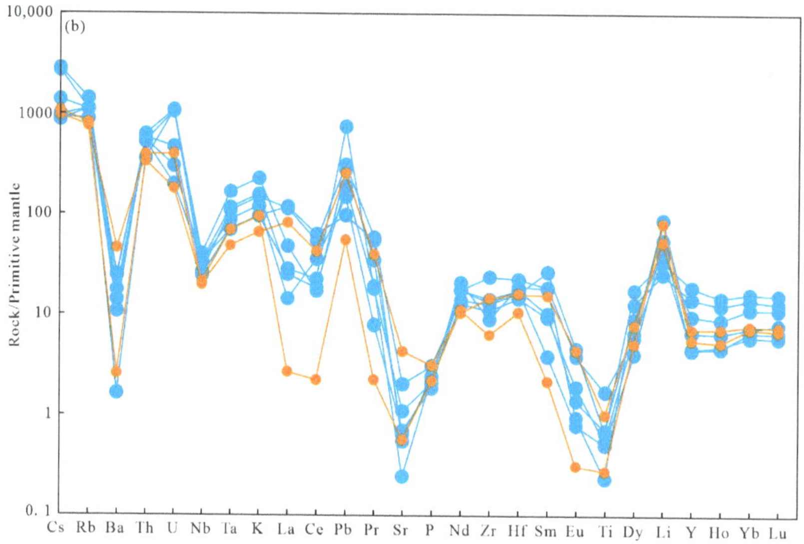

Figure 5. (**a**) Chondrite-normalized REE and (**b**) N-MORB-normalized spider patterns of the Rizhao granites (normalizing factors are from [49]; N-MORB compositions are from [49]).

4.2. Geochronology

The zircons used in this age test had good idiomorphism, mainly existing a long columnar shape, and were relatively complete overall. Age tests were conducted at zircon locations where cracks and inclusions were not developed. The locations tested were mostly in magmatic zircons with developed oscillation rings [50–52] (Figure 6). The ages measured according to the zircons are all less than 1.0 Ga, as suggested by the zircon $^{206}Pb/^{238}U$ age [53], and the test method is ICP-MS. Zircon test data are shown in Table S2.

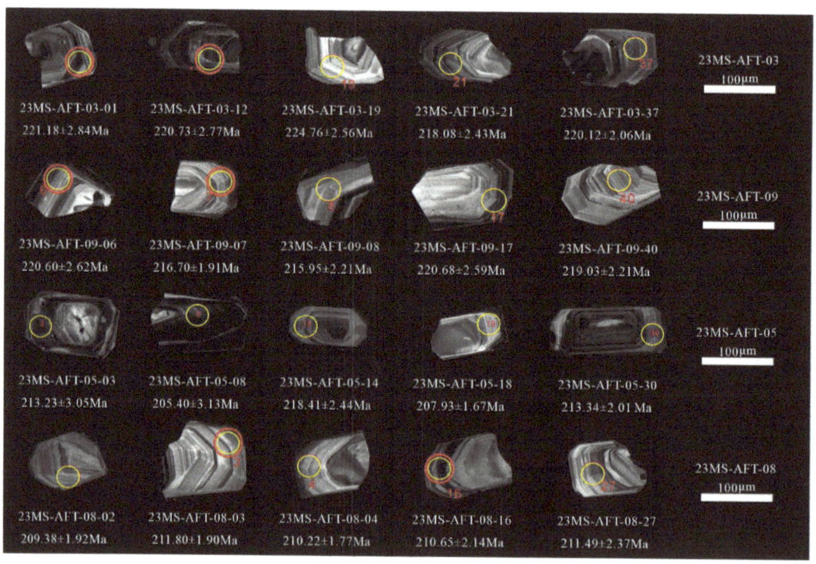

Figure 6. Cathodoluminescence (CL) images of zircons for the Mengshan granitoid complex.

The zircon Pb content in the medium–fine-grained biotite monzogranite (23MS-AFT-03) is 50.30 × 10^{-6}–366.17 × 10^{-6}, the Th content is 653.87 × 10^{-6}–3004.36 × 10^{-6}, the U content is 975.78 × 10^{-6}–8013.64 × 10^{-6}, and the Th/U ratio is 0.37–0.76. The zircon age of the sample is from 214.69 ± 1.95 Ma to 225.68 ± 3.20 Ma, with a weighted average age of 220.7 ± 1.0 Ma and a MSWD value of 0.83 (Figure 7a). The zircon Pb content of sample 23MS-AFT-09 is 49.42 × 10^{-6}–429.63 × 10^{-6}, the Th content is 721.51 × 10^{-6}–3100.52 × 10^{-6}, the U content is 914.49 × 10^{-6}–9460.35 × 10^{-6}, and the Th/U ratio is 0.33–0.81. The zircon age of the sample is from 215.21 ± 2.09 Ma to 220.89 ± 2.13 Ma, with a weighted average age of 217.98 ± 0.76 Ma and a MSWD value of 0.76 (Figure 7b).

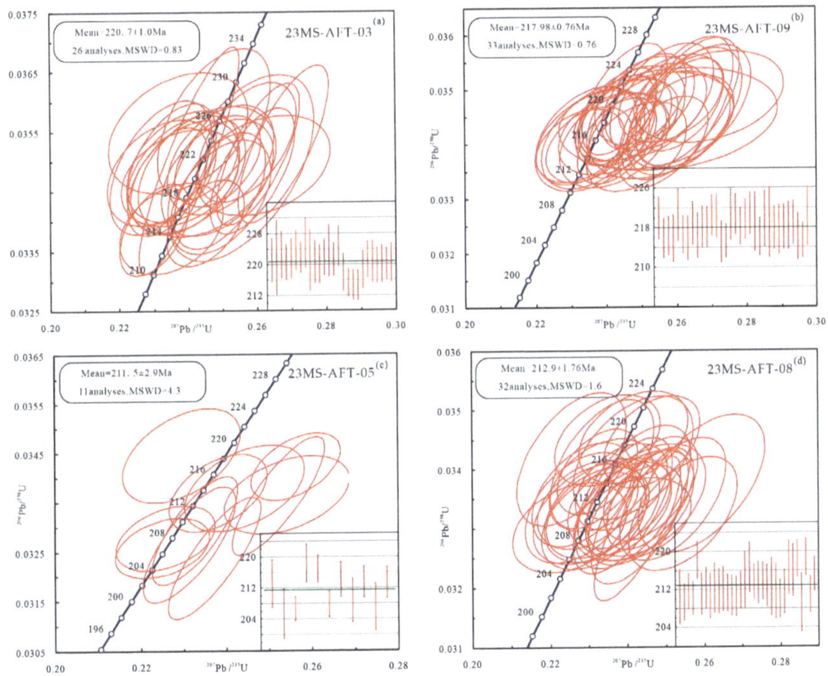

Figure 7. U–Pb concordia diagram for the Mengshan granitoid complex: (**a**) 23MS-AFT-03, (**b**) 23MS-AFT-09, (**c**) 23MS-AFT-05, and (**d**) 23MS-AFT-08.

The zircon Pb content of fine-grained monzogranite (23MS-AFT-05) is 30.92 × 10^{-6}–150.48 × 10^{-6}, the Th content is 379.99 × 10^{-6}–1220.79 × 10^{-6}, the U content is 429.48 × 10^{-6}–3063.13 × 10^{-6}, and the Th/U ratio is 0.25–0.91. The zircon age of the sample is from 205.40 ± 3.13 Ma to 218.41 ± 2.44 Ma, with a weighted average age of 211.5 ±2.9 Ma (MSWD = 4.3, Figure 7c). The zircon Pb content of sample 23MS-AFT-08 is 52.06 × 10^{-6}–222.29 × 10^{-6}, the Th content is 785.98 × 10^{-6}–5085.20 × 10^{-6}, the U content is 1114.67 × 10^{-6}–5293.28 × 10^{-6}, and the Th/U ratio is 0.40–1.70. The zircon age of the sample is from 209.27 ± 3.10 Ma to 218.86 ± 1.98 Ma, with a weighted average age of 212.9 ± 1.0 Ma and a MSWD value 1.6 (Figure 7d). All the granites tested were formed in the Triassic.

4.3. *Hf Isotopic Compositions*

The $^{176}Lu/^{177}Hf$ ratios of the medium–fine-grained biotite monzogranite (23MS-AFT-03) are 0.000819–0.001644, and the $^{176}Lu/^{177}Hf$ ratio is less than 0.002, indicating that the zircon only had a tiny amount of radiogenic Hf accumulation after its formation. Therefore, the initial $^{176}Hf/^{177}Hf$ ratio can represent the Hf isotope composition at the time of formation [54]. The zircon Hf isotope ratios of $^{176}Yb/^{177}Hf$ are 0.022454–0.044441, the $^{176}Hf/^{177}Hf$ ratios are 0.282524–0.282640, the $f_{Lu/Hf}$ values are from −0.95 to −0.98, the zircon $\varepsilon_{Hf}(t)$ values are from −4.1 to 0, the zircon Hf single-stage model ages (T_{DM}) are from 866 Ma to 1049 Ma, and the two-stage model ages ($T_{DM}{}^{C}$) are from 1682 Ma to 2058 Ma (Table S3; Figure 8).

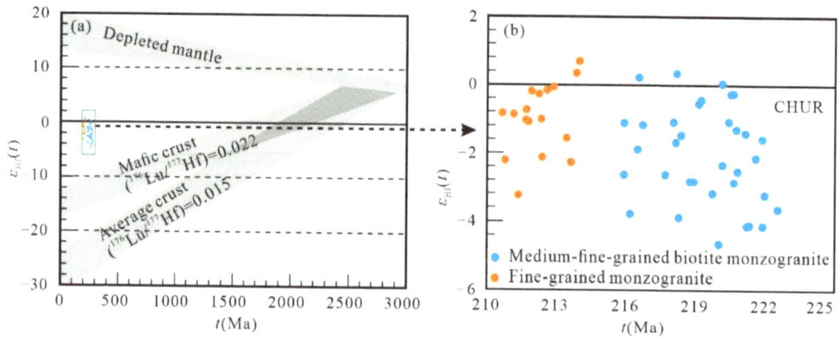

Figure 8. $\varepsilon_{Hf}(t)$ versus age diagram (**a**,**b**) (base image based on [54]).

The $^{176}Lu/^{177}Hf$ ratios of medium–fine-grained biotite monzogranite (23MS-AFT-09) are 0.000785–0.003383. Except for a few points, the $^{176}Lu/^{177}Hf$ ratios are less than 0.002, indicating that the zircons only had a small amount of radiogenic Hf accumulation after formation. Therefore, the initial $^{176}Hf/^{177}Hf$ ratio can represent the Hf isotope composition at the time of formation [54]. The zircon Hf isotope ratios of $^{176}Yb/^{177}Hf$ are 0.021857–0.095409, the $^{176}Hf/^{177}Hf$ ratios are 0.282518–0.282649, the $f_{Lu/Hf}$ values are from −0.90 to −0.98, the zircon $\varepsilon_{Hf}(t)$ values are from −4.7 to 0.3, the zircon Hf single-stage model ages (T_{DM}) are 850–1110 Ma, and the two-stage model ages ($T_{DM}{}^{C}$) are 1653–2103 Ma (Table S3; Figure 8).

The $^{176}Lu/^{177}Hf$ ratios of the fine-grained monzogranite (23MS-AFT-08) are between 0.000704 and 0.002538. Except for a few points, the $^{176}Lu/^{177}Hf$ ratios are less than 0.002, indicating that zircons only had a tiny amount of radiogenic Hf accumulation after formation. Therefore, the initial $^{176}Hf/^{177}Hf$ ratio can represent the Hf isotope composition at the time of formation [54]. The zircon Hf isotope ratios of $^{176}Yb/^{177}Hf$ are 0.019312–0.069183, the $^{176}Hf/^{177}Hf$ ratios are 0.282559–0.282662, the $f_{Lu/Hf}$ values are from −0.92 to −0.98, the zircon $\varepsilon_{Hf}(t)$ values are from −3.2 to 0.7, the zircon Hf single-stage model ages (T_{DM}) are 835–1023 Ma, and the two-stage model ages ($T_{DM}{}^{C}$) are 1619–1970 Ma (Table S3; Figure 8).

5. Discussion

5.1. Chronological Significance

The Mengshan granitoid complex is a multi-stage composite rock mass, and the age of the rock mass was tested to varying degrees. The isotopic age determined using the biotite K-Ar method is 174 Ma, and it is believed that the rock mass was formed in the Jurassic [55]. Since the Mengshan granitoid complex was formed via multiple magmatic intrusions and the biotite K-Ar system has a low closure temperature, it is easily affected by later thermal events, resulting in a K-Ar age younger than the actual formation age of the rock mass. The zircon U-Pb system has the characteristics of a high closure temperature and strong anti-interference ability. The zircon U-Pb age measured using the in situ technology can accurately reflect the formation time of the rock mass. In recent years, a large number of new zircon SHRIMP and LA-ICP-MS U-Pb age data have shown that the formation time of the Indosinian granite in South China is mainly 210–240 Ma [26,27,31,32,56,57]. It can be seen that the time during which Mengshan granite magma activity occurred is consistent with the intrusion time of other Indosinian granites in South China. The Indosinian peraluminous granites mainly include garnet-bearing granites, cordierite-bearing granites, biotite/two-mica granites, and muscovite-granites, which are closely related to mineralization, such as that of tungsten, tin, niobium, and tantalum [58–61].

According to previous research results, the first phase of the Mengshan granitoid complex is currently divided into three stages, namely 236–223 Ma (white medium–coarse-grained porphyritic biotite granite in the first stage), 222–220 Ma (flesh-red biotite granite in the second stage), and 218–217 Ma (gray-white fine-grained granite in the third stage). Because of the complexity of the magmatic activity of the Mengshan granitoid complex, this work selected representative samples for zircon LA-ICP-MS age testing and obtained four isotopic ages, which are 220.7 ± 1.0 Ma, 217 ± 0.76 Ma, 212.9 ± 1.76 Ma, and 211.5 ± 2.9 Ma (Figure 7). This work also obtained granite from three stages. Based on a combination of the field geological characteristics and the latest age data, the first phase of magmatism in the Mengshan granitoid complex is suggested to be divided into three stages. The first stage is the medium–coarse-grained porphyritic biotite granite with an age of ca. 236–223 Ma; the second stage is the medium–fine-grained biotite monzogranite, with an age of ca. 222–217 Ma; and the third stage is the fine-grained monzogranite, with an age of ca. 212–211 Ma.

5.2. Petrogenesis and Magmatic Source

Granite can be divided into the I type, S type, M type, and A type, according to the differences in source area and tectonic environment. S-type granite is produced via the partial melting and crystallization of crustal sediments [1,62,63]. Mineralogical signs are reliable signs for distinguishing between granite types. For example, magmatic muscovite, cordierite, garnet, corundum, and tourmaline are essential signs of S-type granite [62,64–66]. In the medium–fine-grained biotite monzogranite, panidiomorphic lamellar muscovite in sizes of 200–500 μm is generally present in the form of poikilitic with quartz, plagioclase, and potassium feldspar (Figure 3e–g). In the fine-grained monzogranite, there is hypidiomorphic lamellar muscovite (Figure 3i,k) in the form of poikilitic with quartz, plagioclase, and potassium feldspar, or panidiomorphic lamellar muscovite wrapped in plagioclase (Figure 3l), indicating that there is a large amount of primary muscovite of magmatic origin in these acidic rocks in the Mengshan area. In addition, previous researchers have also proposed that the color of primary biotite of magmatic origin is also related to the origin of granite. Biotite in S-type granite is usually rich in Fe^{2+} and presents a reddish-brown color, while biotite in I-type granite is richer in Fe^{3+} and Mg and presents a brownish-green color [2,62,63]. The primary biotite in the medium–fine-grained biotite monzogranite and the fine-grained monzogranite is of a significant reddish-brown color (Figure 3c,h). The above mineralogical evidence shows that the granites of the Mengshan granitoid complex have the typical characteristics of S-type granite. The 10,000 Ga/Al ratio of the granite is less than the lower limit of 2.6 for A-type granite [1,67], indicating that it does not have the characteristics of A-type granite. In the K_2O+Na_2O and $w(FeO^*)/w(MgO)$ versus 10,000 Ga/Al diagrams, almost all of the sample spots fall into the I- or S-type granite range (Figure 9a,b). However, the P_2O_5 content of the rock mass is 0.04%–0.07% (Table S1), which is different from the typical content in I-type granite. The standard mineral calculation (CIPW) of the rock mass contains a large number of standard corundum molecules (1.76–8.37), and the A/CNK is 1.15–2.38, showing its strong peraluminous characteristics, similar to those of S-type granite (Figure 4d). The samples in the ACF diagram (Figure 9c) are projected to the side of S-type granite. Therefore, the geochemical analysis of the granite in this study also shows that the granites of the Mengshan granitoid complex belong to S-type granite.

S-type granite is a type of granite that is produced via the partial melting and crystallization of crustal sediments. S-type granite is closely associated with W, Sn, and light rare earth element (LREE) deposits regarding genesis. W-rich granite's ideal magma source area is usually rich in significant ion lithophile and ore-forming elements [68–71]. Biotite and muscovite in crustal rocks are the main minerals that host highly incompatible significant ion lithophile elements such as W, Cs, and U. Partial melting of metamorphic sedimentary rocks will form melts with high U content and low Th/U ratios [5,72]. Therefore, the W-rich initial magma produced via the partial melting of enriched crustal materials will also be rich in U and Cs and show characteristics of high U and a low Th/U ratio. The granites of the Mengshan granitoid complex are rich in Cs, Rb, U, Sn, and Li to varying degrees, and the Th/U ratio is low (Table S2; Figure 5b), indicating that the magma of the Mengshan granitoid complex comes from the partial melting of enriched crustal materials. "Nb/Ta = 5" is considered to be the geochemical distinguishing mark for normal crystallization differentiation and magma–hydrothermal interaction, and "Zr/Hf = 26" is considered to be the geochemical boundary-distinguishing mark for highly differentiated and ultra-differentiated granites [3]. In the diagram of Nb/Ta vs. Zr/Hf that characterizes the origin of granite (Figure 9d), both types of granitoid complexes show a high degree of differentiation and strong melt–fluid interaction close to granite areas associated with mineralization, such as that in W-Sn rare metals, suggesting that the Mengshan granitoid complex has the potential for W-Sn rare metal mineralization; for example, it may be related to W mineralization (Taizibi W deposit; Figure 2) around the granitoid complex.

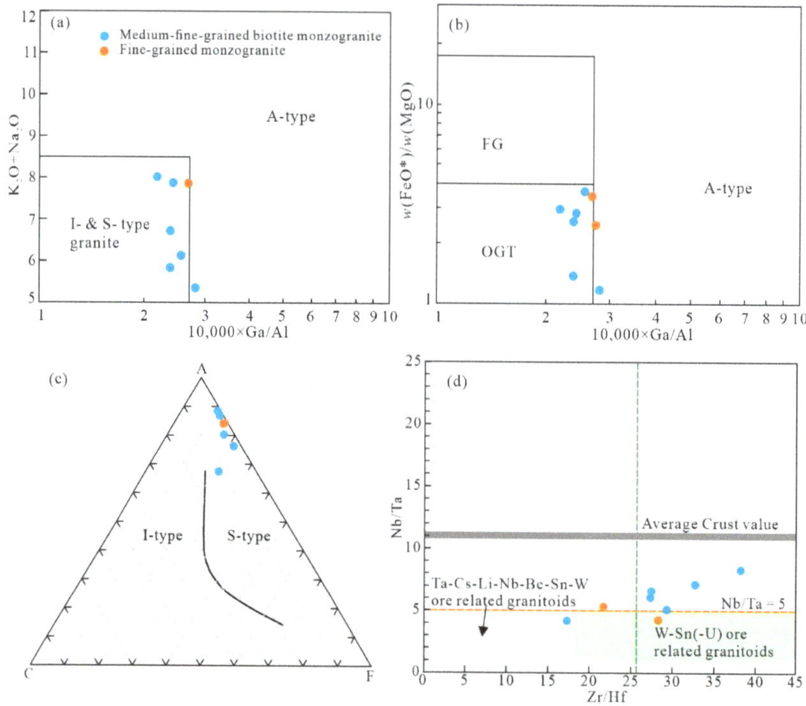

Figure 9. (**a**) K_2O+Na_2O vs. 10,000 Ga/Al diagram; (**b**) $w(FeO^*)/w(MgO)$ vs. 10,000 Ga/Al diagram; (**c**) ACF diagram [73]; (**d**) Nb/Ta vs. Zr/Hf diagram [1,2].

In this study, zircon Hf isotope tests were conducted on representative samples, and the $f_{Lu/Hf}$ values of the samples were all less than −0.90. The zircon $\varepsilon_{Hf}(t)$ values of the medium–fine-grained biotite monzogranite samples 23MS-AFT-03 and 23MS-AFT-09 were from −4.1 to 0 and −4.7 to 0.3, and the zircon $\varepsilon_{Hf}(t)$ values of fine-grained monzogranite sample 23MS-AFT-08 were from −3.2 to 0.7 (Figure 8). The zircon $\varepsilon_{Hf}(t)$ values of the magmatic rocks of these two periods were mainly negative, and there was also a small number of positive points, indicating that the source of these magmas is mainly ancient crustal materials and that a small amount of mantle materials are mixed in. The two-stage model ages ($T_{DM}{}^C$) of the zircons from the biotite monzogranite samples 23MS-AFT-03 and 23MS-AFT-09 were 1682–2058 Ma and 1653–2103 Ma, and the two-stage model ages (T_{DM}^{C}) of the zircons from monzogranite sample 23MS-AFT-08 were 1619–1970 Ma. The results show that ancient crustal material in the magma source area was formed in the early Mesoproterozoic period. Combined with the previous research results, it is believed that the Mengshan granite is an S-type granite, and the magma source area is mainly composed of Early Mesoproterozoic crustal materials [27,31,32].

5.3. Tectonic Setting

High-field-strength elements (HFSE), such as Nb, Ta, Th, Zr, Hf, HREE, etc., can effectively indicate the tectonic environment of rocks [74]. In the Rb–Y+Nb diagram of granite (Figure 10a), the samples are plotted in the syn-collision area, close to the boundary with the plate. In the Rb–Yb+Ta diagram (Figure 10b), all the samples are plotted in the syn-collision area. The above structural discrimination diagrams show that the samples are projected in the intra-plate tectonic and syn-collision tectonic environments, indicating that the granite has the geochemical characteristics of intracontinental and syn-collision granite.

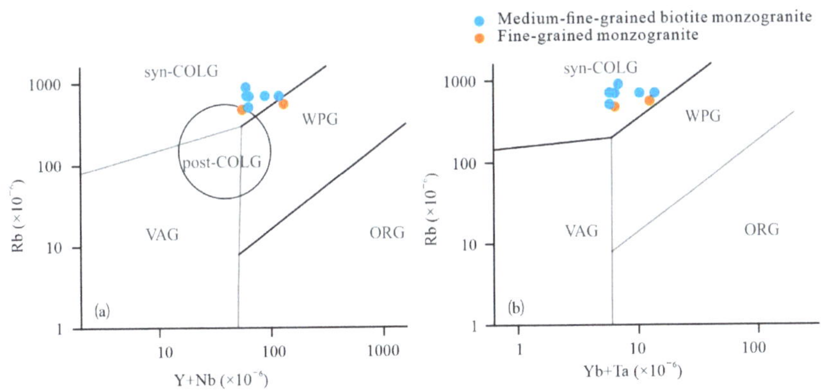

Figure 10. Tectonic discrimination diagrams for magmatic rocks (base image according t [75]). (**a**) Rb vs. (Y+Nb) diagram; (**b**) Rb vs. (Yb+Ta) diagram. Syn-COLG, syn-collision granite; VAG, volcanic arc granite; WPG, within plate granite; ORG, ocean ridge granite; post-COLG, post-collision granite.

The South China Plate is located at the intersection of multiple tectonic domains of the North China Plate, the Indosinian Plate, and the Paleo-Pacific Plate in the Triassic. Tectonic deformation events and magmatic activities are widely distributed in the South China Plate. The collision and convergence of the north and south margins of the South China Plate led to the thickening of the crust in South China, forming a large-scale Early Mesozoic fold-nappe system, a large ductile shear zone, and a series of S-type granites [76,77]. Studies have shown that the initial collision between the South China Plate and the North China Plate occurred in the Early Triassic [78]. The subduction–accretion and collision orogeny of the North China Plate and the South China Plate along the Mianlue suture in the Middle Triassic led to the development of the NW/NWW-trending fold-thrust tectonic belt in the Qinling–Dabie orogenic belt and its foreland [77,79,80]. The related deformation is not limited to the Qinling–Dabie orogenic belt and its foreland but affects the Jiangnan orogenic belt and the interior of the Cathaysia block [81–83], resulting in strong intracontinental compression deformation in Jiangxi, and the development of NE-trending fold-nappe structures in the strata before the Early Triassic period. Ref. [84] pointed out through the study of A-type granites in South China that the South China plate completed its final collision and amalgamation with the Indian and North China plates around 230 Ma. Some scholars also believe that 240–220 Ma was the main collision period between the Yangtze block and the North China plate [85,86]. Ref. [56] proposed that the formation time of the Indosinian granite in South China has excellent temporal coupling with the

Indosinian movement the Dabie orogenic belt in the north, indicating that the formation of the Indosinian granite in South China is related to the strong subduction/collision orogeny around the South China plate. Ref. [6] proposed the flat-slab subduction model of the Pacific plate to explain the tectonic evolution of South China since the early Mesozoic. From the early Mesozoic period (about 265 Ma), the ancient Pacific plate began to subduct gently under the Eurasian continent along the NW direction, resulting in large-scale magmatic activity in South China. Refs. [87,88] suggested that there is a lack of conclusive evidence of collision between the South China plate and the Indosinian plate in the Triassic period and that the tectonic-magmatic thermal event in South China during the Indosinian period was related to the NW subduction of the ancient Pacific Plate.

In summary, the driving force of the Indosinian tectonic–magmatic thermal event in South China should have come from the reduplicated remote effects of the collision between the North China Plate and the South China Plate in the north and of the subduction of the Paleo-Pacific Plate to the South China Plate (Figure 11). The collision of the North China Plate with the South China Plate and the subduction of the Pacific Plate caused the compression of the South China Plate, forming a series of NE-trending thrust structures. The thickening of the Earth's crust facilitated crustal delamination, underplating of mantle-derived magma, and crustal heating, triggering intense partial melting of the lithosphere and magma enrichment. As a result, peraluminous S-type granite was formed in the Mengshan area, and its formation environment was an intracontinental subduction environment under the convergent tectonic system.

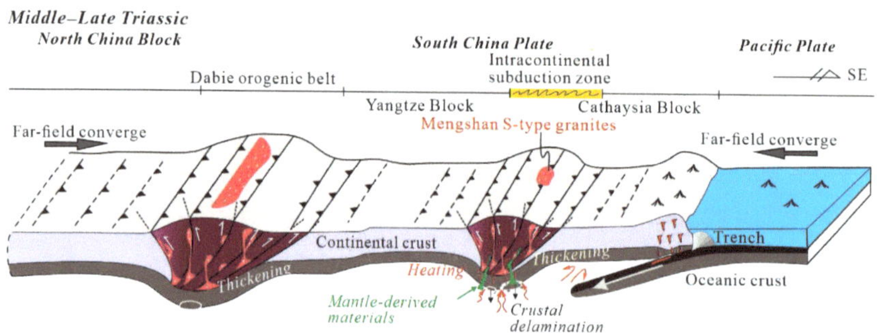

Figure 11. A dynamic model of the Mesozoic intracontinental subduction of the South China Plate (non-scaled for crust and lithosphere).

6. Conclusions

(1) The medium–fine-grained biotite monzogranite and the fine-grained monzogranite in the Mengshan granitoid complex mainly formed at 220.7 ± 1.0 Ma to 218.0 ± 0.8 Ma and 211.5 ± 2.9 Ma to 212.9 ± 1.0 Ma, respectively. They are the products of the magmatic evolution of the Mengshan complex up to a late stage. They belong to the category of S-type granites and are characterized by strongly peraluminous, high K calc-alkaline content, high silica content, high Cs and U concentrations, and low Th/U ratios. They were derived from the partial melting of Proterozoic continent basement rocks, and minor mantle materials were involved during their generation.

(2) Combined with the systematic field investigation in this study and previous data, three stages of magmatism are proposed for the formation of the Mengshan granitoid complex, comprising medium–coarse-grained porphyritic biotite granite formed at ca. 236–223 Ma, medium–fine-grained biotite monzogranite formed at ca. 222–217 Ma, and fine-grained monzogranite formed at ca. 212–211 Ma.

(3) The presence of the Early Mesozoic Mengshan granitoid complex reflects a reduplicated far-field converge effect of the collision of the North China and South China blocks and the subduction of the Palaeo-Pacific plate into the South China block. The thickening of the Earth's crust facilitated crustal delamination, the underplating of

mantle-derived magma, and crustal heating, triggering the intense partial melting of the lithosphere and magma enrichment.

Supplementary Materials: The following supporting information can be downloaded at https://www.mdpi.com/article/10.3390/min14090854/s1. Table S1: Major element (wt%), REE element (ppm), and trace element (ppm) compositions of the Mengshan granitoid complex; Table S2: LA-ICP-MS zircon U-Pb data for the Mengshan granitoid complex; Table S3: LA-ICP-MS Lu-Hf analysis results of the Mengshan granitoid complex zircons [89–91].

Author Contributions: Conceptualization, J.W., Y.O., X.H. and Y.Z.; methodology, J.W., J.Z., R.Z. and Y.O.; software, R.Z.; validation, J.Z. and D.Z.; formal analysis, J.W., Y.Z. and X.H.; investigation, J.W., X.Z., X.H., T.Z., S.S. and D.Z.; resources, S.S., T.Z. and X.Z.; data curation, J.W. and Y.Z.; writing—original draft preparation, J.W. and Y.O.; writing—review and editing, X.H. and Y.Z.; visualization, S.S.; supervision, T.Z.; project administration, X.H., J.W., D.Z. and Y.O.; funding acquisition, X.H. and Y.Z. All authors have read and agreed to the published version of the manuscript.

Funding: This study was supported by a Jiangxi Geological Bureau Young Science and Technology Leader Training Programme Project (2023JXDZKJRC05 and 2022JXDZKJRC02), a National Natural Science Foundation of China (42302099), a project of Science and Technology Department of Hunan Province (2023JJ40624), a Natural Science Foundation of Jiangxi Province (20224BAB203037), a Geological survey project funded by the Provincial Department of Finance of Jiangxi Province (20240022), a Key R&D Project of Science and Technology Department of Jiangxi Province (20212BBG73045), and a Science and Technology Programme Project of Yingtan City (20233-185656 and 2024SYD017).

Data Availability Statement: The data are contained within the article and Supplementary Materials.

Acknowledgments: We are grateful to Youguo Deng, Lin He, Xianghui, Zeng and Qi Chen of the Tenth Geological Brigade of Jiangxi Geological Bureau, and Shuxun Wang, Mingjian Yang, and Mengying Cai of the China University of Geosciences (Beijing) for their help with fieldwork and indoor mapping. Special thanks go to the reviewers for their constructive comments on this paper.

Conflicts of Interest: The authors declare no conflicts of interest.

References

1. Whalen, J.B.; Currie, K.L.; Chappell, B.W. A-type granites: Geochemical characteristics, discrimination and petrogenesis. *Contrib. Mineral. Petrol.* **1987**, *95*, 407–419. [CrossRef]
2. Chappell, B.W. Aluminium saturation in I-and S-type granites and the characterization of fractionated haplogranites. *Lithos* **1999**, *46*, 535–551. [CrossRef]
3. Ballouard, C.; Poujol, M.; Boulvais, P.; Branquet, Y.; Tartèse, R.; Vigneresse, J.L. Nb-Ta fractionation in peraluminous granites: A marker of the magmatic-hydrothermal transition. *Geology* **2016**, *44*, 231–234. [CrossRef]
4. Moghadam, H.S.; Hoernle, K.A.; Hauff, K.A.; Hauff, F.; Chiaradia, M.; Schönberg, D.G.; Esquivel, T.O.; Bindeman, I.N.; Karsli, O.; Ghorbani, G.; et al. Middle-Late Miocene to Pleistocene Post-Collisional Magmatism in the Arabia-Eurasia Collision Zone, an Example from Northwest Iran. *J. Petrol.* **2023**, *64*, 081. [CrossRef]
5. Förster, H.J.; Tischendorf, G.; Trumbull, R.B.; Gottesmann, B. Late-collisional granites in the Variscan Erzgebirge, Germany. *J. Petrol.* **1999**, *40*, 1613–1645. [CrossRef]
6. Li, Z.X.; Li, X.H. Formation of the 1300km wide intracontinental orogen and postorogenic magmatic province in Mesozoic South China: A flat-slab subduction model. *Geology* **2007**, *35*, 179–182. [CrossRef]
7. Zheng, Y.F.; Chen, R.X.; Xu, Z.; Zhang, S.B. The transport of water in subduction zones. *Sci. China Earth Sci.* **2016**, *59*, 651–682. [CrossRef]
8. Rong, W.; Zhang, S.B.; Zheng, Y.F. Back-reaction of Peritectic Garnet as an Explanation for the Origin of Mafic Enclaves in S-type Granite from the Jiuling Batholith in South China. *J. Petrol.* **2017**, *58*, 569–598. [CrossRef]
9. Li, X.H.; Liu, D.Y.; Sun, M.; Li, W.X.; Liang, X.R.; Liu, Y. Precise Sm-Nd and U-Pb isotopic dating of the supergiant Shizhuyuan polymetallic deposit and its host granite, SE China. *Geol. Mag.* **2004**, *141*, 225–231. [CrossRef]
10. Li, X.H.; Li, Z.X.; Li, W.X.; Liu, Y.; Yuan, C.; Wei, G.J.; Qi, C.S. U–Pb zircon, geochemical and Sr–Nd–Hf isotopic constraints on age and origin of Jurassic I-and A-type granites from central Guangdong, SE China: A major igneous event in response to foundering of a subducted flat-slab? *Lithos* **2007**, *96*, 186–204. [CrossRef]
11. Brown, M. Granite: From genesis to emplacement. *Geol. Soc. Am. Bull.* **2013**, *125*, 1079–1113. [CrossRef]
12. He, X.L.; Yu, Q.Y.; Liu, S.Y.; Yang, M.J.; Zhang, D. Origin of the Erdaohe Ag–Pb–Zn deposit, central Great Xing'an Range, northeast China: Constraints from fluid inclusions, zircon U-Pb geochronology, and stable isotopes. *Ore Geol. Rev.* **2021**, *107*, 10439. [CrossRef]

13. He, X.L.; Zhang, D.; Di, Y.J.; Wu, G.G. Evolution of the magmatic–hydrothermal system and formation of the giant Zhuxi W-Cu deposit in South China. *Geosci. Front.* **2022**, *13*, 101–278. [CrossRef]
14. Vatuva, A.; He, X.L.; Zhang, X.M.; Zhang, D.; Feng, H.B.; Yuan, Y.; Wang, S.; Yi, J.J.; Di, Y.J. Genesis of Makeng-type Fe-polymetallic deposits in SE China: New constraints by geochronological and isotopic data from the Dapai–Makeng metallogenic system. *Geosci. Front.* **2023**, *14*, 101614. [CrossRef]
15. Wan, L.; Jin, W.; Kusky, T.; Tian, Y.; Wang, J.; Ke, X.Z.; Long, W.G.; Yang, J.; Sun, X.M.; Chen, C. Late Jurassic granitoids in Mufushan complex and their significance for the Mesozoic tectonic evolution of eastern South China. *Lithos* **2024**, *482–483*, 107707. [CrossRef]
16. Sajona, F.G.; Maury, R.C.; Pubellier, M.; Leterrier, J.; Bellon, H.; Cotten, J. Magmatic source enrichment by slab-derived melts in a young post-collision setting, Central Mindanao (Philippines). *Lithos* **2000**, *54*, 173–206. [CrossRef]
17. Xiao, W.J.; Windley, B.; Sun, S.; Li, J.L.; Huang, B.C.; Han, C.M.; Yuan, C.; Sun, M.; Chen, H.L. A Tale of Amalgamation of Three Permo-Triassic Collage Systems in Central Asia: Oroclines, Sutures, and Terminal Accretion. *Annu. Rev. Earth Planet. Sci.* **2015**, *43*, 477–507. [CrossRef]
18. McGee, L.E.; Smith, I.E. Interpreting chemical compositions of small scale basaltic systems: A review. *J. Volcanol. Geotherm. Res.* **2016**, *325*, 45–60. [CrossRef]
19. Takach, M.K.; Bohrson, W.A.; Spera, F.J.; Viccaro, M. The Role of Crustal Contamination throughout the 1329–2005 CE Eruptive Record of Mt. Etna Volcano Italy. *J. Petrol.* **2024**, *65*, 028. [CrossRef]
20. Cawood, P.A.; Buchan, C. Linking accretionary orogenesis with supercontinent assembly. *Earth-Sci. Rev.* **2007**, *82*, 217–256. [CrossRef]
21. Maruyama, S.; Hasegawa, A.; Santosh, M.; Kogiso, T.; Omori, S.; Nakamura, K.; Kawai, K.; Zhao, D. The dynamics of big mantle wedge, magma factory, and metamorphic-metasomatic factory in subduction zones. *Gondwana Res.* **2009**, *16*, 414–430. [CrossRef]
22. Wakita, K.; Pubellier, M.; Windley, B. Tectonic processes, from rifting to collision via subduction, in SE Asia and the western Pacific: A key to understanding the architecture of the Central Asian Orogenic Belt. *Lithosphere* **2013**, *5*, 265–276. [CrossRef]
23. Curry, A.; Caricchi, L.; Lipman, P.W. Magmatic Evolution of Zoned and Unzoned Ignimbrites: Evidence for a Complex Crustal Architecture Feeding four Rapid-sequence, Caldera-forming Eruptions in the San Juan Mountains. *Colorado J. Petrol.* **2021**, *62*, 006. [CrossRef]
24. He, X.L.; Zhang, D.; Chen, G.H.; Di, Y.J. Genesis of Zhuxi Copper-Tungsten Deposit, Jiangxi Province: Insights from mineralogy and chronology. *J. Jilin Univ. Earth Sci. Ed.* **2018**, *48*, 1050–1070.
25. He, X.L.; Zhang, D.; Wu, G.G.; Di, Y.J. Control of interaction between stress and fluid in tectonic transition background on metallogenesis of giant Zhuxi W-Cu deposit, South China. *Miner. Depos.* **2021**, *40*, 1135–1159.
26. Wang, X.; Hu, Z.; Yu, X.; Chen, G.; Li, Y.; Zhan, T.; Chen, S.; Liu, S.; Cheng, X.; Yang, S. Geological Characteristics and Prospecting Significance of the Shizhushan Superlarge Wollastonite Deposit in Mengshan, West Jiangxi Province. *Acta Geosci. Sin.* **2019**, *40*, 259–264.
27. Yang, Y.; Pan, X.; Hou, Z.; Deng, Y.; Ouyang, Y.; Meng, D.; Xie, T. Petrogenesis, Redox State, and Mineralization Potential of Triassic Granitoids in the Mengshan District, South China. *Front. Earth Sci.* **2021**, *9*, 657618. [CrossRef]
28. Yang, G.Q.; Zhang, Y.Y.; Liu, K.; Zhou, Y.; Wang, S.X.; Huo, H.L. Geochemistry and Zircon U–Pb Geochronology of the Wugongshan Granites in the Northwestern Jiangxi Area, China: Implications for the Paleozoic Tectonic Development of South China. *Minerals* **2023**, *13*, 1427. [CrossRef]
29. Zhang, Y.Y.; Liu, K.; He, Q.C.; Hao, M.Y.; Guo, C.B.; Bai, J.J.; Yu, T.X. Zircon U-Pb ages, Hf isotopic characteristics, and geological significance of the Mesozoic granites in Wugong Mountains area, Jiangxi. *Geol. Rev.* **2022**, *68*, 1301–1319.
30. Mu, J.; Zhao, S.; Brzozowski, M.; Li, H.; Wu, C.; Li, W. Geology, geochemistry and genesis of the world-class Shizhushan wollastonite deposit, Mengshan area, South China. *Ore Geol. Rev.* **2023**, *158*, 105469. [CrossRef]
31. Zhong, Y.F.; Ma, C.Q.; She, Z.B.; Xu, H.J.; Wang, S.M.; Wang, L.X. U-Pb-Hf Isotope of Zircons, Geochemistry and Genesis of Mengshan Granitoids in Northwestern Jiangxi Province. *Earth Sci.-J. China Univ. Geosci.* **2011**, *36*, 703–720.
32. Sun, J.D.; Li, H.L.; Lu, F.; Xu, M.C.; Wu, S.Z.; Zhang, Z.K. Geochemistry, Zircon U-Pb ages, and Hf isotopes of the Mengshan rock mass in Western Jiangxi Province and their geologic implication. *Geol. Explor.* **2022**, *58*, 96–106.
33. Zhang, Y.Y.; Liu, K.; Tong, J.; He, Q.C.; He, X.L.; Jia, W.H.; Zhang, H.R.; Wang, S.X. Characteristics and Heat Source Mechanism of Geothermal Resources in Qianshan Area of Ji'an, Jiangxi Province. *Acta Geosci. Sin.* **2022**, *1*, 39–52.
34. Hu, Z. *The Formation Conditions and Matallogenic Regularity of Zhuxi Tungsten Polymentallic Deposit in Northeast of Jiangxi Province*; Chengdu University of Technology: Chengdu, China, 2016.
35. Wei, X.; Shi, G.H.; Zhang, X.C.; Zhang, J.J.; Shih, M. A New Nephrite Occurrence in Jiangxi Province, China: Its Characterization and Gemological Significance. *Minerals* **2024**, *14*, 432. [CrossRef]
36. Liao, M. Geological Characteristics and Genesis of Non-Metallic and Polymetallic Deposits of Mengshan Region, Jiangxi Province. *Adv. Geosci.* **2012**, *2*, 211–216.
37. Ye, Y.F.; Zhang, H.H. Study on Rock Petrogenesis and Tectomic Background of Mengshan Rock Mass in Western Jiangxi Province. *Jiangxi Coal Sci. Technol.* **2022**, *1*, 125–129.
38. Faure, M.; Sun, Y.; Shu, L.S.; Monié, P.; Charvet, J. Extensional Tectonics within a Subduction-Type Orogen; The Case Study of the Wugongshan Dome (Jiangxi Province, Southeastern China). *Tectonophysics* **1996**, *263*, 77–106. [CrossRef]

39. Faure, M.; Lepvrier, C.; Nguyen, V.V.; Vu, T.V.; Lin, W.; Chen, Z.C. The South China Block Indochina Collision: Where, When, and How. *J. Asian Earth Sci.* **2014**, *79*, 260–274. [CrossRef]
40. Lou, F.S.; Shen, W.Z.; Wang, D.Z.; Shu, L.S.; Yu, J.H. Zircon U-Pb Isotopic Chronology of the Wugongshan Dome Compound Granite in Jiangxi Province. *Acta Geol. Sin.* **2005**, *79*, 636–644.
41. Wang, D.Z.; Shu, L.S. Late Mesozoic basin and range tectonics and related magmatism in Southeast China. *Geosci. Front.* **2012**, *3*, 109–124. [CrossRef]
42. Liu, X.Y.; Yang, X.H.; Nie, L.M.; Lei, T.H. Basic Characteristics of the Magma Core Complex at Wugongshan, Jiangxi. *Acta Geol. Sin.* **2016**, *90*, 468–474.
43. Shu, L.S.; Wang, D.Z.; Shen, W.Z. Nd-Sr Isotopic Compositions of Granitic Rocks of the Mesozoic Metamorphic Core Complex in the Wugongshan Area, Jiangxi Province. *J. Nanjing Univ. (Nat. Sci.)* **2000**, *36*, 306–311.
44. Hou, K.J.; Li, Y.H.; Tian, Y.R. In situ U-Pb zircon dating using laser ablation-multi ion counting-ICP-MS. *Miner. Depos.* **2009**, *28*, 481–492.
45. Liu, Y.S.; Hu, Z.C.; Zong, K.Q.; Gao, C.G.; Gao, S.; Xu, J.; Chen, H.H. Reappraisement and refinement of zircon U-Pb isotope and trace element analyses by LA-ICP-MS. *Chin. Sci. Bull.* **2010**, *55*, 1535–1546. [CrossRef]
46. Griffin, W.L.; Pearson, N.J.; Belousova, E.; Jackson, S.E.; Achterbergh, E.V.; O'Reilly, S.Y.; Shee, S.R. The Hf isotope composition of cratonic mantle: LAM-MC-ICPMS analysis of zircon megacrysts in kimberlites. *Geochim. Cosmochim. Acta* **2000**, *64*, 133–147. [CrossRef]
47. Griffin, W.L.; Wang, X.; Jackson, S.E.; Pearson, N.J.; O'Reilly, S.Y.; Xu, X.S.; Zhou, X.M. Zircon chemistry and magma mixing, SE China: In situ analysis of Hf isotopes, Tonglu and Pingtan igneous complexes. *Lithos* **2002**, *61*, 237–269. [CrossRef]
48. Middlemost, E.A.K. Naming materials in the magma/igneous rock system. *Earth Sci. Rev.* **1994**, *37*, 215–224. [CrossRef]
49. Sun, S.S.; Mcdonough, W.F. *Chemical and Isotopic Systematics of Ocean Basalts: Impications for Mantle Composition and Processes*; Geological Society of London: London, UK; Blackwell Scientific Publications: London, UK, 1989; pp. 313–345.
50. Rubatto, D.; Gebauer, D. Use of cathodoluminescence for U-Pb zircon dating by IOM Microprobe: Some examples form the western Alps. In *Cathodoluminescence in Geoscience*; Springer-Verlag: Berlin/Heidelberg, Germany, 2000; pp. 373–400.
51. Cleasson, S.; Vetrin, V.; Bayanova, T.; Downes, H. U-Pb zircon ages from a Devonian carbonatite dyke, Kola peninsula, Russia: A record of geological evolution from the Archaean to the Palaeozoic. *Lithos* **2000**, *51*, 95–108. [CrossRef]
52. Belousova, E.A.; Griffin, W.L.; O'Reilly, S.Y.; Fisher, N.I. Igneous zircon: Trace element composition as an indicator of source rock type. *Contrib. Mineral. Petrol.* **2002**, *143*, 602–622. [CrossRef]
53. Griffin, W.L.; Belousova, E.A.; Shee, S.R. Crustal Evolution in the northern Yilarn Craton: U-Pb and Hf-isotope evidence from detrital zircons. *Precambrian Res.* **2004**, *131*, 213–282. [CrossRef]
54. Wu, F.Y.; Li, X.H.; Zheng, Y.F.; Gao, S. Lu-Hf isotopic systematics and their applications in petrology. *Acta Petrol. Sin.* **2007**, *23*, 185–220.
55. TGRGNP (The Granitoid Research Group of Nanling Project); Ministry of Geology and Mineral Resources. *Geology of Granitoids of Nanling Region and Their Petrogenesis and Mineralization*; Geological Publishing House: Beijing, China, 1989; pp. 18–66.
56. Wang, Y.J.; Fan, W.M.; Sun, M.; Liang, X.Q.; Zhang, Y.H.; Peng, T.P. Geochronological geochemical and geothermal constraints on petrogenesis of the Indosinian peraluminous granites in the South China block: A case study in the Hunan Province. *Lithos* **2007**, *96*, 475–502. [CrossRef]
57. Zhou, X.M.; Sun, T.; Shen, W.Z.; Shu, L.S. Petrogenesis of Mesozoic granitoids and volcanic rocks in South China: A response to tectonic evolution. *Episodes* **2006**, *29*, 26–33. [CrossRef] [PubMed]
58. Wang, Y.J.; Fan, W.M.; Zhang, G.W.; Zhang, Y.H. Phanerozoic tectonics of the South China Block: Key observations and controversies. *Gondwana Res.* **2013**, *23*, 1273–1305. [CrossRef]
59. Li, X.F. Spatial and temporal distributions and the geological setting of the W-Sn-Mo-Nb-Ta deposits at the northeast Guangxi, South China. *Acta Geol. Sin.* **2012**, *86*, 1713–1725.
60. Mao, J.W.; Chen, Y.B.; Cheng, M.H.; Franco, P. Major types and time-space distribution of Mesozoic ore deposits in South China and their geodynamic settings. *Miner. Depos.* **2013**, *48*, 1019–1045.
61. Feng, M.; Feng, Z.H.; Kang, Z.Q.; Fu, W.; Qing, Y.; Hu, R.G.; Cai, Y.F.; Feng, Y.Y.; Wang, C.Z. Establishing an Indosinian geochronological framework for episodic granitic emplacement and W-Sn-Nb-Ta mineralization in Limu mining district, South China. *Ore Geol. Rev.* **2019**, *107*, 1–13. [CrossRef]
62. Chappell, B.W.; White, A.J.R. Two contrasting granite types. *Pac. Geol.* **1974**, *8*, 173–174.
63. Chappell, B.W.; White, A.J.R. Two contrasting granite types: 25 years later. *Aust. J. Earth Sci.* **2001**, *48*, 489–499. [CrossRef]
64. Deng, J.F.; Liu, C.; Di, Y.J.; Feng, Y.F.; Su, S.G.; Xiao, Q.H.; Zhao, G.C.; Dai, M.; Duan, P.X. Crustal convergent and accretional consumption zones, and continent-continent collisional orogenes and subduction aceretional orogenes: Records from the igneous petrotectonic assemblages. *Earth Sci. Front.* **2016**, *23*, 034–041.
65. Sylvester, P.J. Post-collisional strongly peraluminous granites. *Lithos* **1998**, *45*, 29–44. [CrossRef]
66. Wu, F.Y.; Li, X.H.; Yang, J.H.; Zheng, Y.F. Discussions on the petrogenesis of granites. *Acla Petrol. Sin.* **2007**, *23*, 1217–1238.
67. Collins, W.J.; Beams, S.D.; White, A.J.R.; Chappell, B.W. Nature and origin of A-type granites with particular reference to southeastern Australia. *Contrib. Miner. Petrol.* **1982**, *80*, 189–200. [CrossRef]
68. Pitcher, W.S. Granite type and tectonic environment. In *Mountain Building Processes*; Hsu, K., Ed.; Academic Press: London, UK, 1983; pp. 19–40.

69. Huang, L.C.; Jiang, S.Y. Highly fractionated S-type granites from the giant Dahutang tungsten deposit in Jiangnan Orogen, Southeast China: Geochronology, petrogenesis and their relationship with W-mineralization. *Lithos* **2014**, *202*, 207–226. [CrossRef]
70. Romer, R.L.; Kroner, U. Sediment and weathering control on the distribution of Paleozoic magmatic tin-tungsten mineralization. *Miner. Depos.* **2015**, *50*, 327–338. [CrossRef]
71. Romer, R.L.; Kroner, U. Phanerozoic tin and tungsten mineralization-Tectonic controls on the distribution of enriched protoliths and heat sources for crustal melting. *Gondwana Res.* **2016**, *31*, 60–95. [CrossRef]
72. Bea, F.; Montero, P. Behavior of accessory phases and redistribution of Zr, REE, Y, Th, and U during metamorphism and partial melting of metapelites in the lower crust: An example from the Kinzigite Formation of Ivrea-Verbano, NW Italy. *Geochim. Cosmochim. Acta* **1999**, *63*, 1133–1153. [CrossRef]
73. Nakada, S.; Takahashi, M. Regional variationin chemistry of the Miocene intermediate to felsic magmas in the Outer Zone and the Setouchi Province of Southwest Japan. *Geol. Soc. Jpn.* **1979**, *85*, 571–582. [CrossRef]
74. Pearce, J.A.; Cann, J.R. Tectonic setting of basic volcanic rocks determined using trace element analyses. *Earth Planet. Sci. Lett.* **1973**, *19*, 290–300. [CrossRef]
75. Pearce, J.A.; Harris, N.B.W.; Tindle, A.G. Trace element discrimination diagrams for the tectonic interpretation of granitic rocks. *J. Petrol.* **1984**, *25*, 956–983. [CrossRef]
76. Zhang, Y.Q.; Xu, X.B.; Jia, D.; Shu, L.S. Deformation Record of the Change from Indosinian Collision Related Tectonic System to Yanshanian Subduction-Related Tectonic System in South China during the Early Mesozoic. *Earth Sci. Front.* **2009**, *16*, 234–247.
77. Zhang, Y.Q.; Dong, S.W.; Li, J.H.; Cui, J.J.; Shi, W.; Su, J.B.; Li, Y. The New Progress in the Study of Mesozoic Tectonics of South China. *Acta Geosci. Sin.* **2012**, *33*, 257–279.
78. Ma, Q.L.; Yang, J.H.; Du, Y.S.; Dai, X.D.; Chai, R.; Guo, G.; Xu, Y.J. Early Triassic Initial Collision between the North China and South China Blocks in the Eastern Qinling Orogenic Belt. *Tectonophysics* **2021**, *814*, 228965. [CrossRef]
79. Dong, Y.P.; Santosh, M. Tectonic Architecture and Multiple Orogeny of the Qinling Orogenic Belt, Central China. *Gondwana Res.* **2016**, *29*, 1–40. [CrossRef]
80. Shi, W.; Dong, S.W.; Zhang, Y.Q.; Huang, S.Q. The Typical Large-Scale Superposed Folds in the Central South China: Implications for Mesozoic Intracontinental Deformation of the South China Block. *Tectonophysics* **2015**, *664*, 50–66. [CrossRef]
81. Lin, W.; Faure, M.; Sun, Y.; Shu, L.S.; Wang, Q.C. Compression to Extension Switch during the Middle Triassic Orogeny of Eastern China; The Case Study of the Jiulingshan Massif in the Southern Foreland of the Dabieshan. *J. Asian Earth Sci.* **2001**, *20*, 31–43. [CrossRef]
82. Shu, L.S. An Analysis of Principal Features of Tectonic Evolution in South China Block. *Geol. Bull. China* **2012**, *31*, 1035–1053.
83. Wang, Y.J.; Wang, Y.; Zhang, Y.Z.; Qian, X. Indosinian Deformation in the South China Block and Interaction with the Adjoining Blocks. *Geotecton. Metallog.* **2022**, *46*, 399–415.
84. Xia, Y.; Xu, X.H. The Epilogue of Paleo-Tethyan Tectonics in the South China Block: Insights from the Triassic Aluminous A-Type Granitic and Bimodal Magmatism. *J. Asian Earth Sci.* **2020**, *190*, 104129. [CrossRef]
85. Faure, M.; Lin, W.; Schärer, U.; Shu, L.S.; Sun, Y.; Arnaud, N. Continental subduction and exhumation of UHP rocks: Structural and geochronological insights from the Dabieshan (East China). *Lithos* **2003**, *70*, 213–241. [CrossRef]
86. Weislogel, A.L.; Graham, S.A.; Chang, E.Z.; Wooden, J.L.; Gehrels, G.E.; Yang, H.S. Detrital zircon provenance of the Late Triassic Songpan-Ganzi complex: Sedimentary record of collision of the North and South China blocks. *Geology* **2006**, *34*, 97–100. [CrossRef]
87. Carter, A.; Clift, P.D. Was the Indosinian orogeny a Triassic mountain building or a thermotectonic reactivation event? *C. R. Geosci.* **2008**, *340*, 83–93. [CrossRef]
88. Ferrari, O.M.; Hochard, C.; Stampfli, G.M. An alternative plate tectonic model for the Palaeozoic-Early Mesozoic Palaeotethyan evolution of Southeast Asia (northern Thailand-Burma). *Tectonophysics* **2008**, *451*, 346–365. [CrossRef]
89. Blichert-Toft, J.; Chauvel, C.; Albarède, F. Separation of Hf and Lu for high-precision isotope analysis of rock samples by magnetic sector-multiple collector ICP-MS. *Contrib. Mineral. Petrol.* **1997**, *127*, 248–260. [CrossRef]
90. Goolaerts, A.; Mattielli, N.; Jong, J.D.; Weis, D.; Scoates, J.S. Hf and Lu isotopic reference values for the zircon standard 91500 by MC-ICP-MS. *Chem. Geol.* **2004**, *206*, 1–9. [CrossRef]
91. Soderlund, U.; Patchett, P.J.; Vervoort, J.D.; Isachsen, C.E. The 176Lu decay constant determined by Lu-Hf and U-Pb isotope systematic of Precambrian mafic intrusions. *Earth Planet. Sci. Lett.* **2004**, *219*, 311–324. [CrossRef]

minerals

MDPI

Article

Tectonic Inversion in Sediment-Hosted Copper Deposits: The Luangu Area, West Congo Basin, Republic of the Congo

Hongyuan Zhang [1,2,*], Shenghong Cheng [1,3,*], Gongwen Wang [1], William F. Defliese [2] and Zhenjiang Liu [1]

1 School of Earth Sciences and Mineral Resources, China University of Geosciences, Beijing 100083, China; gwwang@cugb.edu.cn (G.W.); lzj@cugb.edu.cn (Z.L.)
2 School of the Environment, University of Queensland, Brisbane, QLD 4072, Australia; w.defliese@uq.edu.au
3 Societe de Recherche et D'Exploitation Miniere (SOREMI), Pointe-Noire n°11B 247, Congo
* Correspondence: zhanghongyuan@cugb.edu.cn (H.Z.); shenghong.cheng@soremi.net (S.C.)

Abstract: Complex Neoproterozoic tectonic processes greatly affected the West Congo Basin, resulting in a series of dispersed copper deposits in the Niari Sub-basin, Republic of the Congo. Structural observation and analysis can help in understanding both the transportation pathways for copper accumulation and the detailed tectonic evolution processes. This study examines cases from four copper mine sites in the Luangu region of the Niari Basin, using a set of codes that consider the three regional tectonic regimes (extension, extrusion, and contraction) and three deformation criteria (maximum effective moment criterion, tensile fracture criterion, and the Coulomb criterion). By combining these two aspects, nine new codes are introduced: the extension maximum effective moment criterion (EM), extension tensile fracture criterion (ET), extension Coulomb criterion (EC), strike-slip maximum effective moment criterion (SM), strike-slip tensile fracture criterion (ST), strike-slip Coulomb criterion (SC), compression maximum effective moment criterion (CM), compression tensile fracture criterion (CT), and compression Coulomb criterion (CC). By analyzing and applying these codes to the selected sites, we show that the new codes can present a geometric coordination catering to an exhumation-related inversion process from extension, strike-slipping, to contraction. The existence of SM- and CM-related structures that occurred during regional extrusional and contractional events may indicate a deeper level of exhumation for layers related to copper deposits in the field sites. A new tectonic evolution model is presented, considering the hypothesis of vertical principal stress changes while the two horizontal principal stresses remain relatively constant during copper mineralization affected by the Western Congo Orogen. The application of the nine codes facilitates the determination of interrelations between different tectonic regimes.

Keywords: new structural codes; Luangu copper deposits; basin inversion model; West Congo Basin

1. Introduction

Previous studies have revealed sediment-hosted stratiform copper deposits (SSC), supergene-enriched copper deposits, and vein-type copper deposits within the Central African Copper Belt [1,2], with total estimated resources of more than 150 million tons of copper and 8 million tons of cobalt [3]. The formation mechanisms of sediment-hosted deposits are still debated; however, it is likely that structural factors play a significant role.

Deformation is localized within sediment-hosted deposits in the Central African Copper Belt [2], but the main phase of stratiform mineralization occurred during burial after the Lufilian Orogeny [1], following the diagenetic phase and the formation of siderite. This may be related to Pan-African orogenic superimposition. In tradition, researchers believed that sediment-hosted deposits did not exhibit early structural movements, but are controlled by extensional tectonic systems [3]. Recent studies reveal the mineralization of the sediment-hosted copper deposits may also be related to tensile fractures which act as pathways for fluid flow and metal transport in orogenic-related fold-thrust tectonics [4–6].

Tensile fractures are governed by the tensile criterion and have no direct relationship with the tectonic regime, as they can occur in multiple settings. Extension refers to a tectonic regime defined by extensional structures, especially normal faults. However, tensile fractures with vertical occurrences can locally belong to extensional structures. Sometimes, vertical tensile fractures appear as a set of en echelon fractures, indicating a general strike-slip movement. Horizontal tensile fractures result from contraction. Therefore, mapping outcrop structures is a key to determining copper mineralization mechanisms when gravity and aeromagnetic surveys do not support the presence of such deep-seated bodies.

Fractures in sedimentary rocks include tension fractures (brittle), shear fractures (brittle), and ductile (local plasticity) shear zones, which are respectively controlled by the tensile criterion (T, the tensile strength of the rock, e.g., [7,8]), the Coulomb criterion (C, e.g., [7,8]), and the maximum effective moment criterion (M, e.g., [9]). Different criteria can be combined in Mohr space as shown in Figure 1. The conjugate angle 2α of these three can be used to differentiate the conditions easily in outcrops, being approximately 0° (T), 60° (C), and 110° (M), respectively [8]. Similarly, there are three regional stress regime types—extension, strike-slip, and contraction [10]—coexisting within the same ore deposit fields (Table 1). Each stress regime exhibits typical structures that can be mapped using the three different faulting criteria mentioned above.

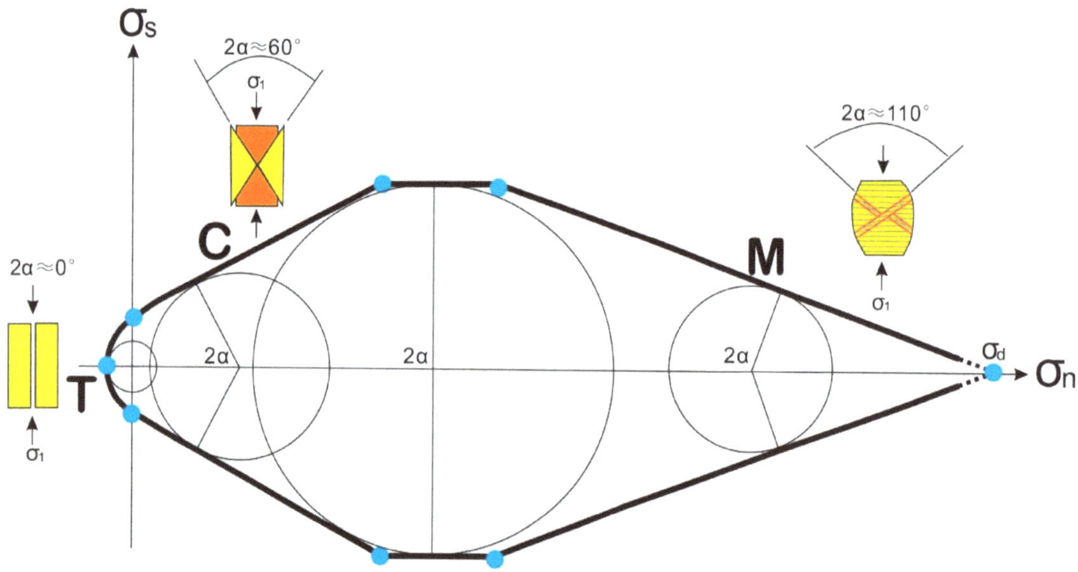

Figure 1. Three different fracture criteria combined in Mohr space (after [7–9]). Three styles of fracturing are depicted as tensile fractures (T), Coulomb criterion-related shear fractures (C), and maximum effective moment criterion-related ductile shear bands (M). The transition criteria among these three types are not displayed. The point (σ_d) refers to the lithostatic stress with zero differential stress.

Table 1. Lithostratigraphic units in the Luangu area and its surrounding region corresponding to Figure 2. The green dashed line in the table represents the studied copper unit.

Craton Texture	Unit	Rock Assemblage (and Thickness)	Time	Tectonic Setting	References
	Mpioka Subgroup	Alternating sandstone-clayey sediments	~540 Ma, ~1000 m	Late orogenic related	[11]
	Schisto-Calcaire Subgroup	I–V carbonate-dominated subgroups	~630–580 Ma, ~1100 m	Related to the rifting of the basin.	[11]
	Upper diamictite formation	Diamictite	~640–635 Ma; ~150 m	Related To the Macaúbas Rift II: Cryogeonian	[12–14]
	Haut Shiloango Subgroup	III clast-supported conglomerates and breccias	~678–640 Ma; ~1050 m		
		II nodular wackestone			
		I alternating limestones and claystones			
	Lower diamictite formation	Diamictite with basalt sheets	~694–678 Ma; ~400 m		
	Sansikwa Subgroup	Continental rift basin fill intruded with dolerite sills	Earlier than 694 Ma ~1650 m		
		Nonconformity			
Basement (Before Early Neoproterozic)		Lufu granite	917 ± 14 Ma	Related To the Macaúbas Rift I: Tonian	[13]
		Gangila Meta basalt	920 ± 8 Ma		
		Inga metarhyolite	924 ± 25 Ma		
		Yelala metaconglomerate	~930 Ma		
		Kimezian Basement	~2.1 Ga		[15,16]

This paper presents a case study on tectonic regime changes in the Luangu area of the Niari Sub-basin in the Neoproterozoic West Congo Basin, Republic of the Congo (Figure 2). Regionally, copper deposits are distributed along sinistral strike-slip shear zones that terminate at the thrust front of the fold-and-thrust belt of the West Congo Orogen (Figure 2C; [17]). However, they are also sediment-hosted and relevant to regional extension processes (Table 2; e.g., [12,13]), which presents a challenge in clarifying the various ore-controlling models affected by different regimes. By selecting a north–south rectangular area across the Luangu region, we provide observational results from the field to regional scales by combining deformation criteria theory with structural analysis. Structural features in different stress fields at the outcrop scale delineate ore-controlling systems for mineralization. A copper formation model controlled by tectonic regimes is discussed.

Table 2. Copper mine locations and deformation features.

Mines	GPS	Ductile Shearing	Extensile Fractures	Shear Fractures
L1, Northwest Luangu	+13°55′50″–04°19′30″	Layer-parallel sliding; stretch lineations	Veinlet stockworks,	Silication alteration in wall rocks, iron stained
L2, Central West Luangu	+13°55′50″–04°20′15″	Layer-parallel sliding; folding	Fissures with malachite veins	Silication alteration in wall rocks, iron stained
L3, Southwest Luangu	+13°56′00″–04°21′15″	Layer-parallel sliding; folding	Disseminated	Silication alteration in wall rocks
LE, Northeast Luangu	+13°59′00″–04°19′25″	Layer-parallel sliding; folding	Stockworks	Fault zone with silication alteration

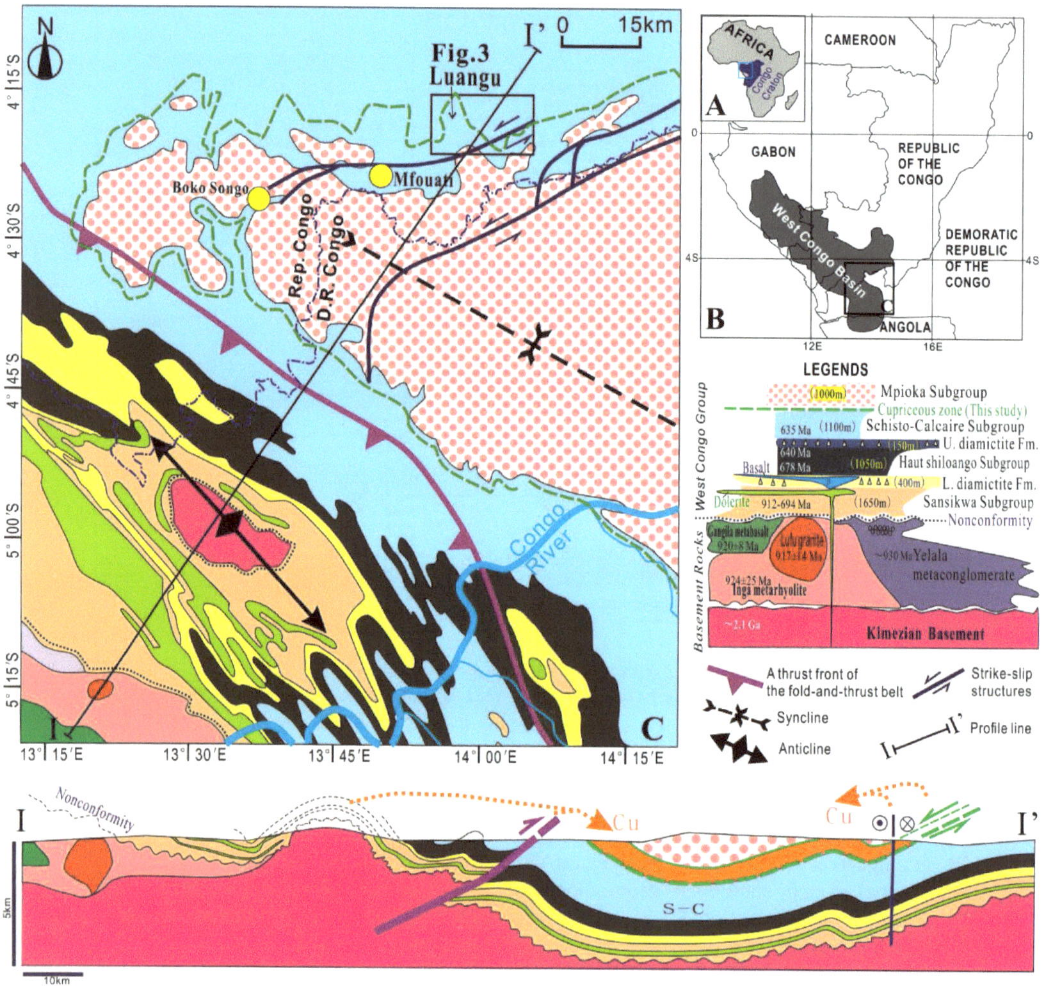

Figure 2. (**A**,**B**) Tectonic locations of the Congo Craton and the West Congo Basin (after [11]). (**C**) Tectonic framework in the area southeast of the West Congo Basin after [18–21]. Orange color dashed lines with arrows in the I-I′ profile show some possible movement directions of copper (Cu). The orange region in the profile indicates the current study layer which had been deformed by extensional, strike-slipping, and thrusting and folding structures.

2. Geological Setting

The study area is located in the Niari sub-basin, a southeast part of the West Congo Basin rifted upon the southwest boundary of the Proterozoic Congo Craton (Figure 2A,B; [11,22–25]). The Paleoproterozoic basement is distributed in the west of the study area and is locally intruded by younger granites dated at 1.0 Ga (Figure 2; [26]). This basement is capped by 5–6 km of thick basalts and rhyolites, dated between 912 $\pm$ 7 Ma and 920 $\pm$ 8 Ma (U-Pb zircon dates; [18]). The Neoproterozoic cover rocks are carbonate platforms and foreland metasediments of the West Congolian Group [27,28]. With a nonconformable relationship to the basement, the West Congolian Group includes, upwards, a continental rift basin fill (Sansikwa Subgroup) intruded with dolerite sills, as well as marine carbonate successions with interbeds of glacial deposits and basalt sheets (Table 1; [11–14]).

The Luangu deposit is hosted by the West Congo Group within a cyclical transgressive–regressive zone, situated between the underlying basement complex rocks and the overlying terrigenous red and tan siltstones and mudstones (Table 1). The timing of mineralization is not well constrained. Mineralization either occurred between 1000 and 700 Ma during the diagenetic period of the covering sedimentary system [29,30], or coincided with the Pan-African orogeny in the Neoproterozoic Ediacaran period or slightly later (635–541 Ma for the Schisto-Calcaire/Lukala Group [11]; Mpioka Group, about 540 Ma [11]; see Table 1). Regionally, there is an approximately east–west-trending copper mineralization zone in the Luangu region (see Figure 3). Various cupriferous deposit locations within the Luangu region have been verified to be situated in the upper section of the West Congolian Group, between the carbonate Schisto-Calcaire Subgroup and the terrigenous Mpioka Subgroup (as indicated by the green dashed line in Table 1). The cupriferous units have also been significantly deformed by layer-parallel detachments, strike-slip faults, and reverse faults (see Figure 3). A series of north–south-trending fold-and-thrust structures, along with some northeast–southwest- and northwest–southeast-trending conjugate strike-slip structures, can also be mapped to superpose upon the mineralization zone into north–south-trending stripes (see Figure 3). In general, three stages of evolution after the formation of the basement can be identified:

Rift Basin Period (about 0.93 Ga to about 0.54 Ga [11,12,31,32]): Characterized by extensional detachment, new crustal growth, and the formation of the Luangu copper mineralization zone.

Pan-African Orogeny (Neoproterozoic to Early Paleozoic): This was a notable period of tectonic, magmatic, and metamorphic activity that contributed to the formation of Gondwanaland [32–35]. Regional strike-slip faults and related reverse faulting and folding structures serve as the main ore-controlling structures of the region and may overlie some mineral bodies.

Gondwana Continental Rifting Period (Mesozoic): This period was marked by rifting related to the formation of the Atlantic Ocean [36–39].

New structural features from outcrops of some copper mines—L1, L2, L3, and LE—are summarized and listed in Table 2. Due to alteration, there is significant variation in the lithology of sedimentary wall rocks, necessitating the use of original sedimentary structures for identification. Further studies on ore-controlling structures in oxide ores could be beneficial for better understanding the mineralization process.

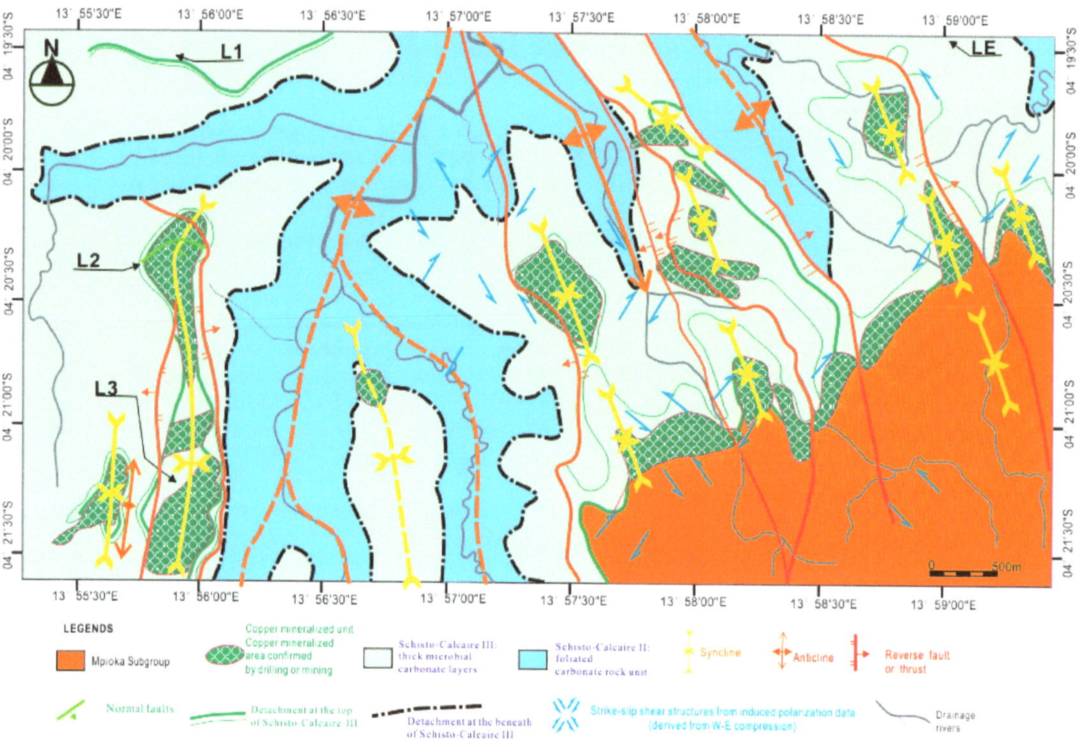

Figure 3. Structural map of the Luangu Area.

3. New Structural Codes Based on Both Structural Regimes and Faulting Criteria

Nine new structural codes are defined to help perform structural analysis in the Luangu region in this paper. In this section, the codes are explained, including how the nine codes are formulated and defined.

There are three types of regional deformation regimes including extension, strike-slipping, and contraction [10]. Here, we will provide a new but more general interpretation of geometric relationships during the superposition of different tectonic regimes in an inversion process. When observing stress field patterns, if the directions of the maximum and minimum horizontal compressions (σH, σh) across the aforementioned regimes are constant, and only the magnitude of the vertical principal stress changes with time or space, this suggests the presence of an ore-forming system (see Figure 4). It is important to study vertical or regional variations in the stress field based on signature structures characteristic of the tectonic stress field. The inversion system logically encompasses tectonic processes from early to later stages, such as rifting, strike-slip faulting, reverse faulting, and folding (e.g., [7,40,41]), similar to the evolution of the West Congo Basin.

Figure 4 illustrates the paleostress field sequence during basin inversion processes, highlighting strain relationships. From the deformation perspective, an ore-forming system that combines the extension regime and the strike-slip regime should align with the maximum principal strain axis; the strike-slip regime and the contraction regime (including fold-thrust and reverse faults) should align with the minimum principal strain axis. Additionally, the medium principal strain axis (y) in the contraction regime should be parallel to the maximum principal strain axis (x) in the extension regime. Extensional structures focus primarily on investigating interlayer sliding, combinations of high and low angles, and attitude relationships, among others. Extrusion structures including X-shaped shear fracture combinations are typically observed on the original bedding plane (S0).

Contraction structures including reverse faults and related folds are caused by the stress field when both the axes of the minimum principal stress and the maximum principal stress are horizontal. It is crucial to analyze the relationships between structures formed during the three different regimes above.

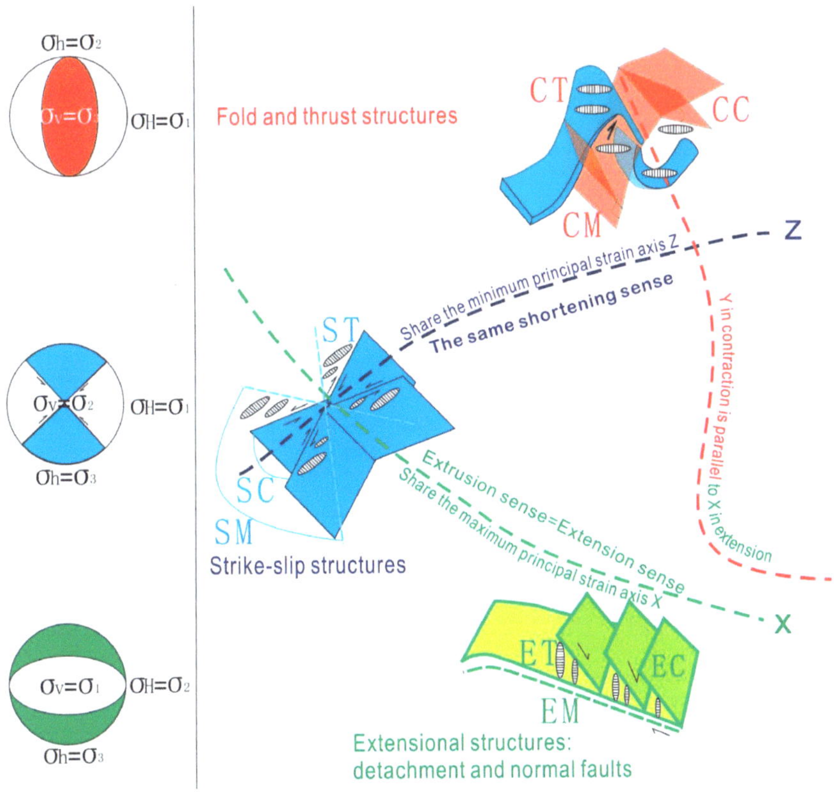

Figure 4. This diagram shows the superposition of three types of structures caused by a decrease in stress solely in the vertical direction over time; or the superposition of three types of structures caused by a decrease in stress solely in the vertical direction from bottom to top in space. Note: EM refers the extension maximum effective moment criterion; ET refers the extension tensile fracture criterion; EC refers the extension Coulomb criterion; SM refers the strike-slip maximum effective moment criterion; ST refers the strike-slip tensile fracture criterion; SC refers the strike-slip Coulomb criterion; CM refers the compression maximum effective moment criterion; CT refers the compression tensile fracture criterion; CC refers the compression Coulomb criterion. X refers to the maximum principal strain axis; Y refers to the medium principal strain axis; Z refers to the minimum principal strain axis. Extrusion sense here refers to the extruded out direction, parallel to the *X* axis; extension sense is the extended direction, parallel to the *X* axis. In contraction, the vertical axis is the *X* axis; the *Y* axis is the medium undeformed axis, and is parallel to the *X* axis in the extension regime in the whole inversion system.

There are also three types of structural criteria available for mapping in outcrops: the tensile fracture criterion, the Coulomb criterion, and the maximum effective moment criterion [9]. These three criteria can form a unified deformation criterion to explain natural shear planes and cracks (Figure 1; [8]).

If we consider three stress regimes, which are extension, strike-slipping, and compression, and the three deformation criteria, there will be nine types, including extension maximum effective moment criterion (EM); extension tensile fracture criterion (ET); extension Coulomb criterion (EC); strike-slip maximum effective moment criterion (SSM); strike-slip tensile fracture criterion (SST); strike-slip Coulomb criterion (SSC); compression maximum effective moment criterion (CM); compression fracture criterion (CT); and compression Coulomb criterion (CC) (Table 3). Therefore, considering both the tectonic regime and geometric factors, nine common classifications have been identified from outcrop studies.

Table 3. Classification of nine codes for different stress regime structures in copper deposits.

Stress Regime	Related Structures in Outcrops	Codes and $2\alpha_f$		Related Reports
Extensional	Detachment or layer-parallel sliding plane; normal faults, extensional veins/fractures, vein structures/tensile fractures	EC	<90°	[42,43]
		ET	~0°	
		EM	~110°	
Strike-Slip or Extrusion	Strike-slip faults, releasing/restraining bends, dilational jogs, vein structures/tensile fractures	SC	<90°	[44]
		ST	~0°	
		SM	~110°	
Compressional	Thrust or reverse faults, folds, duplexes, imbricate thrust systems; sulfide-filled foliation boudinage structures, vein structures/tensile fractures	CC	<90°	[5,6]
		CT	~0°	
		CM	~110°	

Note: Codes definition and explanation are the same as in Figure 4. $2\alpha_f$ is the conjugate angle between shear planes containing σ_1.

4. Results

4.1. Primary and Weathering Structures in the Cupriferous Units

Field surveys, drill core analyses, and geophysical techniques on copper mineralization have been conducted, revealing a new cupriferous microbial dolomite unit in the western Luangu region (see Figures 5 and 6). Thick dolomite with stromatolites occurred in a warm, shallow water environment within a Neoproterozoic rifting basin [11]. On the way from L2 to L3, particularly near the outcrop close to L3, one can observe an exposed outcrop of inclined bedding surfaces running north–south (see Figures 5 and 6). From a distance, it exhibits directional distribution characteristics. The diameter of the layered stones is generally 10–30 cm. The curved layers display alternating bright and dark patterns (see Figure 6). The ore rock types here have been mostly proven to be weathering-related (Figure 7). For example, chessboard-like structures are commonly observed where only silicified veins remain, while other parts of the primary sandstone rocks have weathered away. Larger ore bodies are usually confined to fault planes. In oxide ore zones, weathered ores are not conducive to diagenetic analysis; however, in many cases, the siliceous framework remains intact, with neogenic veinlets, malachite, azurite, etc., attached to the framework (see Figure 3; Table 2).

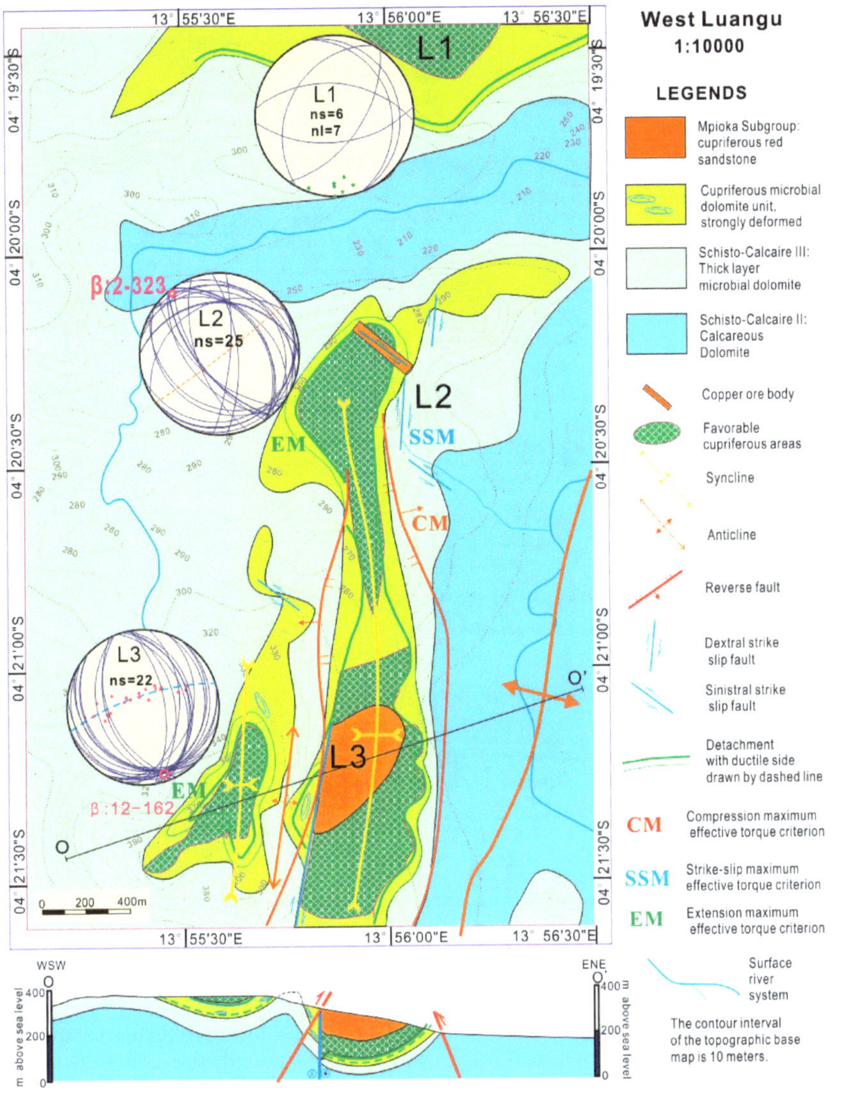

Figure 5. Regional geology with structures of three tectonic regimes in the western Luangu region. ns is the number of bedding orientation datapoints; nl refers to the number of stretching lineations.

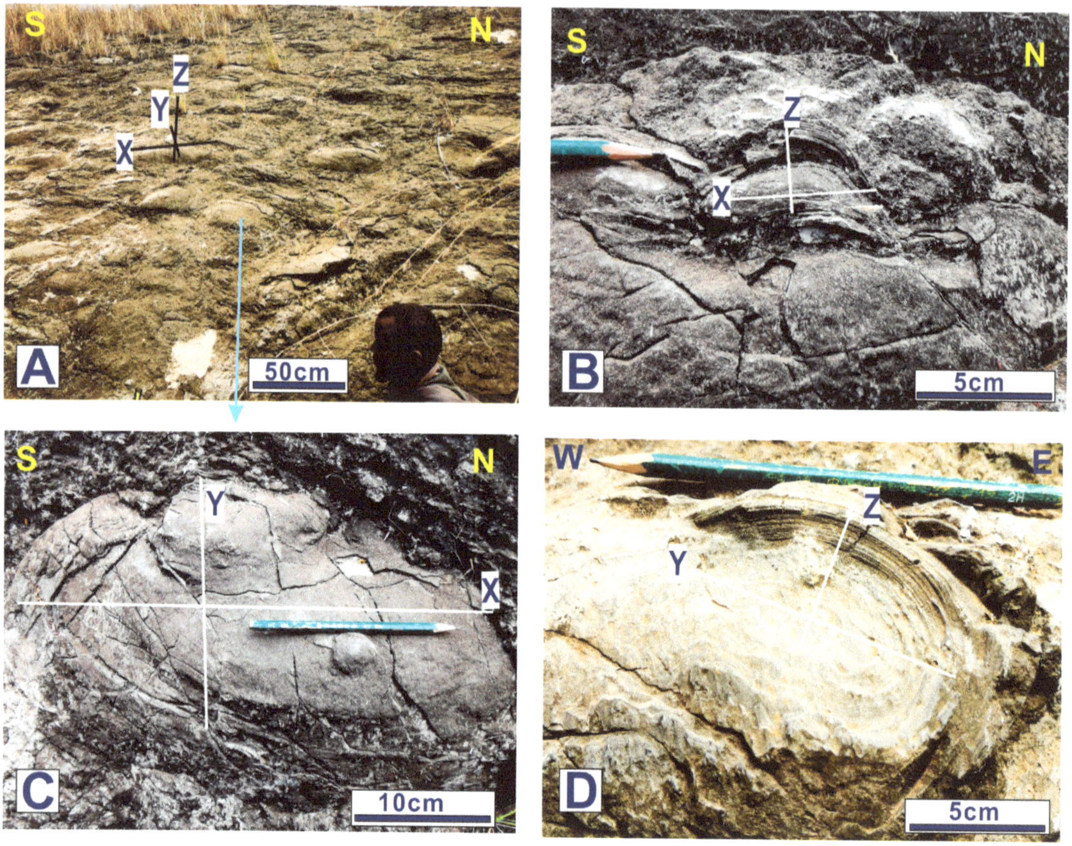

Figure 6. Outcrop of stromatolite at L3: (**A**) dip surface; (**B**) XZ plane; (**C**) XY plane; (**D**) YZ plane. X refers to the maximum principal strain axis; Y refers to the medium principal strain axis; Z refers to the minimum principal strain axis.

4.2. Results and Analyses on Ore-Controlling Structures

Outcrop structures from four mine sites in the Luangu copper deposits are carefully studied and depicted in this section. Table 4 contains the results of observations of the new codes and the relations between the different tectonic regimes experienced.

Table 4. Coded results and geometric relationships between different regimes.

Mine Site	Coded Results	Contraction and Extrusion Share the Z Axis or Not?	Extension and Extrusion Share the X Axis or Not?	Is Y Axis in Contraction Parallel the X Axis in Extension?
L1, Northwest Luangu	EM; SC; CT, CC	Deduced YES	Deduced YES	YES by CC and EM from Figure 8F
L2, Central West Luangu	EM, ET, EC; SM; CM	YES from Figure 9E	Deduced YES	Deduced YES
L3, Southwest Luangu	EM; ST, SC; CM	YES from Figure 10C,E	YES from Figure 10B,C	YES from Figure 10B,E
LE, Northeast Luangu	EM; SM, SC; CM, CT, CC	Deduced YES	Deduced YES	YES from Figure 11A,B

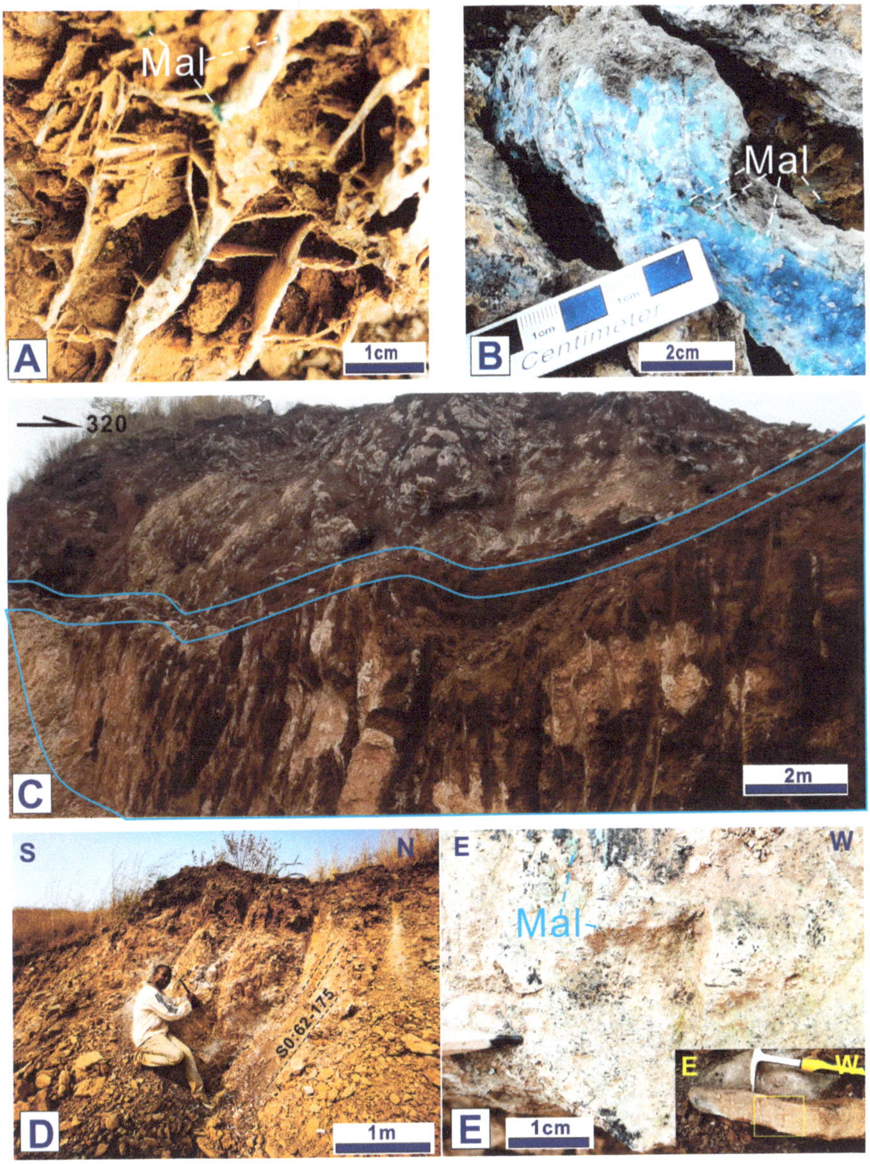

Figure 7. Weathering ore bodies: (**A**,**B**) are from L1; (**C**) is from L2; (**D**,**E**) are from L3. (**A**) Chessboard-like veins that are all silicified (a piece of float in situ). (**B**) Quartz vein (a piece of float in situ). (**C**) The weathering copper ore body is the area marked with blue lines, which is controlled by an upright strike-slip fault. (**D**) The brown hematite-rich sandstone layers have been faulted. (**E**) The silicified margin (chessboard pattern) is higher than the sandstone due to differential weathering. The yellow box shows the outcrop location of (**E**). In photos (**A**,**B**,**E**), 'Mal' refers to malachite.

4.2.1. L1

L1 is currently a mining pit but does not have a well-exposed outcrop profile (Figures 3 and 5; Table 2; Figure 8A). Apart from the oxidized copper ore, primary ore minerals without oxidation cannot be found here (Figure 6A,B).

The euhedral quartz veins can grow in extensile cracks near the Earth's surface environment (possibly coded as CT in contraction), along with oxidized copper minerals attached to their tips (Figure 8B–D). The attitude can be measured in an excavation outcrop (Figure 8E). Layer-parallel extension caused stretching lineations with a mean orientation of 10°–178° (L_M in Figure 8F). L_M is from the low-angle detachment coded as EM. Bedding layers might have also experienced conjugate strike-slip (SC in Figure 8F) and folding processes (CC in Figure 8F). According to Figure 4, the set of structures in L1 is suitable for the extension inversion process caused by a decrease in stress solely in the vertical direction over time. The earlier deformation is characterized by layer-parallel extension controlled by the maximum effective moment criterion. The later, but shallower, deformation involves strike-slip and contraction, both controlled by the Coulomb criterion.

Figure 8. Field outcrop and vein-type ore from L1: (**A**) Field quarry. (**B**) Ore, approximately 20 × 30 × 50 cm in size, featuring developed jointing with quartz crystal clusters forming on the

joint surfaces. (**C**,**D**) depict the tips of the quartz crystal clusters turning dark in color, potentially indicating copper mineralization. (**E**) is the outcrop used for orientation measurement. (**F**) shows an equal-area stereographic projection plotted on the lower hemisphere to display layer surfaces and stretch lineations. The plotted blue dashed circles represent bedding layers that have been transformed by layer-parallel sliding, with stretch lineations on their surfaces.

4.2.2. L2

The copper ore body in the site L2 is NW–SE-trending, controlled by regional strike-slip structures (Figure 5; Table 4; Figures 6C and 9A). In the field, it appears as steep-standing blocks, specifically lenses of ore bodies within the strike-slip fault systems (Figure 9B).

The cupriferous shear zone, controlled by EM, ET, and EC, is marked in Figure 9C,D, indicating that it may have formed on extensional detachment fault surfaces on top of the SC-III unit (Figures 3 and 5).

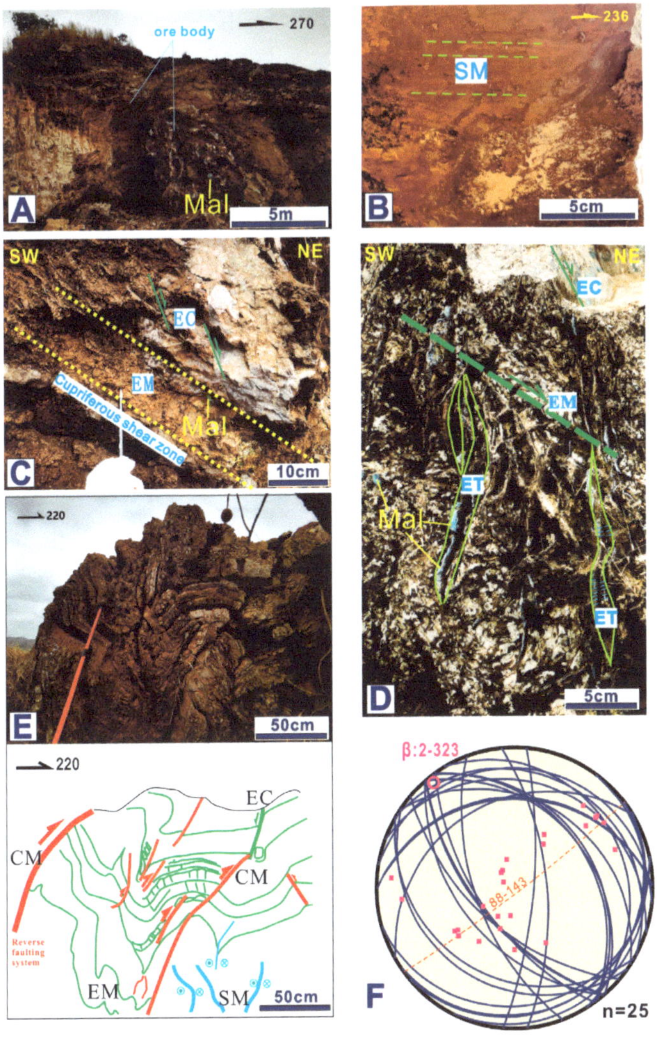

Figure 9. Deformation structures from L2. (**A**) The NW–SE-trending and vertical ore body, approximately 5 m wide in the field. (**B**) Strike-slip-related slickenside lineations coded SM. (**C**) The cupriferous

shear zone is parallel to the bedding layers marked EM, with high-angle shear fractures marked EC. (**D**) In addition to the EM-EC structures, extensional cracks filled with cupriferous veins are observed and marked as ET. (**E**) An excellent complex folding and reverse faulting outcrop, outlined and compared as different regimes with defined codes EM, EC, SM, and CM. (**F**) A stereographic projection plotted on the lower hemisphere with data measured on the bedding layer at the top of the (**D**) profile.

The latest contraction structures can also be observed in well-exposed outcrops in the field from mine L2 (Figure 9D,E). The fold-thrust structures align with the projection results. Statistical analysis of 25 bedding planes results in a beta axis of or y-axis orientation of 2–323, representing the medium strain orientation during the final contraction process.

Analysis indicates that extensional structures have been modified by both strike-slip and thrust regimes. From early to late, there is layer-parallel shearing (EM) with high-angle normal faulting (EC), strike-slip faulting (SM), and high-angle reverse faulting (CM).

4.2.3. L3

The site L3 is located in the southwestern part of the Luangu study area, some several meters under the top surface of the Schisto-Calcaire carbonate rock unit and the below surface of the Mpioka red sandstone unit, where malachite mineralization is observed near the surface but inside the Schisto-Calcaire unit (Figure 10A). In the field, strain ellipsoid markers and iron mineralization are visible.

The large extensional lineations in the outcrop also indicate that the maximum strain axis is 165° (see Figure 10B). A closer observation also shows small strike-slip structures. The en echelon cracks on the surface containing large lineations can be coded as ST (Figure 10C). The ST cracks opened at the larger shear plane are marked as SC. Sinistral and dextral kinematic senses can be determined by a set of SC-ST geometric systems (Figure 10C). An arcuate anticline with sub-kink folds (CM) and sub-reverse blind faults (CM) has been identified, with a β range of 12°–162° (see Figure 10D–F). It is inferred that this fold has a medium strain axis orientation of about 162°. Therefore, the medium strain axis in the contraction system is parallel to the earlier extensional maximum strain axis, which can be explained by the method shown in Figure 4.

4.2.4. LE

The cupriferous shearing zone between the thick dolomite (Schisto-Calcaire III) and the red sandstone layers (Mpioka Subgroup) at two pits from the northeastern Luangu (LE) mine site was observed and measured (Figure 3; Figure 11). The natural erosion in the north pit is much deeper than that of the south mining pit and has caused the cupriferous shearing zone to be lost.

The cupriferous shearing zone with malachite and limonite can be observed in the south pit of the EL site. The extensional detachment surface is nearly horizontal with smooth undulations (Figure 11A). There are some smaller sheath folds that can be found in the shallower dip planes. The kinematic sense of the detachment zone (EM) is roughly with the top plate moving to the south, which is consistent with that in other sites of Luangu. Extrusion-related shear planes (SC) can also be found, cutting through the detachment surfaces.

The overall dip of the strata is gentle, but becomes steeper as it nears the faulting locations. Extrusion-related strike-slip faulting (SM) with cataclastics occurred in the north pit of the EL site, as well as contraction-related kink folding (CM), layer-parallel cracks and veins (CT), and reverse faulting (CC) (Figure 11B).

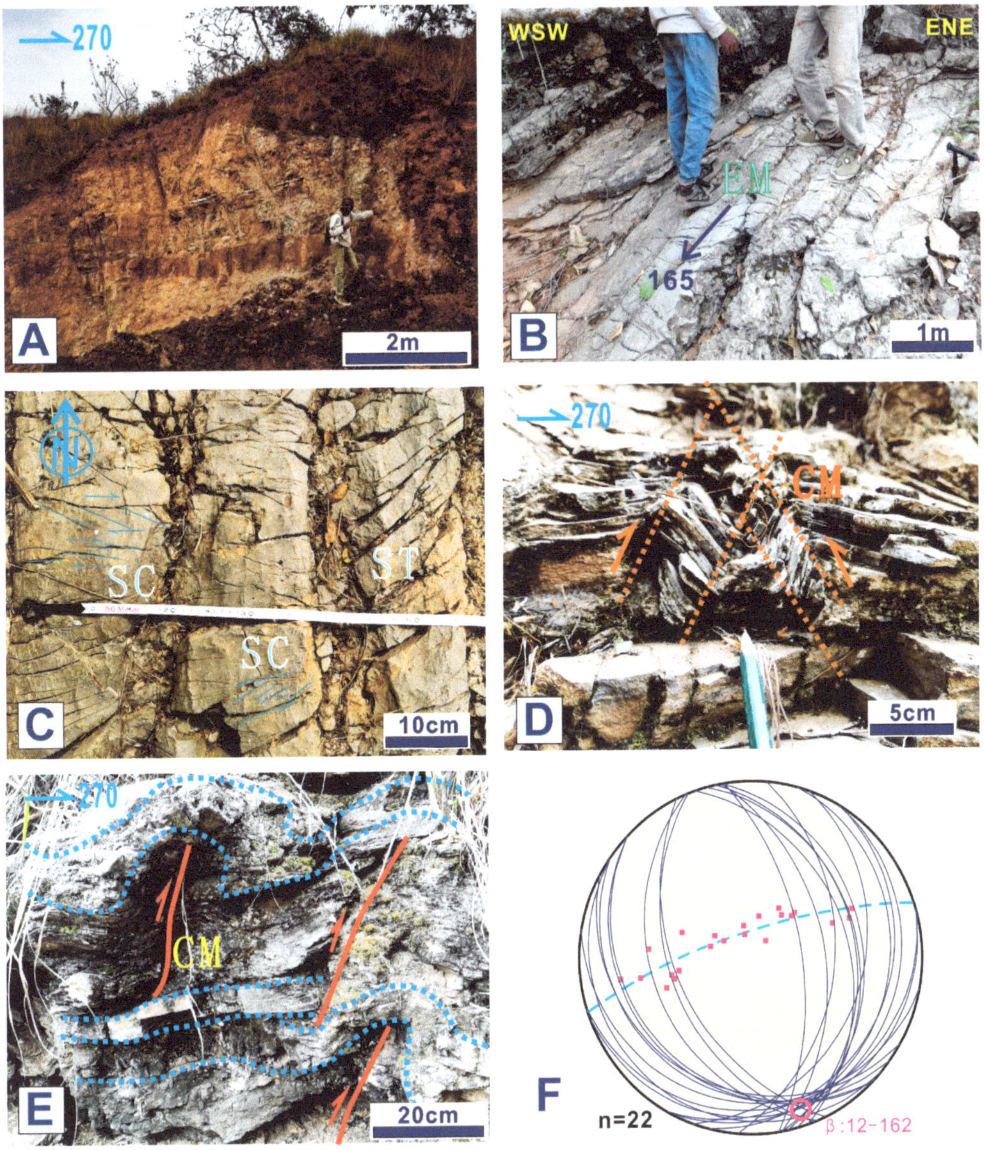

Figure 10. Field outcrop structures and elemental studies at L3. (**A**) Boundary between the Mpioka Subgroup and Schisto-Calcaire-III, showing red sandstone layers above the unconformity. (**B**) Depicts the fold hinge area and the initial stretching lineation bands (coded EM) when facing north. (**C**) Shows extensile cracks (ST), sinistral and dextral strike-slip shear fracture sets (SC). (**D**,**E**) Illustrates thrust-and-fold structures when facing south (CM). The dotted lines in (**D**) illustrate two reverse kink zones, and the dotted lines in (**E**) outline curved bedding surfaces. (**F**) Displays equal-area stereographic projections of plane elements with a beta axis of 12–162°.

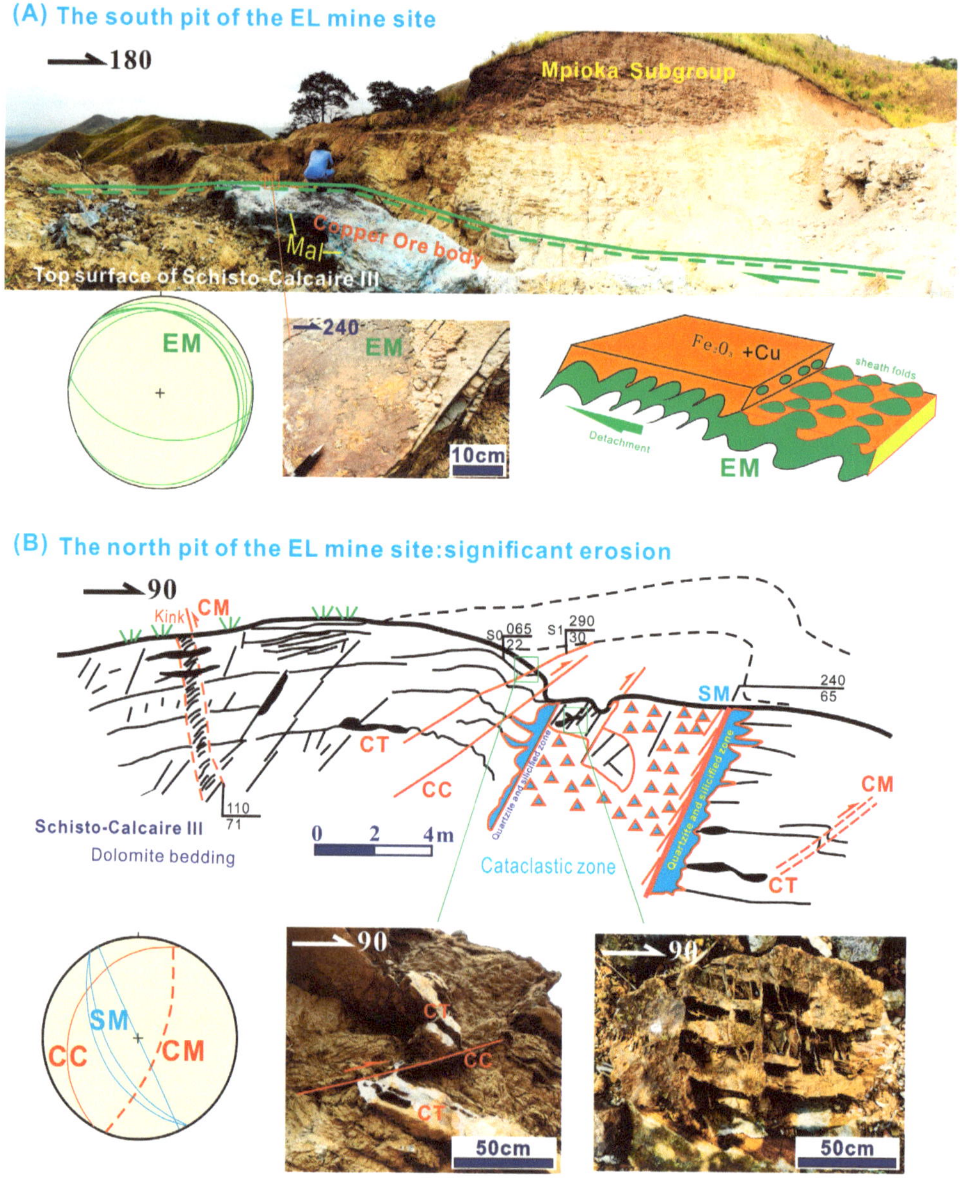

Figure 11. Presents the outcrop observation results in the eastern part of Luangu (EL): (**A**) Illustrates a copper ore body within layer-parallel sliding structures (EM), situated below a detachment surface (the green solid/dotted line) in the south pit of the EL mine site. The cupriferous unit, indicated by malachite, is located between the red sandstones (Mpioke Subgroup) and the dolomite bedding (Schisto-Calcaire III). (**B**) Shows evidence of significant erosion following strike-slip sliding (SM), a kink zone (CM), layer-parallel extensional fractures with veins (CT), and reverse faulting (CC). Stereographic projections are plotted on the lower hemisphere with codes corresponding to those found in the outcrop figures.

According to the geometric relations revealed by the stereographic projection in Figure 11, the maximum extensional principal strain axis is nearly parallel to the fold axis and the strike of the reverse faults, which reflects an inversion process.

5. Discussion

In the previous section of this paper, we analyzed complex outcrop structures from four mine sites using our new codes (Table 4; Figures 8–11). These observations are dynamically related to both the tectonic regimes and possible deformation criteria. However, the results remain too isolated to provide a comprehensive understanding of the tectonic evolution processes. Our results and analysis from four mine sites in Luangu clearly indicate that three different regimes were experienced. Here, we present a further discussion on regional evolution by comparing various sites of the Luangu copper deposits.

5.1. The Significance of the New Codes Series Used for Structural Analysis

Previous structural analysis only differentiated three regimes qualitatively according to reverse, strike-slip, and normal faults, and could only roughly determine a major tectonic regime on site. However, through our nine new codes, it is possible to uncover more details about the transitions between different tectonic regimes. Moreover, a quantitative determination of the coordination relations can be made and deduced from these codes recorded at outcrops, allowing us to establish a regime change order.

It is easy to understand that a location point should be controlled by a single tectonic regime at any given time. However, extension at one point could change to strike-slip at another time and eventually to contraction. The final stage should depend on regional tectonic conditions. In other words, when the basin rift is relatively localized, if the regional tectonic condition is strike-slip, then the final stage will be strike-slip; if the regional tectonic condition is contraction, then the final stage will be contraction. Some previous studies believe that regional copper deposits are more obviously controlled by a set of ENE–WSW strike-slip faults (Figure 2C; [11,17]) in a regional extrusion regime. However, structures associated with contraction are also shown to be prevalent in our study. From all pits, upright veins and horizontal veins can be observed to form due to fracturing that occurred under tensile criteria during different regimes (see the lower right picture in Figure 11B).

By classifying structures related to different regimes and levels at the field scale, a changing process for different erosion levels of each tectonic regime can be easily deduced (Table 5). The tendency of these changes can be more easily understood in Figure 12, which compares different evolutionary routes in physical style. EM is likely the most prominent during a deep ductile extension stage, which is clearly related to the formation of the Luangu copper deposit (Figure 13). In L2 and LE, the copper grade reaches up to 3%–4% within the dolomite layer detachment zone (Figures 9 and 11). However, other types of structures may hinder preservation conditions due to higher erosion levels. For the L1 site, this may indicate good preservation of the copper deposit, as no SM and CM structures have been observed there. For the L2, L3, and LE sites, this suggests limited preservation conditions for the copper deposit, as CM structures have been observed there (Table 5).

5.2. Models for Regional Distribution and Evolution of the Luangu Deposits

From the geological background in the Section 2, three stages of tectonic development can be concluded: basement rift and West Congo Basin formation, Pan-African orogeny, and the assembly and rifting of Gondwanaland. Large sediment-hosted copper deposits often experienced superimposed tectonic systems [1,3]. Since the four study sites are primarily located at the same tectonic level within a relatively small region, it is plausible to determine the superposition order in time, beginning with extension, followed by strike-slip, and finally contraction. The regime changes correspond to an inversion process during exhumation. This study further demonstrates the geometric, kinematic, and dynamic relationships resulting from tectonic inversion.

Table 5. Deformation history of each site using the new codes.

Mine Site	Results	Extension		Strike-Slip		Contraction		Preservation Assessment
		Ductile	Brittle	Ductile	Brittle	Ductile	Brittle	
		EM	ET/EC	SM	ST/SC	CM	CT/CC	
L1, Northwest Luangu	EM; SC; CT, CC	√	-	-	√	-	√	Good
L2, Central West Luangu	EM, ET, EC; SM; CM	√	√	√	-	√	-	Limited
L3, Southwest Luangu	EM; ST, SC; CM	√	-	-	√	√	-	Limited
LE, Northeast Luangu	EM; SM, SC; CM, CT, CC	√	-	√	√	√	√	Limited
Ore preservation assessment		Best	Best	Good	Better	Bad	Good	-

Note: The tick mark '√' indicates that structures related to different regimes and levels at the field scale have been found at the corresponding mine site.

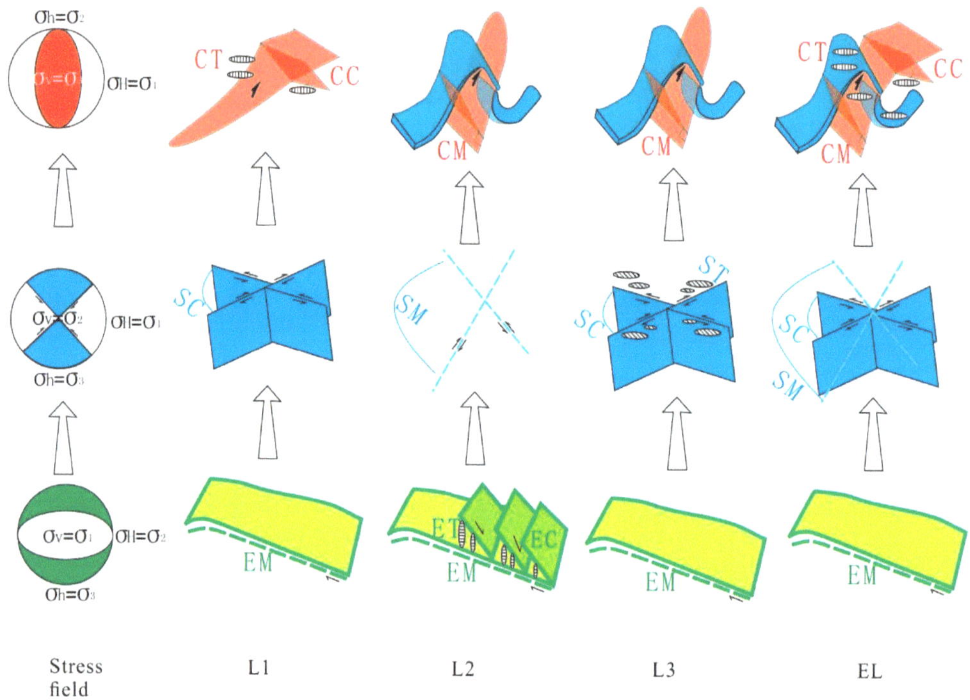

Figure 12. Evolution routes in physical style based on our structural observations for different sites. Single geometric styles are mostly after [7,45].

Stage A: Rift valley basin period

This time period is a transition from N–S contraction to N–S extension; that is, from the Lufilian orogeny to the rift valley, resulting in the sedimentation of the West Congo Group (Figure 14A). Previous studies have suggested the presence of a regional Neoproterozoic coastal environment during the extensional regime [11,46]. Our research has also revealed the presence of abundant stromatolitic structures between L1, L2, and L3, indicating a lacustrine depositional setting. Within these formations, sedimentary copper deposits are found, as illustrated in field photographs (Figures 8–11).

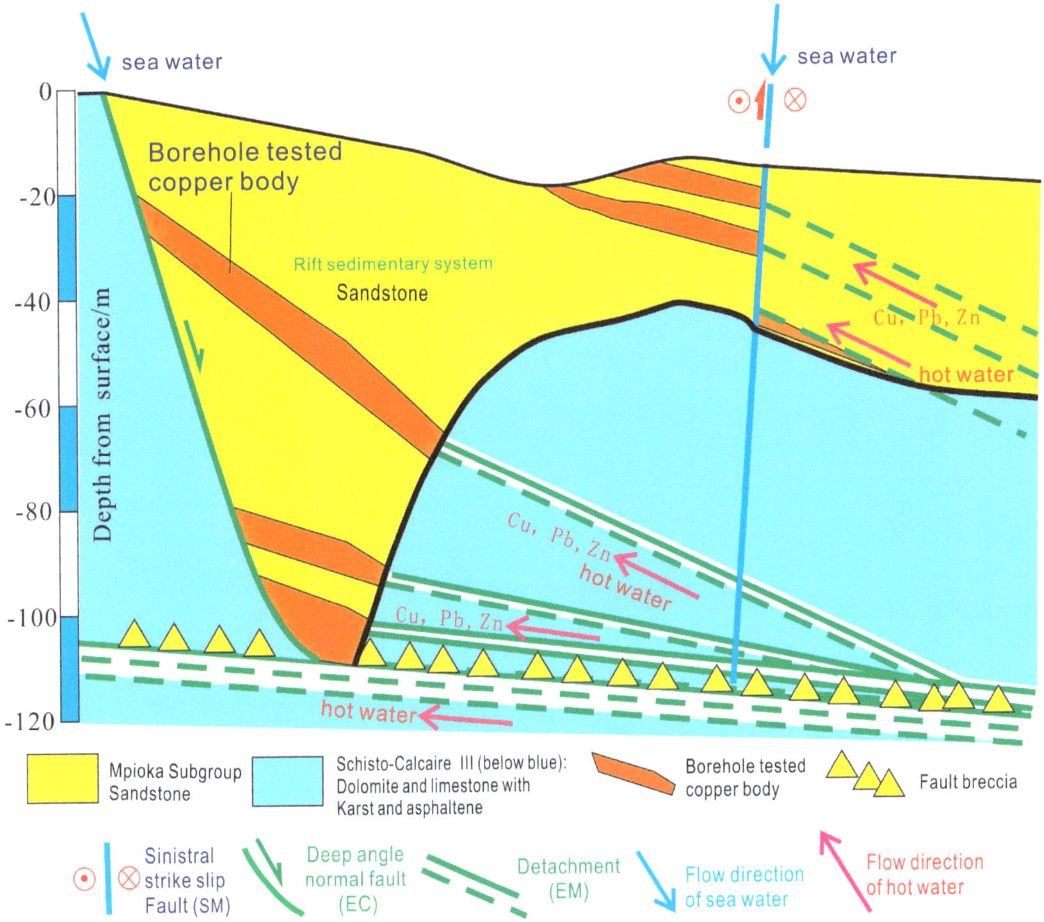

Figure 13. A mineralization model related to extension and strike-slip.

Stage B: Pan-African orogenic period

This is a period characterized by E–W contraction but occasionally by extension due to deep tectonic heat upwelling, resulting in the formation of copper deposits in the study area via hydrothermal processes along fracture zones (Figure 14B). The major superimposed tectonic systems transformed the boundary between the red bed sediments of the Mpioka unit and the underlying Schisto-Calcaire unit, resulting in isolated copper-bearing units in the study area, although they remain within a defined zone on the regional map (Figure 2C).

Systems of extensional, extrusive, and thrust structures coexist in field observations, indicating an evolving inversion process at each site. Field studies indicate that the extensional-related EM, ET, and EC systems are major mineralization factors. However, the strike-slip-related SM, ST, and SC systems, as well as the contraction-related CM, CT, and CC systems, may also have ore control significance.

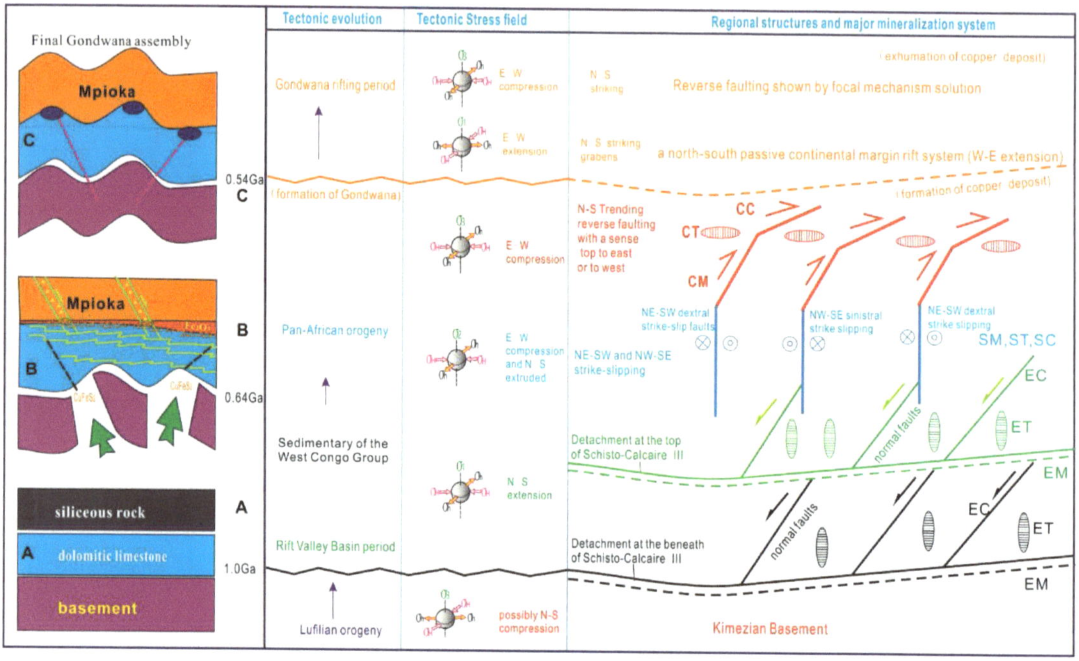

Figure 14. Copper-polymetallic mineralization in the basin inversion system showing the main copper bodies formed during E–W compression at the end of the Pan-African Orogen. (**A**) illustrates the marine sedimentary cover above the basement; (**B**,**C**) shows the deformation of cover layers with extension-related structures and contraction-related thrust and fold structures, respectively. Inversion occurs from (**B**) to (**C**), as indicated by the normal fault plane represented by black dashed lines in (**B**) and the reverse fault plane represented by red dashed lines in (**C**).

Stage C: Assembly and rifting of Gondwanaland

This is a period characterized by E–W extension and contraction processes, resulting in the exhumation of copper deposits in the study area (Figure 14B). The extension is related to Gondwana continental rifting and the formation of the Atlantic. The current predominant stress system is east–west contraction [47]. Although the copper mineralization is more intense near the oceanic spreading center [48], it is difficult to determine whether this significantly affects the formation of ore deposits within the continent.

6. Conclusions

(1) This study introduced and defined nine codes (EM, ET, EC, SM, ST, SC, CM, CT, CC) considering both tectonic regimes and deformation criteria for structural analysis. The significance of the new code series used for structural analysis makes the deformation history easier to understand for each mine site.

(2) Taking the Luangu deposits as an example, analysis results from four mine sites are provided using the new codes and discussed through comparisons with related physical diagrams. The key geometric relationships between different regimes, as observed and measured directly, prove that contraction and extrusion share the Z axis at L2 and L3, extension and extrusion share the X axis at L3, and the Y axis in contraction is parallel to the X axis in extension at L1, L3, and LE.

(3) A further basin inversion model for regional copper deposits in the West Congo Basin is established based on the previous tectonic settings. The E–W compression at the end of the Pan-African Orogen may represent the primary stress field for the formation of significant copper bodies.

Author Contributions: H.Z. was responsible for field data collection and writing the paper; S.C. organized geological field investigations, reviewed figures, and contributed to discussions; G.W. participated in project initiation, conducted field geological investigations, and engaged in discussions; W.F.D. contributed to writing, discussion, and revision of the paper; Z.L. contributed to the idea discussion and proofreading. All authors have read and agreed to the published version of the manuscript.

Funding: The study was supported by the Major Research Plan of the National Natural Science Foundation of China (Grant No. 92062219) and by the fund titled 'Analysis and Prediction of Copper Polymetallic Control Conditions in the Sorimi Mining Rights Area of Congo (Brazzaville) and Selection of Exploration Target Areas (Year 2023).

Data Availability Statement: The data presented in this study are available on request from the corresponding author. The data are not publicly available due to Privacy.

Acknowledgments: We extend our gratitude to Sheng Jin, Gaofeng Ye, and Yi Cao from the China University of Geosciences (Beijing) for their insightful contributions during our seminars in 2023. We also thank Grant Dawson from the University of Queensland and Thomas Blenkinsop from Cardiff University for their valuable suggestions on the title and research method. Fieldwork was supported by Deng Pan and Qing Lei, along with two local workers. We would like to thank Gideon Rosenbaum for his support regarding H.Z.'s research conditions at the University of Queensland. We would also like to thank the anonymous reviewers and editors for their valuable feedback.

Conflicts of Interest: Shenghong Cheng is an employes of Societe de Recherche et D'Exploitation Miniere. The paper reflects the views of the scientists and not the company.

References

1. Dewaele, S.; Muchez, P.; Vets, J.; Fernandez-Alonzo, M.; Tack, L. Multiphase origin of the Cu–Co ore deposits in the western part of the Lufilian fold-and-thrust belt, Katanga (Democratic Republic of Congo). *J. Afr. Earth Sci.* **2006**, *46*, 455–469. [CrossRef]
2. De Oliveira, S.B.; Bertossi, L.G. 3D structural and geological modeling of the Kwatebala Cu–Co deposit, Tenke-Fungurume district, Democratic Republic of Congo. *J. Afr. Earth Sci.* **2023**, *199*, 104825. [CrossRef]
3. Misra, K.C. Sediment-Hosted Stratiform Copper (SSC) Deposits. In *Understanding Mineral Deposits*; Misra, K.C., Ed.; Springer: Dordrecht, The Netherlands, 2000; pp. 539–572.
4. Curzi, M.; Aldega, L.; Billi, A.; Boschi, C.; Carminati, E.; Vignaroli, G.; Viola, G.; Bernasconi, S.M. Fossil chemical-physical (dis)equilibria between paleofluids and host rocks and their relationship to the seismic cycle and earthquakes. *Earth-Sci. Rev.* **2024**, *254*, 104801. [CrossRef]
5. Williams, B.J.; Blenkinsop, T.G.; Lilly, R.; Thompson, M.P.; Ila'ava, P. Foliation boudinage structures in the Mount Isa Cu system. *Aust. J. Earth Sci.* **2023**, *70*, 972–989. [CrossRef]
6. Williams, B.J.; Blenkinsop, T.G. Strain in sulphide filled foliation boudinage structures at the Mount Isa Cu deposit, Australia. *J. Struct. Geol.* **2024**, *179*, 105034. [CrossRef]
7. Fossen, H. *Structural Geology*, 2nd ed.; Cambridge University Press: Cambridge, UK, 2016; pp. 1–524.
8. Hou, Q.L.; Cheng, N.N.; Shi, M.Y.; Lu, X. The union of various rock deformation criteria at different structural levels and its further development. *Acta Petrol. Sin.* **2018**, *34*, 1792–1800. Available online: http://www.ysxb.ac.cn/search (accessed on 18 October 2024). (In Chinese with English Abstract).
9. Zheng, Y.D.; Wang, T.; Ma, M.B.; Davis, G.A. Maximum effective moment criterion and the origin of low-angle normal faults. *J. Struct. Geol.* **2004**, *26*, 271–285. [CrossRef]
10. Curzi, M.; Aldega, L.; Bernasconi, S.M.; Berra, F.; Billi, A.; Boschi, C.; Franchini, S.; Van der Lelij, R.; Viola, G.; Carminati, E. Architecture and evolution of an extensionally-inverted thrust (Mt. Tancia Thrust, Central Apennines): Geological, structural, geochemical, and K–Ar geochronological constraints. *J. Struct. Geol.* **2020**, *136*, 104059. [CrossRef]
11. Mfere, A.P.A.; Delpomdor, F.; Proust, J.-N.; Boudzoumou, F.; Callec, Y.; Préat, A. Facies and architecture of the SCIc Formation (Schisto-Calcaire Group), Republic of Congo, in the Niari-Nyanga and Comba subbasins of the Neoproterozoic West Congo Basin after the Marinoan glaciation event. *J. Afr. Earth Sci.* **2020**, *166*, 103776. [CrossRef]
12. Delpomdor, F.; Schröder, S.; Préat, A.; Lapointe, P.; Blanpied, C. Sedimentology and chemostratigraphy of the late Neoproterozoic carbonate ramp sequences of the Hüttenberg Formation (northwestern Namibia) and the C5 Formation (western central Democratic Republic of Congo): Record of the late post-Marinoan marine transgression on the margin of the Congo Craton. *S. Afr. J. Geol.* **2018**, *121*, 23–42. [CrossRef]
13. Souza, M.E.; de Souza Martins, M.; Queiroga, G.N.; Leite, M.; Oliveira, R.G.; Dussin, I.; Pedrosa-Soares, A.C. Paleoenvironment, sediment provenance and tectonic setting of Tonian basal deposits of the Macaúbas basin system, Araçuaí orogen, southeast Brazil. *J. S. Am. Earth Sci.* **2019**, *96*, 102393. [CrossRef]

14. Delpomdor, F.; Kant, F.; Préat, A. Neoproterozoic uppermost Haut-Shiloango Subgroup (West Congo Supergroup, Democratic Republic of Congo): Misinterpreted stromatolites and implications for sea-level fluctuations before the onset of the Marinoan glaciation. *J. Afr. Earth Sci.* **2014**, *90*, 49–63. [CrossRef]
15. Danderfer Filho, A.; Lana, C.C.; Nalini Júnior, H.A.; and Costa, A.F.O. Constraints on the Statherian evolution of the intraplate rifting in a Paleo-Mesoproterozoic paleocontinent: New stratigraphic and geochronology record from the eastern São Francisco craton. *Gondwana Res.* **2015**, *28*, 668–688. [CrossRef]
16. De Castro, M.P.; Queiroga, G.; Martins, M.; Alkmim, F.; Pedrosa-Soares, A.; Dussin, I.; Souza, M.E. An Early Tonian rifting event affecting the São Francisco-Congo paleocontinent recorded by the Lower Macaúbas Group, Araçuaí Orogen, SE Brazil. *Precambrian Res.* **2019**, *331*, 105351. [CrossRef]
17. Nikis, N.; Putter, T. Recherches Géo-Archéologiques Dans Les Zones Cuprifères Du Bassin Du Niari En République Du Congo. *Nyame Akuma* **2015**, *84*, 142–153. (In French) [CrossRef]
18. Tack, L.; Wingate, M.T.D.; Liégeois, J.P.; Fernandez-Alonso, M.; Deblond, A. Early Neoproterozoic magmatism (1000–910 Ma) of the Zadinian and Mayumbian Groups (Bas-Congo): Onset of Rodinia rifting at the western edge of the Congo craton. *Precambrian Res.* **2001**, *110*, 277–306. [CrossRef]
19. Garzanti, E.; Vermeesch, P.; Vezzoli, G.; Andò, S.; Botti, E.; Limonta, M.; Dinis, P.; Hahn, A.; Baudet, D.; De Grave, J.; et al. Congo River sand and the equatorial quartz factory. *Earth-Sci. Rev.* **2019**, *197*, 102918. Available online: https://hdl.handle.net/1854/LU-8641851 (accessed on 18 October 2024). [CrossRef]
20. De Putter, T.; Nikis, N. Le cuivre du Niari, une ressource ancienne et prisée. Études géologique et archéologique des mines de cuivre-plomb-zinc du Bassin du Niari (République du Congo). *Sci. Connect.* **2016**, *50*, 35–39. (In French) [CrossRef]
21. Eeckhout, S.; De Grave, J.; Van den Haute, P. Geology and petrology of the felsic intrusions in the metasedimentary and metavolcanic rocks of the early Neoproterozoic Zadinian Group around Matadi (Lower-Congo region, D.R. Congo). In *A State Affair: Privatising Congo's Copper Sector*; The Carter Center: Atlanta, GA, USA, 2017; pp. 1–104. Available online: https://lib.ugent.be/en/catalog/rug01:002163635 (accessed on 18 October 2024).
22. Kampunzu, A.B.; Kapenda, D.; Manteka, B. Basic magmatism and geotectonic evolution of the Pan African belt in central Africa: Evidence from the Katangan and West Congolian segments. *Tectonophysics* **1991**, *190*, 363–371. [CrossRef]
23. De Waele, B.; Johnson, S.P.; Pisarevsky, S.A. Palaeoproterozoic to Neoproterozoic growth and evolution of the eastern Congo Craton: Its role in the Rodinia puzzle. *Precambrian Res.* **2008**, *160*, 127–141. [CrossRef]
24. Begg, G.C.; Griffin, W.L.; Natapov, L.M.; O'Reilly, S.Y.; Grand, S.P.; O'Neill, C.J.; Hronsky, J.M.A.; Djomani, Y.P.; Swain, C.J.; Deen, T.; et al. The lithospheric architecture of Africa: Seismic tomography, mantle petrology, and tectonic evolution. *Geosphere* **2009**, *5*, 23–50. [CrossRef]
25. Fernandez-Alonso, M.; Cutten, H.; De Waele, B.; Tack, L.; Tahon, A.; Baudet, D.; Barritt, S.D. The Mesoproterozoic Karagwe-Ankole Belt (formerly the NE Kibara Belt): The result of prolonged extensional intracratonic basin development punctuated by two short-lived far-field compressional events. *Precambrian Res.* **2012**, *216–219*, 63–86. [CrossRef]
26. De Wit, M.J.; Linol, B. Precambrian Basement of the Congo Basin and Its Flanking Terrains. In *Geology and Resource Potential of the Congo Basin*; de Wit, M.J., Guillocheau, F., de Wit, M.C.J., Eds.; Springer: Berlin/Heidelberg, Germany, 2015; pp. 19–37. [CrossRef]
27. Ball, S.H.; Shaler, M.K. A Central African Glacier of Triassic Age. *J. Geol.* **1910**, *18*, 681–701. Available online: https://www.jstor.org/stable/30081323 (accessed on 18 October 2024). [CrossRef]
28. Frimmel, H.E.; Tack, L.; Basei, M.S.; Nutman, A.P.; Boven, A. Provenance and chemostratigraphy of the Neoproterozoic West Congolian Group in the Democratic Republic of Congo. *J. Afr. Earth Sci.* **2006**, *46*, 221–239. [CrossRef]
29. Vicat, J.-P.; Gioan, P.; Caby, R. Occurrence of greenschist facies metamorphism in the Bouenza sequence of the south-eastern border of the Chaillu Massif, Congo. *Comptes Rendus De L'académie Des Sci. Série II Mécanique Phys. Chim. Astron.* **1991**, *312*, 517–522. Available online: https://www.researchgate.net/publication/313104753 (accessed on 18 October 2024).
30. Rademakers, F.W.; Nikis, N.; De Putter, T.; Degryse, P. Copper Production and Trade in the Niari Basin (Republic of Congo) During the 13th to 19th Centuries ce: Chemical and Lead Isotope Characterization. *Archaeometry* **2018**, *60*, 1251–1270. [CrossRef]
31. Hanson, R.E. Proterozoic geochronology and tectonic evolution of southern Africa. *Geol. Soc. Lond. Spec. Publ.* **2003**, *206*, 427–463. [CrossRef]
32. Loemba, A.P.R.; Ferreira, A.; Ntsiele, L.J.E.P.; Opo, U.F.; Bazebizonza Tchiguina, N.C.; Nkodia, H.M.D.V.; Klausen, M.; Boudzoumou, F. Archean crustal generation and Neoproterozoic partial melting in the Ivindo basement, NW Congo craton, Republic of Congo: Petrology, geochemistry and zircon U–Pb geochronology constraints. *J. Afr. Earth Sci.* **2023**, *200*, 104852. [CrossRef]
33. Kröner, A.; Stern, B. Pan-African Orogeny. In *Encyclopedia of Geology*; Elsevier: Amsterdam, The Netherlands, 2004; Volume 1, pp. 1–12. [CrossRef]
34. Monié, P.; Bosch, D.; Bruguier, O.; Vauchez, A.; Rolland, Y.; Nsungani, P.; Buta Neto, A. The Late Neoproterozoic/Early Palaeozoic evolution of the West Congo Belt of NW Angola: Geochronological (U-Pb and Ar-Ar) and petrostructural constraints. *Terra Nova* **2012**, *24*, 238–247. [CrossRef]
35. Delvaux, D.; Maddaloni, F.; Tesauro, M.; Braitenberg, C. The Congo Basin: Stratigraphy and subsurface structure defined by regional seismic reflection, refraction and well data. *Glob. Planet. Chang.* **2021**, *198*, 103407. [CrossRef]
36. White, R.; McKenzie, D. Magmatism at rift zones: The generation of volcanic continental margins and flood basalts. *J. Geophys. Res. Solid Earth* **1989**, *94*, 7685–7729. [CrossRef]

37. Haapala, I.; Frindt, S.; Kandara, J. Cretaceous Gross Spitzkoppe and Klein Spitzkoppe stocks in Namibia: Topaz-bearing A-type granites related to continental rifting and mantle plume. *Lithos* **2007**, *97*, 174–192. [CrossRef]
38. Pérez-Díaz, L.; Eagles, G. Constraining South Atlantic growth with seafloor spreading data. *Tectonics* **2014**, *33*, 1848–1873. [CrossRef]
39. Bah, B.; Lacombe, O.; Beaudoin, N.E.; Zeboudj, A.; Gout, C.; Girard, J.-P.; Teboul, P.-A. Paleostress evolution of the West Africa passive margin: New insights from calcite twinning paleopiezometry in the deeply buried syn-rift TOCA formation (Lower Congo basin). *Tectonophysics* **2023**, *863*, 229997. [CrossRef]
40. Sieberer, A.K.; Willingshofer, E.; Klotz, T.; Ortner, H.; Pomella, H. Inversion of extensional basins parallel and oblique to their boundaries: Inferences from analogue models and field observations from the Dolomites Indenter, European eastern Southern Alps. *Solid Earth* **2023**, *14*, 647–681. [CrossRef]
41. Curzi, M.; Viola, G.; Zuccari, C.; Aldega, L.; Billi, A.; van der Lelij, R.; Kylander-Clark, A.; Vignaroli, G. Tectonic Evolution of the Eastern Southern Alps (Italy): A Reappraisal From New Structural Data and Geochronological Constraints. *Tectonics* **2024**, *43*, e2023TC008013. [CrossRef]
42. Japas, M.S.; Gómez, A.L.R.; Rubinstein, N.A. Unravelling the hydro-mechanical evolution of a porphyry-type deposit by using vein structures. *Ore Geol. Rev.* **2021**, *133*, 104074. [CrossRef]
43. Tassara, S.; Ague, J.J. A role for crustal assimilation in the formation of copper-rich reservoirs at the base of continental arcs. *Econ. Geol.* **2022**, *117*, 1481–1496. [CrossRef]
44. Andersson, J.B.H.; Bauer, T.E.; Lynch, E.P. Evolution of structures and hydrothermal alteration in a Palaeoproterozoic supracrustal belt: Constraining paired deformation–fluid flow events in an Fe and Cu–Au prospective terrain in northern Sweden. *Solid Earth* **2020**, *11*, 547–578. [CrossRef]
45. Twiss, R.J.; Moores, E.M. *Structural Geology*, 2nd ed.; W. H. Freeman: New York, NY, USA, 2007; pp. 1–736.
46. Trompette, R.; Boudzoumou, F. Palaeogeographic significance of stromatolitic buildups on late proterozoic platforms: The example of the West Congo basin. *Palaeogeogr. Palaeoclimatol. Palaeoecol.* **1988**, *66*, 101–112. [CrossRef]
47. Nkodia, H.M.D.V.; Miyouna, T.; Kolawole, F.; Boudzoumou, F.; Loemba, A.P.R.; Bazebizonza Tchiguina, N.C.; Delvaux, D. Seismogenic Fault Reactivation in Western Central Africa: Insights From Regional Stress Analysis. *Geochem. Geophys. Geosystems* **2022**, *23*, e2022GC010377. [CrossRef]
48. Syverson, D.D.; Awolayo, A.N.; Tutolo, B.M. Seafloor spreading and the delivery of sulfur and metals to Earth's oceans. *Geology* **2023**, *51*, 1168–1172. [CrossRef]

minerals

MDPI

Article

Geochronology and Genesis of the Shuigou Gold Deposit, Qixia-Penglai-Fushan Metallogenic Area, Jiaodong Peninsula, Eastern China: Constraints from SHRIMP U-Pb, $^{40}Ar/^{39}Ar$ Age, and He-Ar Isotopes

Zhenjiang Liu [1], Shaobo Cheng [2,3,*], Changrong Liu [4], Benjie Gu [5] and Yushan Xue [6]

1 School of Earth Sciences and Resources, China University of Geosciences, Beijing 100083, China; lzj@cugb.edu.cn
2 International Mining Research Center, China Geological Survey, Beijing 100083, China
3 China Mining News, Beijing 100083, China
4 Shandong Provincial 6th Exploration Institute of Geology and Mineral Resources, Weihai 264200, China; lcr0631@163.com
5 School of Earth Resources, China University of Geosciences, Wuhan 430074, China; 18612698826@163.com
6 Xi'an Northwest Geological Institute of Nonferrous Metals Co., Ltd., Xi'an 710054, China; ysxuecugb@126.com
* Correspondence: chengshaobo0631@sina.com

Abstract: The Jiaodong Peninsula is renowned for its significant gold reserves, which exceed 4500 tons. In this study, we conducted zircon SHRIMP U-Pb dating, $^{40}Ar/^{39}Ar$ geochronology, electron probe microanalysis (EPMA) analysis, and He-Ar isotope analysis on samples from the Shuigou gold deposit located in the Qixia-Penglai-Fushan metallogenic area of central Jiaodong. This quartz vein-type gold deposit is characterized by three mineralization stages: (I) the quartz-pyrite stage, (II) the quartz-polymetallic sulfide stage, and (III) the calcite stage. In stages I and II, gold primarily exists as native gold or electrum. Preliminary analysis suggests that the deposit contains rare critical metals, including bismuth (Bi), tellurium (Te), and antimony (Sb). The Sb is found as pyrargyrite in stage III, while the other critical elements occur as isomorphisms or nanoparticles within host minerals such as pyrite, native gold, and electrum. Geochronology data indicate that the pre-mineralization Guojialing monzogranite formed around 126 $\pm$ 1.6 Ma, the syn-mineralization muscovite formed at approximately 125 Ma, and the post-mineralization diorite porphyrite formed at 120.4 $\pm$ 1.8 Ma. The $^{3}He/^{4}He$ ratios of fluid inclusions in the main-stage pyrite range from 0.26 to 1.26 Ra, and the $^{40}Ar/^{36}Ar$ ratios vary from 383 to 426.6. These findings suggest that the Shuigou gold deposit formed during the destruction of the North China Craton (NCC), similar to other super-large gold deposits in the Jiaodong Mesozoic gold metallogenic province. Gold mineralization has been influenced by mantle, crustal, and meteoric fluids.

Keywords: Shuigou Au deposit; ore genesis; geochronology; He-Ar isotope; rare critical metals; Jiaodong

1. Introduction

The Jiaodong Peninsula, located in the eastern part of the North China Craton (NCC) (Figure 1a), is the most important gold metallogenetic province in China. Preliminary estimates suggest that proven gold reserves totaling 4500 tons, of which 2700 tons come from deep deposits at depths between 600 and 2000 m [1,2]. Jiaodong is also recognized as the only region in the world where significant gold accumulation occurred billions of years after the host rocks were formed [3]. The gold deposits in this area are primarily found in the Northwest Jiaodong, Qixia-Penglai-Fushan, and Mouping-Rushan Metallogenic Areas (Figure 1b). These deposits exhibit typical characteristics controlled by faults (Figure 1b), mainly influenced by NNE–NE trending faults or interlayer detachment structures, and are classified as "gold-only" type deposits [4]. Recent comprehensive research has confirmed that the critical metals such as Cr, Co, Cd, Rb, Nb, Ta, Hf, W, Sn, REEs, Sb, Bi, and As in the gold deposits exhibit varying degrees of enrichment compared to the average crustal abundances in NCC, with Te, Co, and Cd that can be directly recycled in the Xiadian and Dayingezhuang deposits [5,6].

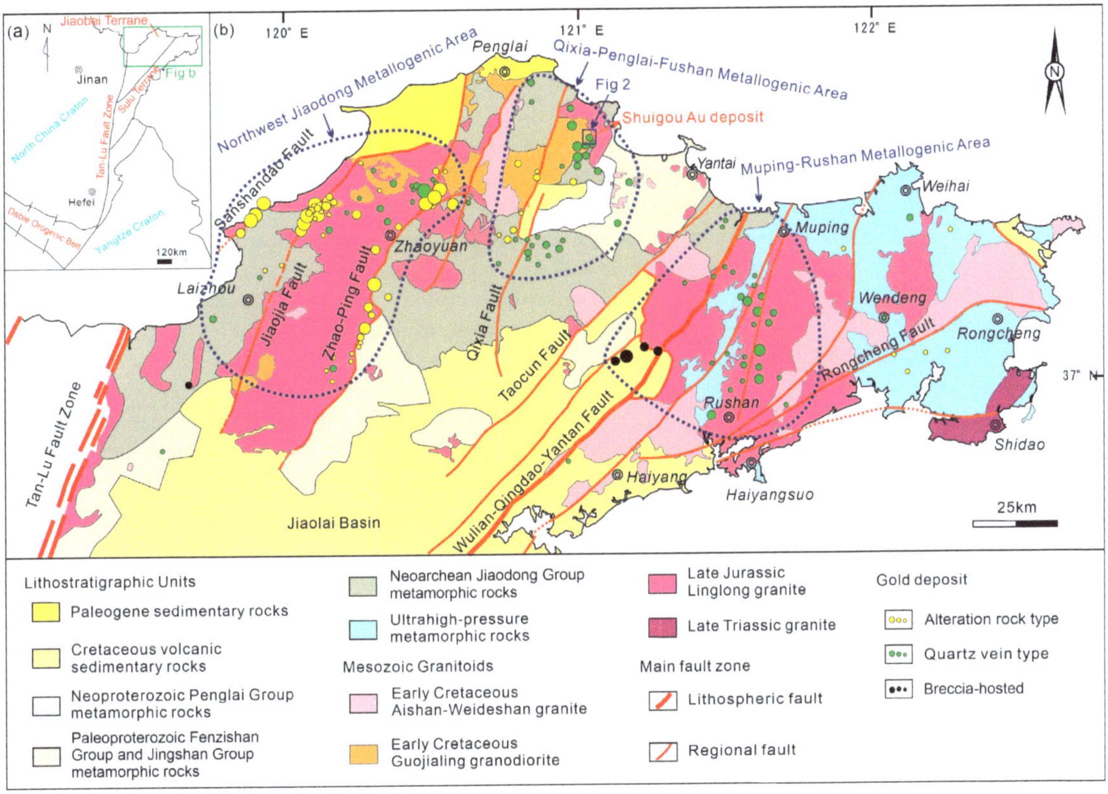

Figure 1. (**a**) Tectonic location of the Jiaodong Peninsula; (**b**) Simplified geological map of the Jiaodong Peninsula, showing the distribution of major tectonic units, Mesozoic igneous rocks, faults, and Au deposits (modified from [7]). The division of the Northwest Jiaodong, Qixia-Penglai-Fushan, and Mouping-Rushan Metallogenic Areas is based on [2].

Precise mica $^{40}Ar/^{39}Ar$ isotopic dating results indicate that the Jiaodong gold deposits were primarily formed during the Early Cretaceous period. The earliest age recorded in the Dayingezhuang deposit suggests that Mesozoic gold mineralization may have begun around 130 ± 4 Ma [8]. Gold mineralization peaked between 126 and 120 Ma, marked by the formation of gold deposits, including Cangshang (121.3 ± 0.2 Ma) [9], Jiaojia, Xincheng, and Wangershan (121.0 ± 0.4 Ma~119.2 ± 0.2 Ma) [10], as well as Pengjiakuang (120.9 ± 0.4~119.1 ± 0.2 Ma) [11]. The youngest ages of 109.3 ± 0.3 Ma~107.7 ± 0.5 Ma from the Rushan gold deposit indicate that gold mineralization continued until 120~110 Ma [8]. Stable isotope and fluid inclusion studies showed that the low salinity (6~13 wt% NaCl.eqv) and

CO_2-rich mineralizing fluids are consistent across different regions and depths with $\delta^{18}O_{SMOW}$ values of 0.08–8.85‰ and δD_{SMOW} values ranging from −106.48‰ to −48‰ [12,13]. The He-Ar isotope compositions are distributed in the zone between mantle and crust [1]. Various scholars have proposed multiple genetic models to explain the formation of the Jiaodong gold deposits. Fan et al. suggest that the mineralizing fluids originated from magma and are related to the melting of mafic intrusions formed in the Mesozoic-enriched lithospheric mantle [14]. Li et al. propose that the mineralizing fluids were primarily composed of magmatic-hydrothermal fluids from felsic magma mixed with meteoric water, along with mantle-derived fluids [15]. Some scholars have suggested that the Jiaodong gold deposits represent a new type known as "Jiaodong-type" gold deposits [16]. Goldfarb and Santosh argue that the fluids and mineralizing materials of the Jiaodong gold deposits originate from below the crust, possibly from the lithospheric mantle and subducted oceanic crust sediments [3]. This led to the establishment of an oceanic crust subduction model to explain the formation mechanism of the Jiaodong gold deposits. Deng et al. propose a new model for the remobilization of the metasomatized mantle lithosphere in the Jiaodong gold province, suggesting that the upwelling of the asthenosphere triggered the release of gold and sulfur from an enriched and fertilized mantle lithosphere [17]. This process contributed to the formation of auriferous fluids, leading to widespread gold mineralization in the Jiaodong region. Qiu et al. utilized lithium isotopes to demonstrate a strong genetic connection between carbonate metasomatism in the mantle and gold mineralization in the overlying crust in the Jiaodong. While the pre-enrichment of Au in the mantle is essential for the formation of large gold provinces, the oxidation of the lithospheric mantle plays a crucial role in the mobilization of Au [18]. However, most of genetic models are based on data from Northwest Jiaodong, and Mouping-Rushan Metallogenic Areas, with Qixia-Penglai-Fushan Area less considered. More studies of gold deposits in Qixia-Penglai-Fushan Area are needed for the better understanding of gold mineralization in Jiaodong Peninsula, especially when there are other Mesozoic polymetallic mineralization such as tungsten, molybdenum and antimony in the area [19,20].

The Shuigou gold deposit is a quartz vein-type (Linglong-type) gold deposit located in the Daliuhang area of Penglai, within the Qixia-Penglai-Fushan Metallogenic area of central Jiaodong (Figure 1b). This study described the geology and occurrence of gold and critical elements such as Bi, Te, and Sb in the Shuigou gold deposit. Systematic geochronological and geochemical data have been conducted to provide new insights into regional gold mineralization.

2. Regional Geology

The Jiaodong Peninsula is located at the southeastern edge of the NCC. It is bordered to the west by the trans-lithospheric Tan–Lu (Tancheng–Lujiang) Fault and to the south by the Yangtze Craton (Figure 1a). It primarily consists of Precambrian metamorphic basement rocks and Mesozoic igneous rocks (Figure 1b), with a minor presence of Mesozoic sedimentary-volcanic deposits [21,22].

The Precambrian metamorphic basement rocks are primarily composed of Archean supracrustal rocks, tonalite–trondhjemite–granodiorite (TTG) gneisses, Paleoproterozoic granitoid intrusions, high-grade metamorphic Fenzishan and Jingshan Groups, low-grade metamorphic Zhifu Group, Neoproterozoic low-grade metamorphic Penglai Group, and various lenses or sheets of metamorphic mafic-ultramafic rocks. The Archean supracrustal rocks mainly consist of biotite-plagioclase gneiss, biotite leptite, leucoleptite, amphibolite, and banded iron formation. TTG gneisses are found from northern Qixia to southern Laixi–Laiyang. These gneisses have protolith ages of approximately 2.9 Ga, 2.7 Ga, and 2.5 Ga, undergoing amphibolite to granulite facies metamorphism around 2.5 Ga and

1.86 Ga [23,24]. The Jingshan Group is made up of Al-rich gneisses and marbles, while the Fenzishan Group primarily includes pelitic schists, fine-grained paragneisses, marbles, and minor amphibolites. The Zhifu Group consists of quartzite and quartz schists containing muscovite and tourmaline. The Penglai Group includes quartzite, slate, and limestone [25]. Paleoproterozoic granitoids are characterized by both deformed and undeformed varieties, including alkali feldspar granite, albite granite, gneisses, syenogranite, and pegmatitic granite. These granitoids can be categorized into two groups: those formed around 2.2 to 2.1 Ga, which are pre-tectonic, and those from ca. 1.8 Ga, which are post-tectonic [26].

Mesozoic sedimentary-volcanic rocks are found in the Jiaolai basin (Figure 1b), which consists of three main groups, listed from bottom to top: the Laiyang Group, the Qingshan Group, and the Wangshi Group. The Laiyang Group consists of a sequence of sandstone, siltstone, shale, and conglomerate, with some rare interlayers of volcanoclastic rocks. The Qingshan Group is characterized by eruptive cycles that include both volcanic and clastic rocks. Lastly, the Wangshi Group is made up of red beds, along with mafic and ultramafic lavas, sandstones, and conglomerates [27].

Mesozoic magmatism in the Jiaodong region is primarily characterized by the presence of granitoids and dykes. The Mesozoic granitoids developed during three episodes: (1) the Linglong-Kunyushan-Wendeng suite, with zircon U–Pb ages ranging from 160 to 150 Ma; (2) the Guojialing suite, which has zircon U–Pb ages between 130 to 126 Ma; (3) the Weideshan suite, with zircon U–Pb ages from 120 to110 Ma [19,28,29]. The dykes include lamprophyre, diorite, diorite porphyry, dolerite, granodiorite/granite, and gabbro, with ages ranging from 127 to 114 Ma [30].

3. Ore Deposit Geology

The Shuigou gold deposit is a quartz vein-type gold deposit with a reserve exceeding 5 tons of gold. It is situated in Daliuhang, to the northeast of Penglai City (Figures 1b and 2a). The gold deposit district is characterized by well-developed fault structures that trend NNE–NE, including the Shuigou fault, F_1 fault, and F_2 fault. The primary igneous rocks in the area are the Linglong monzogranite and the Guojialing monzogranite (Figure 2a), which are found on both sides of the Shuigou fault. Additionally, late Yanshanian diorite porphyrite and porphyry, which intersect the gold-bearing quartz veins, are also present.

The ore bodies are primarily located within alteration zones characterized by pyrite-sericite-silicified breccia and quartz veins, which are controlled by NNE–NE trending faults (Figure 2). A total of eleven alteration zones have been identified, with the No. 1 and No. 2 zones being the largest (Figure 2b). The No. 1 alteration zone is controlled by the F_1 fault. It has a surface outcrop length of 250 m and a current measured length of 650 m in the adit. The zone ranges from 0.5 m to 2.0 m in width, strikes between 10° to 20°, and dips to the southeast at 58° to 84°. This alteration zone is composed of pyrite-sericite-silicified breccia, which locally contains pyrite-quartz veins. Both the hanging wall and footwall of this zone are made up of Guojialing monzogranite. Gold mineralization varies inconsistently but is more prominent in areas abundant with pyrite-quartz veins, where local gold grades can exceed 20 g/t. The No. 2 alteration zone is controlled by the F_2 fault, which has a surface outcrop length of 650 m and a width ranging from 0.5 m to 2.0 m. This zone strikes at an angle of 5° to 15° and dips southeast at 65° to 78°. It consists of pyrite-sericite-silicified breccia and locally contains quartz veins. Both the hanging wall and footwall are composed of Guojialing monzogranite. The gold grades in this zone range from 0.05 g/t to 5.59 g/t.

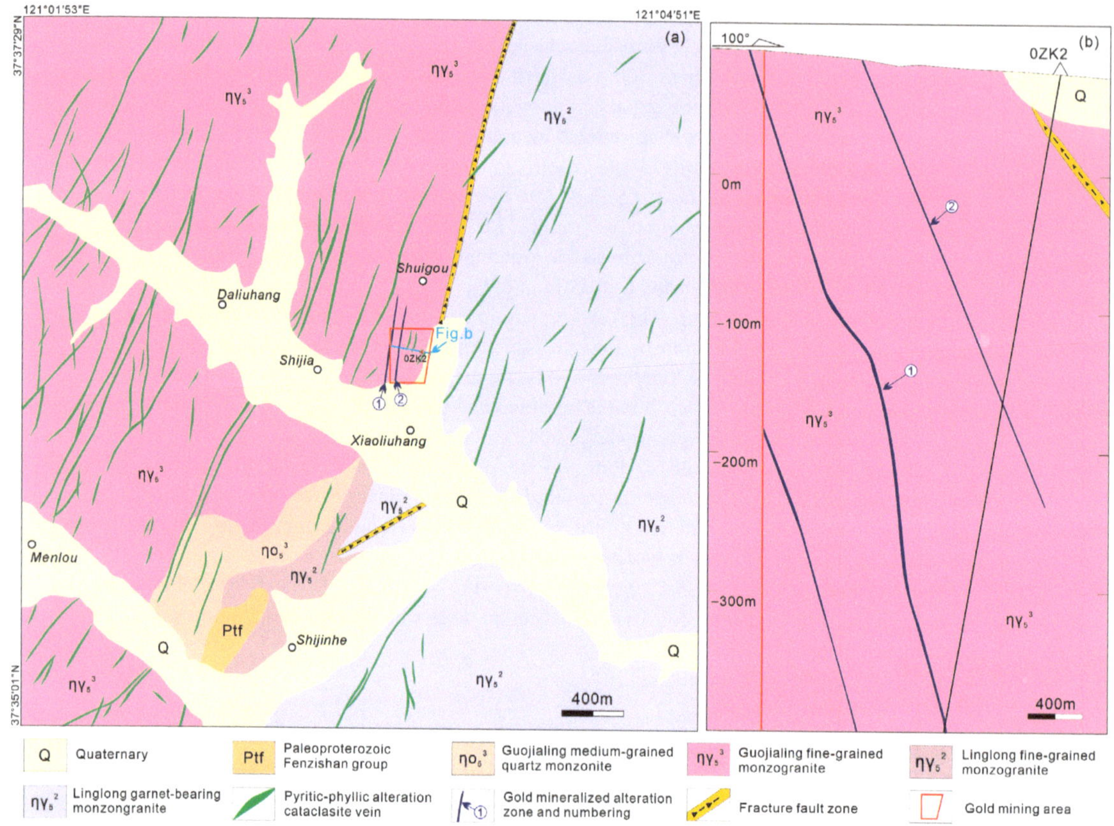

Figure 2. (**a**) Geological sketch map of the Shuigou gold deposit; (**b**) cross-section of Exploration Line No. 0 in the Shuigou gold deposit.

The primary ore minerals found in the Shuigou gold deposit are native gold, electrum, and pyrite, along with minor amounts of pyrrhotite, sphalerite, galena, and chalcopyrite. The gangue minerals primarily consist of quartz, feldspar, sericite, chlorite, and calcite. The main ore textures observed include crushed, euhedral, subhedral-anhedral, replacement-dissolution, inclusion, filling, and colloform textures. Mineralization occurs in three distinct stages: the quartz-pyrite stage (I), the quartz-polymetallic sulfide stage (II), and the calcite stage (III). Gold is mainly precipitated in stages I and II (Figures 3 and 4).

Figure 3. Mine shaft outcrop, hand specimen and microscopic photographs showing the ore texture and structure of the Shuigou Au deposit. (**a**) Gold-bearing quartz vein cutting the beresite and Guojialing monzogranite. (**b**) Hand specimens of Guojialing monzogranite. (**c**) Hand specimens of beresite. (**d**,**e**) Hand specimens of stages I and II. (**f–h**) Photomicrograph of a polished thin section of stage I. (**i**,**j**) Photomicrograph of a polished thin section of stage II. (**k**) Photomicrograph of a polished thin section of stage III. (**l**) Photomicrograph of a polished thin section of stage I epigenetic oxidized ore. Abbreviations: Arn—argentite, Au—native gold or electrum, Ccp—chalcopyrite, Cv—covellite, Gn—galena, Py—pyrite, Po—pyrrhotite, Pyr—pyrargyrite, Q—quartz, Ser—sericite, and Sp—sphalerite. The diameter of the coin is 19 mm.

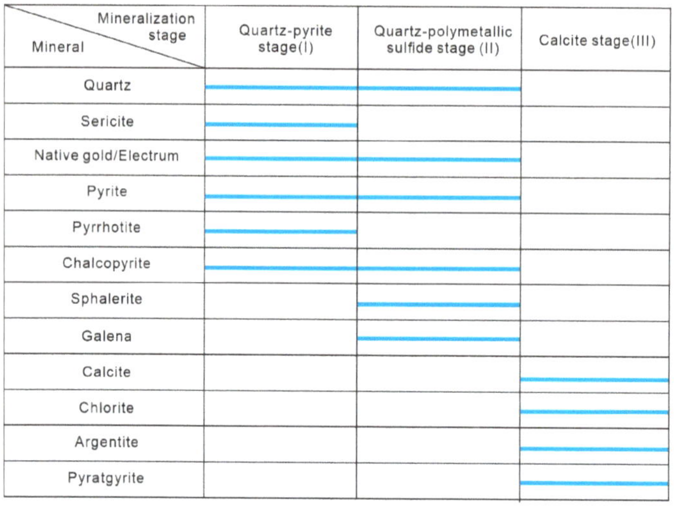

Figure 4. Mineralization stages and paragenetic sequence of the Shuigou gold deposit.

4. Samples and Analytical Methods

All samples in this study were collected from the underground mine of the Shuigou gold deposit. Thin sections of the host rocks and ores were prepared for petrographic and EPMA analyses. Zircons from the Guojialing monzonite granite and post-mineralization diorite porphyrite were separated and prepared for SHRIMP U-Pb dating. Additionally, muscovite and pyrite were separated from samples representing the main mineralization stage for $^{40}Ar/^{39}Ar$ geochronology and He-Ar isotope analysis, respectively.

4.1. EPMA Analysis

Spot analyses and element mapping of samples were performed using a JEOL JXA-8230 electron probe microanalyzer manufactured by Japan at the Key Laboratory of Metallogeny and Mineral Assessment, Chinese Academy of Geological Sciences, Beijing, China. The operating conditions for chosen mineral samples were 15 kV accelerating voltage and a beam current of 50 nA, with variable counting times between 10 and 100 s and between 5 and 50 s in the peak and background, respectively. The beam diameter ranged from 1 to 5 μm. Elements Se, As, Cr, S, Pb, Bi, Sb, Cu, Fe, Zn, Te, Mo, Ag, Au, Co, and Ni were analyzed. The operating conditions for element mapping were 15 kV accelerating voltage with 100 nA beam current, 0.5 μm step size, and 50 ms dwell time for qualitative element maps; Ag, As, Au, Bi, Cu, Fe, S, and Te distribution data were collected.

4.2. SHRIMP Zircon U-Pb Dating

Zircon U–Pb dating was carried out using SHRIMP II manufactured by Australia at the Beijing SHRIMP Center, Chinese Academy of Geological Sciences, Beijing, China. A primary 20–30 μm O^{2-} ion beam of 3–6 nA was used to bombard the surfaces of the zircons. Five scans were made for each analysis after a raster time of 120–200 s. Standard zircons for elemental abundance calibration included 91,500 (U = 91 ppm), SL13 (U = 238 ppm), and M257 (U = 840 ppm) [31,32]. TEMORA with a $^{206}Pb/^{238}U$ age of 417 Ma was used for calibration [33] and was analyzed after every three sample analyses. Common Pb corrections were based on the measured ^{204}Pb contents. Uncertainties for individual analyses are quoted at 1σ, whereas errors for weighted mean ages are quoted at the 95% confidence level.

4.3. $^{40}Ar/^{39}Ar$ Dating

The mineral separate and reference sample ZBH-25 with an age of 132.9 ± 1.3 Ma [34] were irradiated in the high-flux engineering test reactor (HFETR) facility at the Nuclear Power Institute of China. After a cooling period of three months, the $^{40}Ar/^{39}Ar$ analyses were performed at the Beijing Research Institute of Uranium Geology. The decay constant used for calculation is $5.543 \times 10^{-10}\ a^{-1}$.

4.4. He-Ar Isotopes

He and Ar isotopes were analyzed at the Stable Isotope Laboratory, Institute of Mineral Resources, Chinese Academy of Geological Sciences, Beijing, China. The ratios were measured by Helix SFT. The ^{4}He signal was received by a Faraday cup. The ^{3}He signal was received by ion multiplier under the condition of 4.5 kv source voltage, 1218 μA current, 15.56 v trap voltage, and 450 μA trap current. The ^{40}Ar and ^{36}Ar were received with a Faraday cup. The ^{38}Ar was received with a multiplier. The ion source voltage and current were 4.5 kv and 454 μA. The trap voltage and trap current are 15.02 v and 200 μA. The standard air $^{3}He/^{4}He$, $^{40}Ar/^{36}Ar$, and $^{36}Ar/^{38}Ar$ ratios are 1.4×10^{-6}, 295.5, and 5.35, respectively.

5. Analytical Results

5.1. Composition of the Major Ore Minerals

The EPMA analysis data are detailed in Supplementary Table S1, as well as Figures 5 and 6. Major ore minerals measured include pyrite, pyrrhotite, sphalerite, native gold, and electrum.

The stage I pyrite (Py1) and pyrrhotite (Pyr1) samples were analyzed. The Py1 samples (n = 13) contain the following compositions: 45.76–46.69 wt% Fe, 53.07–53.65 wt% S, 0–0.42 wt% As, and negligible amounts of Se, Bi, Te, Co, and Ni. The Pyr1 samples (n = 5) are composed of 59.15–61.11 wt% Fe and 37.99–37.58 wt% S. Additionally, the gold mineral present in stage I is primarily native gold, which consists of 80.49–89.71 wt% Au and 7.54–17.27 wt% Ag.

In stage II, the sphalerite (Sp2) samples (n = 4) contain 59.21–64.29 wt% Zn, 33.22–34.01 wt% S and 2.10–7.12 wt% Fe. The gold mineral in stage II is primarily electrum, comprising 58.26–75.15 wt% Au and 23.00–36.73 wt% Ag.

5.2. Zircon SHRIMP U-Pb Geochronology

Zircon U-Pb results are presented in Table 1 and Figure 7.

The zircon grains from sample DLH2 of the Guojialing monzonite granite are subhedral to euhedral. They can be either stubby or elongated, measuring between 250 and 350 μm in length. In cathodoluminescence (CL) images, these grains display magmatic oscillatory zoning patterns (Figure 7a). A total of 16 analyses were conducted on 16 grains. Thirteen analyses on syn-magmatic domains indicate uranium (U) contents ranging from 202 to 381 ppm, with Th/U ratios between 0.31 and 0.77. These analyses yield a weighted mean $^{206}Pb/^{238}U$ age of 126 ± 1.6 Ma (MSWD = 1.18) (Figure 7b). Additionally, three inherited domains exhibit ages of 134.4 Ma, 150 Ma, and 163.3 Ma (Figure 7a).

Zircon grains from sample DLH5 of diorite porphyrite are subhedral to euhedral in shape, appearing as either stubby or elongated prisms with lengths ranging from 150 to 400 μm. In CL images, these grains display magmatic oscillatory zoning (Figure 7c). A total of 12 analyses were conducted on the syn-magmatic domain across 12 different grains. The results indicate U contents between 193 and 2552 ppm, with Th/U ratios ranging from 0.77 to 2.44. Additionally, the analyses yield a weighted mean $^{206}Pb/^{238}U$ age of 120.4 ± 1.8 Ma (MSWD = 0.77) (Figure 7d).

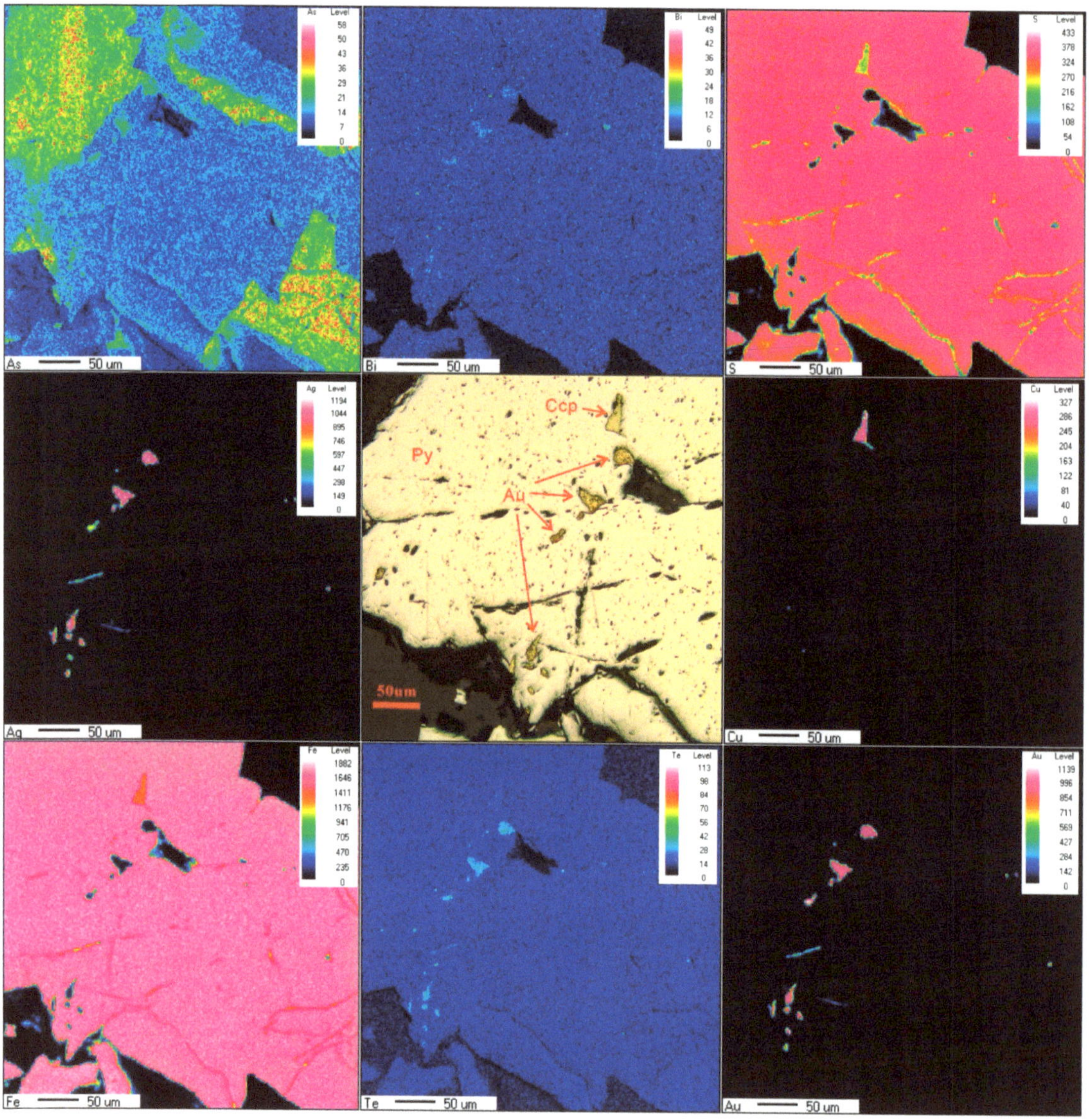

Figure 5. EPMA maps of ore minerals from mineralization stage I. The mapped area is pyrite with gold minerals and chalcopyrite.

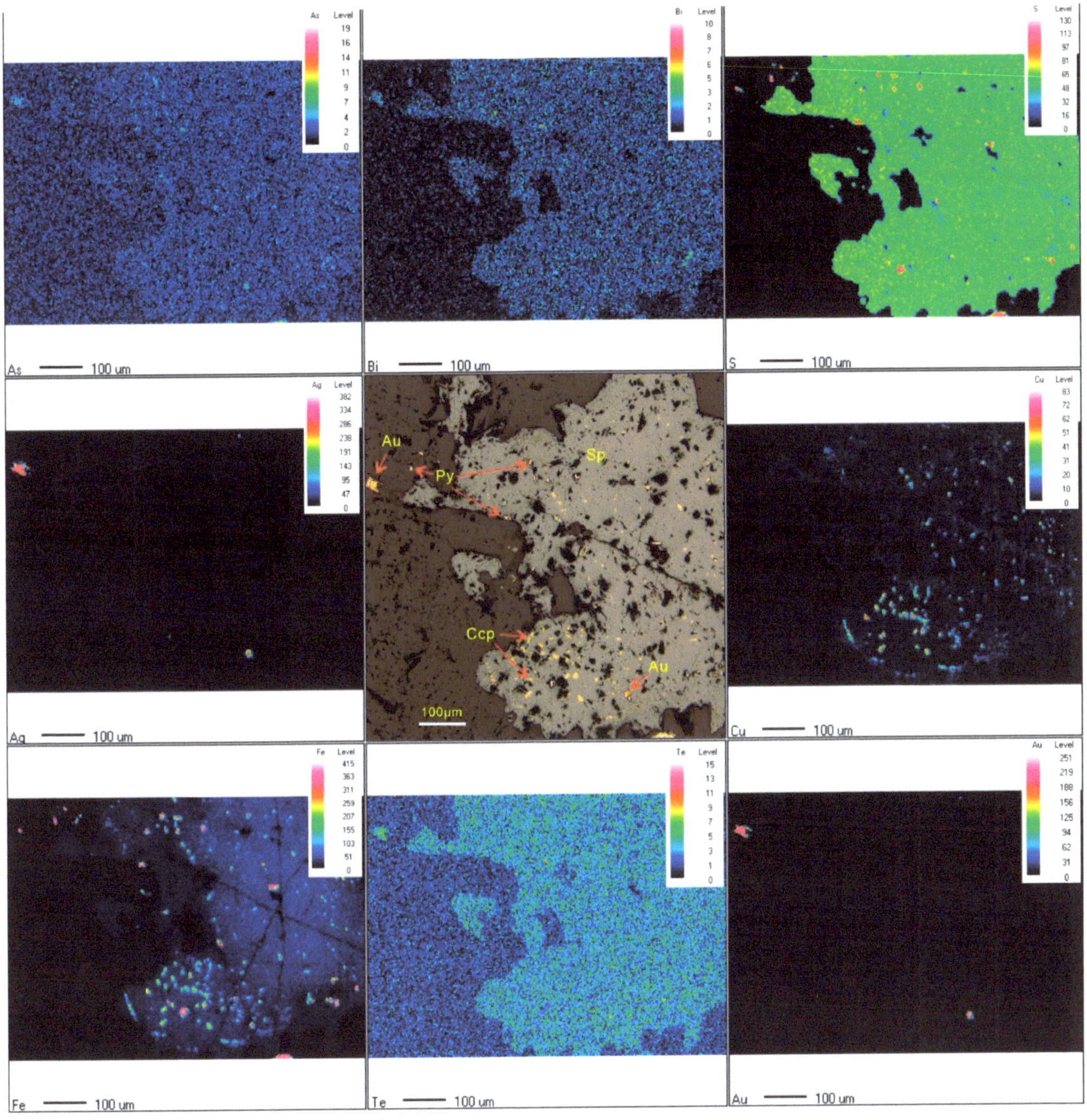

Figure 6. EPMA maps of ore minerals from mineralization stage II. The main mineral is sphalerite with chalcopyrite blebs and pyrite. Gold minerals are isolated or in sphalerite.

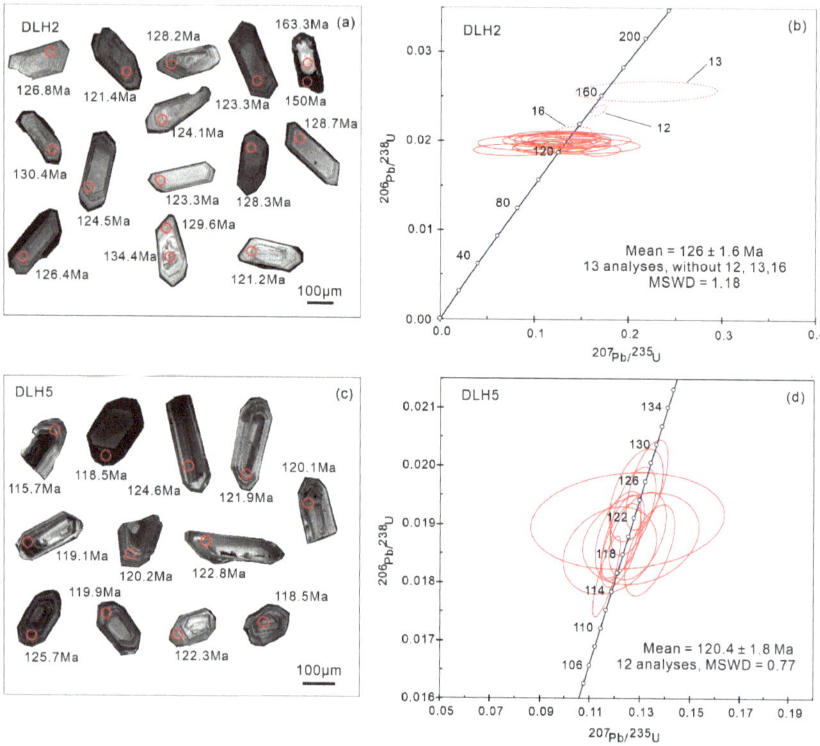

Figure 7. Representative CL images (**a**,**c**)and concordia diagrams (**b**,**d**) of zircon grains from the Shuigou gold deposit. The dashed-line ellipses labeled 12, 13, and 16 are inherited zircons.

Table 1. SHRIMP U-Pb data for zircons from the Shuigou gold deposit.

Spot	U (ppm)	Th (ppm)	Th/U	^{206}Pb * (ppm)	^{207}Pb */^{206}Pb *	±%	^{207}Pb */^{235}U *	±%	^{206}Pb */^{238}U *	±%	^{206}Pb/^{238}U Age
DLH2											
DLH2-1	264	114	0.45	4.57	0.0500	20.0	0.1380	21.0	0.0199	2.4	126.8 ± 3.0
DLH2-2	381	254	0.69	6.49	0.0300	37.0	0.0780	37.0	0.0190	2.3	121.4 ± 2.8
DLH2-3	231	173	0.77	4.09	0.0480	34.0	0.1340	34.0	0.0201	2.9	128.2 ± 3.6
DLH2-4	252	103	0.42	4.28	0.0540	30.0	0.1450	30.0	0.0194	2.9	124.1 ± 3.5
DLH2-5	292	130	0.46	5.28	0.0430	27.0	0.1210	27.0	0.0204	2.3	130.4 ± 3.0
DLH2-6	202	61	0.31	3.43	0.0528	18.0	0.1420	18.0	0.0195	2.4	124.5 ± 2.9
DLH2-7	379	156	0.42	6.40	0.0538	8.1	0.1440	8.3	0.0194	1.9	124.1 ± 2.4
DLH2-8	262	112	0.44	4.66	0.0490	21.0	0.1370	21.0	0.0201	2.5	128.3 ± 3.2
DLH2-10	291	117	0.42	5.12	0.0330	37.0	0.0900	37.0	0.0198	2.7	126.4 ± 3.4
DLH2-11	294	110	0.39	5.02	0.0408	20.0	0.1090	20.0	0.0193	2.2	123.3 ± 2.7
DLH2-14	334	111	0.34	5.40	0.0653	7.6	0.1710	7.8	0.0190	2.0	121.2 ± 2.4
DLH2-17	245	87	0.37	4.35	0.0450	24.0	0.1240	25.0	0.0202	2.3	128.7 ± 3.0
DLH2-18	263	113	0.44	4.70	0.0446	21.0	0.1250	21.0	0.0203	2.3	129.6 ± 2.9
DLH2-12	16,612	1136	0.07	343.00	0.0516	2.5	0.1677	3.1	0.0236	1.7	150 ± 2.6
DLH2-13	152	64	0.44	3.44	0.0620	23.0	0.2190	23.0	0.0257	2.9	163.3 ± 4.6
DLH2-16	269	127	0.49	4.92	0.0484	9.1	0.1410	9.4	0.0211	2.0	134.4 ± 2.7

Table 1. *Cont.*

Spot	U (ppm)	Th (ppm)	Th/U	^{206}Pb * (ppm)	^{207}Pb */^{206}Pb *	±%	^{207}Pb */^{235}U *	±%	^{206}Pb */^{238}U *	±%	^{206}Pb/^{238}U Age
DLH5											
DLH5-1	2552	3262	1.32	39.80	0.0467	1.8	0.1167	3.1	0.0181	2.5	115.7 ± 2.9
DLH5-2	375	345	0.95	6.01	0.0462	10.0	0.1180	10.0	0.0186	2.7	118.5 ± 3.1
DLH5-3	193	382	2.04	3.24	0.0477	4.8	0.1284	5.5	0.0195	2.7	124.6 ± 3.3
DLH5-5	1257	1202	0.99	20.10	0.0476	3.5	0.1224	4.3	0.0186	2.6	119.1 ± 3.0
DLH5-7	302	328	1.12	5.23	0.0476	20.0	0.1250	20.0	0.0191	2.9	121.9 ± 3.5
DLH5-8	419	390	0.96	6.97	0.0480	3.9	0.1272	4.7	0.0192	2.6	122.8 ± 3.2
DLH5-3	404	623	1.59	6.58	0.0470	4.6	0.1220	5.3	0.0188	2.6	120.2 ± 3.1
DLH5-10	394	470	1.23	6.66	0.0489	3.3	0.1328	4.2	0.0197	2.6	125.7 ± 3.2
DLH5-12	279	262	0.97	5.12	0.0960	52.0	0.2500	52.0	0.0192	3.2	122.3 ± 3.9
DLH5-13	357	268	0.77	5.77	0.0498	5.2	0.1292	5.8	0.0188	2.6	120.1 ± 3.2
DLH5-14	332	448	1.39	5.30	0.0523	4.0	0.1339	4.9	0.0186	2.8	118.5 ± 3.3
DLH5-15	519	428	0.85	8.40	0.0473	3.4	0.1225	4.3	0.0188	2.6	119.9 ± 3.1

Note: (1) common lead corrected using ^{204}Pb; (2) ^{206}Pb * is radiogenic lead; (4) age in Ma.

5.3. $^{40}Ar/^{39}Ar$ Geochronology

The two muscovite samples yield well-defined ^{40}Ar/^{39}Ar plateau ages. Sample 13DLH15 has a plateau age of 125.76 ± 0.52 Ma, based on 83.6% of the released ^{39}Ar, while sample 14DLH20 yields a plateau age of 126.8 ± 0.59 Ma based on 84.7% of the released ^{39}Ar. Additionally, the isochron ages for the two samples are 125.55 ± 0.55 Ma and 125.90 ± 0.69 Ma, respectively. These isochron ages are consistent with the plateau ages within the margin of error (Figure 8 and Table 2).

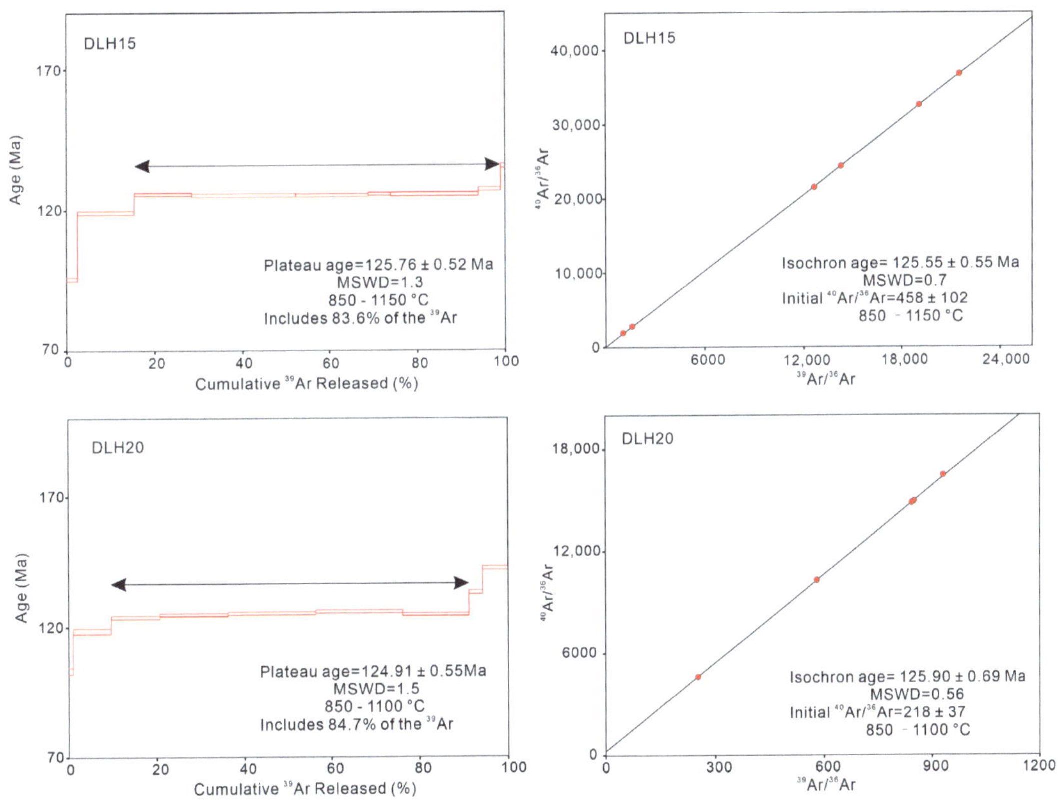

Figure 8. ^{40}Ar/^{39}Ar plateau and isochron ages for muscovite from the Shuigou gold deposit.

Table 2. ^{40}Ar-^{39}Ar step-heating geochronology data for muscovite from the Shuigou gold deposit.

Heating Step	T (°C)	(^{40}Ar/^{39}Ar)$_m$	(^{36}Ar/^{39}Ar)$_m$	(^{37}Ar/^{39}Ar)$_m$	^{40}Ar(%)	F (^{40}Ar*/^{39}Ar)	^{39}Ar (×10^{-14} mol)	^{39}Ar (Cum.) (%)	Age ± 1σ (Ma)
Sample: 14DLH20, weight = 20.4 mg, J = 0.004130									
1	700	22.0330	0.0271	1.1197	64.08	14.1322	0.49	1.07	103.3 ± 1.3
2	800	22.1964	0.0201	0.2500	73.38	16.2913	3.85	8.48	118.6 ± 1.0
3	850	18.2024	0.0040	0.2996	93.59	17.0391	5.03	11.09	123.8 ± 0.63
4	900	17.6757	0.0018	0.3056	97.12	17.1704	7.07	15.59	124.7 ± 0.61
5	950	17.5876	0.0012	0.2303	98.01	17.2404	9.09	20.05	125.2 ± 0.61
6	1000	17.6495	0.0011	0.2812	98.20	17.3357	8.96	19.77	125.9 ± 0.62
7	1050	17.5198	0.0012	0.2513	98.01	17.1753	6.89	15.20	124.8 ± 0.61
8	1100	19.2077	0.0028	0.1427	95.68	18.3804	1.37	3.02	133.2 ± 0.66
9	1200	20.9490	0.0044	0.3891	93.96	19.6893	2.60	5.73	142.4 ± 0.72
Sample: 13DLH15, weight = 20.3 mg, J = 0.004198									
1	700	14.1979	0.0056	2.7976	90.05	12.8152	1.38	2.48	95.4 ± 0.59
2	800	18.7968	0.0093	0.6864	85.64	16.1077	7.14	12.87	119.1 ± 0.71
3	850	17.0325	0.0001	0.0202	99.90	17.0161	7.27	13.11	125.6 ± 0.61
4	900	16.9952	0.0001	0.0311	99.87	16.9739	13.24	23.86	125.3 ± 0.61
5	950	16.9868	0.0001	0.0215	99.86	16.9626	9.08	16.37	125.3 ± 0.61
6	1000	17.0321	0.0001	0.0481	99.91	17.0179	2.83	5.10	125.6 ± 0.61
7	1100	17.1857	0.0006	0.0702	98.94	17.0046	11.16	20.13	125.6 ± 0.61
8	1150	17.5068	0.0010	0.1426	98.42	17.2325	2.78	5.02	127.2 ± 0.62
9	1200	19.1514	0.0033	2.3805	95.94	18.4114	0.59	1.06	135.6 ± 0.77

5.4. He-Ar Isotope Compositions

The He-Ar analysis results indicated that the fluid inclusions in pyrite contain ^{4}He levels ranging from 15.02 to 228.31 × 10^{-8} cm^3 g^{-1} at standard temperature and pressure (STP). The ratios of ^{3}He/^{4}He range from 0.26 to 1.26 Ra. Additionally, the contents of ^{40}Ar varies from 14.69 to 31.62 × 10^{-8} cm^3 STP g^{-1}, with ^{40}Ar/^{36}Ar ratios between 383 and 426.6. All samples demonstrate high F ^{4}He values, ranging from 1459 to 3789 (Table 3) [35].

Table 3. The He-Ar compositions of pyrites from the Shuigou gold deposit.

Sample	^{4}He (10^{-8} cm^3 STP/g)	^{3}He/^{4}He (Ra)	^{40}Ar (10^{-8} cm^3 STP/g)	^{40}Ar/^{36}Ar	^{40}Ar * (10^{-8} cm^3 STP/g)	^{40}Ar */^{4}He	F ^{4}He
DLH11	28.31	0.26	16.57	383.0	3.79	0.13	3789
DLH13	15.02	0.97	14.69	408.7	4.07	0.27	2420
DLH18	18.67	1.26	31.62	426.6	9.72	0.52	1459

Note: (1) 1 Ra = 1.4 × 10^{-6}. (2) ^{40}Ar * is radiogenic ^{40}Ar given all of the Ar come from the fluid inclusions; ^{40}Ar * = ^{40}Ar−295.5 × ^{36}Ar. (3) F ^{4}He = (^{4}He/^{36}Ar sample/(^{4}He/^{36}Ar) air, (^{4}He/^{36}Ar)$_{air}$ = 0.1727.

6. Discussions

6.1. The Timing of Gold Mineralization at the Shuigou Deposit

The formation of gold deposits in the Jiaodong region primarily took place during the Mesozoic period, as mentioned in the introduction. In the Qixia-Penglai-Fushan metallogenic area, most gold deposits are concentrated in the northern Daliuhang ore field and the southern Qixia ore field (Figure 1b). Besides the Shuigou gold deposit analyzed in this paper, over 30 gold deposits have been identified in the Daliuhang area, including Heilangou, Menlou, Ankou, Qijiagou, and Qiangjiagou. Among these, the Heilangou gold deposit is large-scale, while the others are medium to small in scale. The deposits are predominantly of the quartz vein type, with a few classified as alteration rock types. Although the deposits are not large, they feature high ore grades and often contain visible gold [36]. Rb-Sr isotope dating of pyrites from the Penglai gold field indicates an isochron age of 122.3 ± 3.1 Ma for the Hexi deposit and 117.8 ± 6.5 Ma for the Heilangou and Daliuhang deposits [37]. In the southern region, the Qixia gold field generally contains small-scale ore deposits, but recent discoveries include the Hushan large-scale alteration rock-type gold deposit. Monazite U-Pb dating shows that mineralization in this deposit began around 120 Ma [38].

In this study, the ^{40}Ar/^{39}Ar dating of hydrothermal sericite indicates that the Shuigou gold deposit formed around 125 Ma. The mineralization age can be further refined by the chronology of the pre-mineralization Guojialing monzogranite, which is dated at 126 ± 1.6 Ma, and the post-mineralization diorite porphyrite, dated at 120.4 ± 1.8 Ma. When compared to the ages of other large and super-large gold deposits in the Jiaodong region, it can be concluded that the Shuigou gold deposit formed during the regional peak of gold mineralization.

Additionally, based on the crosscutting relationships and ages of ore veins and dykes, a mineralization age between 129.7 Ma and 129.3 Ma has been established for the adjacent Shijia gold deposit, which predates the mineralization observed in this study [39]. We propose that these ages suggest a relatively extended period of mineralization in the Penglai area, which began at ca. 130 Ma, which is similar to the earliest gold event in the Dayingezhuang deposit and continued until the peak mineralization period of the Jiaodong Peninsula.

6.2. Occurrence of Gold and Trace Elements

Previous studies have confirmed that most of the gold in the Jiaodong Peninsula is present as discrete native gold and electrum grains, which are largely sited in fractures at all scales in ore or gangue minerals such as pyrite, galena and quartz [40,41]. The microscopic photographs in this study show that the native gold and electrum grains of the Shuigou deposit mainly occur as (1) fissure gold in the fractures of minerals such as pyrite, quartz, and sphalerite (Figure 3f,g); (2) intergranular gold between minerals such as pyrite, sphalerite, and quartz (Figure 3i); and (3) inclusion gold in pyrite, quartz, sericite, and galena (Figure 3h). Further EPMA results show that the major ore minerals such as pyrite, chalcopyrite, galena, pyrrhotite, and sphalerite barely have any gold, which indicates that little gold is lattice bounded (Supplementary Table S1 and Figures 5 and 6). Combined with the microscopic observation and EPMA analysis, it can be concluded that the gold in the Shuigou deposit shares similar occurrence characteristics to other gold deposits in the Jiaodong Peninsula.

As, Bi, and Te minerals like arsenopyrite, calaverite, hessite, joseite, altaite, and aikinite have been documented in other Jiaodong gold deposits in Northwest Jiaodong and the Mouping-Rushan Metallogenic Areas [42–45], but none were identified in this study.

In contrast, EPMA spot analyses indicate that Py1 and Py2 (n = 14) contain arsenic levels ranging from 0 to 0.417 wt%, Pyr1 (n = 5) shows values between 0.011 and 0.070 wt% As Sp2 (n = 5) demonstrates arsenic content from 0 to 0.034 wt% As (see Supplementary Table S1). These findings suggest that arsenic primarily occurs in pyrite as an isomorphic substitution. Moreover, EPMA mapping of pyrite indicates that As is unevenly distributed within the mineral (Figure 5.). The distribution of Bi and Te aligns closely with that of gold. EPMA mapping reveals that the Bi and Te content increase in areas where native gold and electrum are present (see Figures 5 and 6). The composition of Bi and Te in the gold minerals ranges from 0.249 to 0.628 wt% Bi and from 0 to 0.115 wt% for Te. EPMA analyses indicate that the pyrite, pyrrhotite, and sphalerite from different stages of mineralization contain minimal amounts of Bi and Te (refer to Supplementary Table S1). These findings suggest that Bi and Te may exist as nanoparticles either on the surface or within the fissure of the gold minerals [46].

Au-Sb deposits are typically found in regions where orogenic gold deposits have formed [47]. However, no Sb mineralization has been reported in Northwest Jiaodong and Mouping-Rushan Metallogenic Areas, except for the Chakuang Au-Sb deposit in Yantai. The fault-controlled ore bodies and mineralization age of the Chakuang deposit are consistent with those of other Jiaodong gold deposits, with antimony primarily occurring as stibnite [20]. It is worth noting that the antimony mineral was first confirmed in a gold deposit in the Penglai area. In this study, we confirm the presence of pyrargyrite, which developed in the stage III calcite veins of the Shuigou gold deposit. The observed presence of antimony in both the Chakuang and Shuigou deposits suggests a potential for antimony metallogenesis within the Qixia-Penglai-Fushan area of the central Jiaodong Metallogenic Province.

6.3. Source of Ore-Forming Fluids

Fluid inclusions trapped in pyrite can preserve their noble gas compositions, such as helium (He) and argon (Ar), for up to a hundred million years. This makes them increasingly valuable for tracing ore-forming fluids [48–50]. There are three major reservoirs of helium on Earth: (1) the mantle, which has a $^{3}He/^{4}He$ ratio of 6–9 Ra; (2) the continental crust, with a $^{3}He/^{4}He$ ratio of 0.01–0.05 Ra; and (3) the atmosphere, where the $^{3}He/^{4}He$ ratio is 1 Ra.

Due to the continuous escape of atmospheric helium into space, its abundance in the atmosphere is significantly lower than that in the mantle and crust. As a result, the influence of meteoric water on mantle and crust-derived fluids is usually negligible. Notably, the $^{3}He/^{4}He$ ratios of mantle-derived fluids can be nearly three orders of magnitude higher than those of fluids originating from the crust, making the helium composition a sensitive isotopic indicator of mantle contributions [35].

The high values of F^4 helium ranging from 1459 to 3789 observed in this study indicate that the concentrations of ^{4}He in the ore-forming fluids are three orders of magnitude greater than those found in meteoric water. Assuming that the detected ^{36}Ar originates entirely from meteoric water, there would be only approximately 0.02×10^{-8} cm^3 STP g^{-1} of ^{4}He from meteoric water. This suggests that the influence of meteoric water on the ratios of $^{3}He/^{4}He$ is negligible. The $^{3}He/^{4}He$ ratios from the Shuigou gold deposit range from 3.64×10^{-7} to 1.76×10^{-6} (0.26–1.26 Ra). These ratios fall within the range between the mantle helium and crustal helium lines when plotted on the ^{3}He vs. ^{4}He and $^{3}He/^{4}He$ vs. $^{40}Ar/^{36}Ar$ diagrams (Figure 9a).

The $^{40}Ar/^{36}Ar$ ratios range from 383.0 to 426.6, which are similar to the meteoric water ratios of 295.5. This suggests the involvement of meteoric water (see Figure 9b). By analyzing the He-Ar isotope compositions of the Shuigou gold deposit, we can conclude that the mineralization event was influenced by contributions from mantle, crustal, and meteoric fluids.

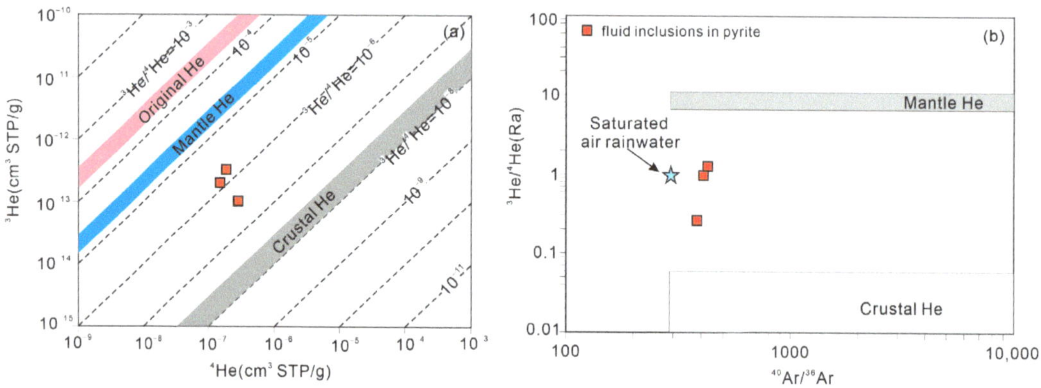

Figure 9. (**a**) Diagrams of ^{3}He vs. ^{4}He and (**b**) $^{3}He/^{4}He$ vs. $^{40}Ar/^{36}Ar$ for fluid inclusions in the main-stage pyrites from the Shuigou gold deposit.

Published isotopic data for He-Ar suggest that most of the gold deposits in the Jiaodong area exhibit $^{3}He/^{4}He$ ratios typical of mantle-crust mixing, along with $^{40}Ar/^{36}Ar$ ratios that are equal to or higher than those found in meteoric water [1,51]. Our findings indicate that the He-Ar isotopic compositions are consistent with those of other gold deposits in the Jiaodong Peninsula.

6.4. Ore Genesis

Based on the geological, geochronological, and isotopic data from this study, it is suggested that the Shuigou gold deposit, along with other gold deposits in the Jiaodong Peninsula, formed in a similar tectonic environment marked by significant crust–mantle interaction [1]. It is generally accepted that the NCC underwent craton destruction during the Mesozoic era [21]. This process resulted in the removal or replacement of over 100 km of lithosphere mantle, leading to intensive magmatism and ductile deformation within the crust [52,53].

The subduction of the Paleo-Pacific (Izanagi) Plate is considered the main tectonic driver behind the Mesozoic geodynamics of the eastern NCC [54]. The onset of this subduction likely began in the Early to Middle Jurassic period, resulting in the metasomatism of the upper NCC subcontinental lithospheric mantle (SCLM) [55]. The rollback of the Paleo-Pacific (Izanagi) Plate is believed to have triggered craton destruction around 140 Ma. An upwelling of the asthenospheric mantle initially caused the thinning of the lithosphere, which was subsequently heated, generating mafic and felsic magmatism [56].

The craton destruction in the Jiaodong Peninsula reached its peak between 130 and 120 Ma, characterized by intense arc-like dykes, volcanic rocks, and the mixing of mafic and felsic granodiorite [19,57,58]. This period was also marked by the replacement of the metasomatized NCC SCLM with the juvenile lithospheric mantle. The ore fluids from this enriched lithospheric mantle ascended along favorable NE- to NNE-trending faults, leading to the formation of the Shuigou gold deposit [59].

7. Conclusions

(1) The Shuigou gold deposit was formed at ca. 125 Ma, aligning with the peak period of gold mineralization in the Jiaodong region.
(2) The He-Ar analysis of fluid inclusions in the main-stage pyrite indicates that the fluids involved in the gold mineralization event originated from the mantle, the crust, and meteoric sources.
(3) The formation of the Shuigou gold deposit occurred in the context of the destruction of the North China Craton (NCC).

Supplementary Materials: The following supporting information can be downloaded at: https://www.mdpi.com/article/10.3390/min15010014/s1, Table S1: EMPA results of significant ore minerals from the Shuigou gold deposit.

Author Contributions: Methodology, S.C.; software, B.G.; validation, Y.X.; formal analysis, S.C. and Z.L.; investigation, Z.L., S.C., C.L. and Y.X.; writing—original draft preparation, S.C. and B.G.; writing—review and editing, Z.L. and S.C.; visualization, Y.X.; supervision, Z.L.; project administration, Z.L.; funding acquisition, Z.L. All authors have read and agreed to the published version of the manuscript.

Funding: This research was jointly funded by the National Natural Science Foundation of China (Grant No. 92062219) and the China Geological Survey project (grant No. DD20230568).

Data Availability Statement: The original contributions presented in the study are included in the article/Supplementary Materials.

Acknowledgments: We thank the chief editor and the anonymous reviewers for their constructive reviews, which have greatly improved our manuscript.

Conflicts of Interest: Author Yushan Xue were employed by the Xi'an Northwest Geological Institute of Nonferrous Metals Co., Ltd. The remaining authors declare that the research was conducted in the absence of any commercial or financial relationships that could be construed as a potential conflict of interest.

References

1. Yang, L.Q.; Deng, J.; Zhang, L.; Yang, W.; Xie, D.; Wang, L.; Qiu, K.F.; Li, D.P. Jiaodong-type gold deposit. *Acta Petrol. Sin.* **2024**, *40*, 1691–1711. [CrossRef]
2. Song, M.C.; Wang, L.; Song, Y.X.; Li, J.; Wang, B.; Wei, X.F.; Zhang, J.J.; Song, G.Z. Geometry and origin of supergiant gold deposits in the Jiaodong gold province, eastern China. *J. Asian Earth Sci.* **2023**, *254*, 105744. [CrossRef]
3. Goldfarb, R.J.; Santosh, M. The dilemma of the Jiaodong gold deposits: Are they unique? *Geosci. Front.* **2014**, *5*, 139–153. [CrossRef]
4. Phillips, G.N.; Powell, R. A practical classification of gold deposits, with a theoretical basis. *Ore Geol. Rev.* **2015**, *65*, 568–573. [CrossRef]
5. Yang, L.Q.; Li, R.H.; Gao, X.; Qiu, K.F.; Zhang, L. A preliminary study of extreme enrichment of critical elements in the Jiaodong gold deposits, China. *Acta Petrol. Sin.* **2020**, *36*, 1285–1314. [CrossRef]
6. Yang, L.Q.; Wei, Y.J.; Wang, S.R.; Zhang, L.; Ju, L.; Li, R.H.; Gao, X.; Qiu, K.F. A preliminary study of reserve estimate and resource potential assessment of critical elements in the Jiaodong gold deposits, China. *Acta Petrol. Sin.* **2022**, *38*, 9–22. [CrossRef]
7. Yang, L.Q.; Deng, J.; Wang, Z.L.; Zhang, L.; Guo, L.N.; Song, M.C.; Zheng, X.L. Mesozoic gold metallogenic system of the Jiaodong gold province, eastern China. *Acta Petrol. Sin.* **2014**, *30*, 2447–2467.
8. Yang, L.Q.; Deng, J.; Goldfarb, R.J.; Zhang, J.; Gao, B.F.; Wang, Z.L. 40Ar/39Ar geochronological constraints on the formation of the Dayingezhuang gold deposit: New implications for timing and duration of hydrothermal activity in the Jiaodong gold province, China. *Gondwana Res.* **2014**, *25*, 1469–1483. [CrossRef]
9. Zhang, X.O.; Cawood, P.A.; Wilde, S.A.; Liu, R.Q.; Song, H.L.; Li, W.; Snee, L.W. Geology and timing of mineralization at the Cangshang gold deposit, north-western Jiaodong Peninsula, China. *Miner. Depos.* **2003**, *38*, 141–153. [CrossRef]
10. Li, J.W.; Vasconcelos, P.M.; Zhang, J.; Zhou, M.F.; Zhang, X.J.; Yang, F.H. 40Ar/39Ar constraints on a temporal link between gold mineralization, magmatism, and continental margin transtension in the Jiaodong Gold Province, eastern China. *J. Geol.* **2003**, *111*, 741–751. [CrossRef]
11. Li, J.W.; Vasconcelos, P.M.; Zhou, M.F.; Zhao, X.F.; Ma, C.Q. Geochronology of the Pengjiakuang and Rushan gold deposits, eastern Jiaodong gold province, northeastern China: Implications for regional mineralization and geodynamic setting. *Econ. Geol.* **2006**, *101*, 1023–1038. [CrossRef]
12. Wen, B.J.; Fan, H.R.; Hu, F.F.; Liu, X.; Yang, K.F.; Sun, Z.F.; Sun, Z.F. Fluid evolution and ore genesis of the giant Sanshandao gold deposit, Jiaodong gold province, China: Constrains from geology, fluid inclusions and H–O–S–He–Ar isotopic compositions. *J. Geochem. Explor.* **2016**, *171*, 96–112. [CrossRef]
13. Deng, J.; Wang, Q.F. Gold mineralization in China: Metallogenic provinces, deposit types and tectonic framework. *Gondwana Res.* **2016**, *36*, 219–274. [CrossRef]
14. Fan, H.R.; Zhai, M.G.; Xie, Y.H.; Yang, J.H. Ore-forming fluids associated with granite-hosted gold mineralization at the Sanshandao deposit, Jiaodong gold province, China. *Miner. Depos.* **2003**, *38*, 739–750. [CrossRef]
15. Li, X.C.; Fan, H.R.; Santosh, M.; Hu, F.F.; Yang, K.F.; Lan, T.G. Hydrothermal alteration associated with Mesozoic granite-hosted gold mineralization at the Sanshandao deposit, Jiaodong Gold Province, China. *Ore Geol. Rev.* **2013**, *53*, 403–421. [CrossRef]
16. Li, L.; Santosh, M.; Li, S.R. The 'Jiaodong type' gold deposits: Characteristics, origin and prospecting. *Ore Geol. Rev.* **2015**, *65*, 589–611. [CrossRef]
17. Deng, J.; Wang, Q.; Santosh, M.; Liu, X.; Liang, Y.; Yang, L.; Zhao, R.; Yang, L. Remobilization of metasomatized mantle lithosphere: A new model for the Jiaodong gold province, eastern China. *Miner. Depos.* **2020**, *55*, 257–274. [CrossRef]
18. Qiu, K.F.; Romer, R.L.; Long, Z.Y.; Williams-Jones, A.E.; Yu, H.C.; Turner, S.; Wang, Q.F.; Li, S.S.; Zhang, J.Y.; Duan, H.R.; et al. The role of an oxidized lithospheric mantle in gold mobilization. *Sci. Adv.* **2024**, *10*, eado6262. [CrossRef]
19. Cheng, S.B.; Lü, Q.T.; Liu, Z.J.; Yang, Y.; Liu, Z.D.; Yan, J.Y.; Zhang, H.; Gong, X.J.; Chen, C.X. Petrogenesis of Mo-associated Mesozoic granitoids on the Jiaodong Peninsula: Implications for crustal architecture and Mo mineralization along the Dabie–Sulu Orogen. *Ore Geol. Rev.* **2022**, *149*, 105015. [CrossRef]
20. Ding, Z.J.; Sun, F.Y.; Liu, F.L.; Liu, J.H.; Peng, Q.M.; Ji, P.; Li, B.L.; Zhang, P.J. Mesozoic geodynamic evolution and metallogenic series of major metal deposits in Jiaodong Peninsula, China. *Acta Petrol. Sin.* **2015**, *31*, 3045–3080.
21. Deng, J.; Wang, C.M.; Bagas, L.; Santosh, M.; Yao, E.Y. Crustal architecture and metallogenesis in the south-eastern North China Craton. *Earth Sci. Rev.* **2018**, *182*, 251–272. [CrossRef]
22. Yang, L.Q.; Deng, J.; Zhang, L.; Zhao, A.H.; Santosh, M.; Yu, X.F.; Yang, W.; Li, D.P.; Shan, W.; Xie, D.; et al. Mantle-rooted fluid pathways and world-class gold mineralization in the giant Jiaodong gold province: Insights from integrated deep seismic reflection and tectonics. *Earth Sci. Rev.* **2024**, *255*, 104862. [CrossRef]
23. Jiang, N.; Guo, J.H.; Fan, W.B.; Hu, J.; Zong, K.Q.; Zhang, S.Q. Archean TTGs and sanukitoids from the Jiaobei terrain, North China craton: Insights into crustal growth and mantle metasomatism. *Precambrian Res.* **2016**, *281*, 656–672. [CrossRef]
24. Wang, W.; Zhai, M.G.; Li, T.S.; Santosh, M.; Zhao, L.; Wang, H.Z. Archean–Paleoproterozoic crustal evolution in the eastern North China Craton: Zircon U–Th–Pb and Lu–Hf evidence from the Jiaobei terrane. *Precambrian Res.* **2014**, *241*, 146–160. [CrossRef]

25. Tam, P.Y.; Zhao, G.C.; Liu, F.L.; Zhou, X.W.; Sun, M.; Li, S.Z. Timing of metamorphism in the Paleoproterozoic Jiao-Liao-Ji Belt: New SHRIMP U–Pb zircon dating of granulites, gneisses and marbles of the Jiaobei massif in the North China Craton. *Gondwana Res.* **2011**, *19*, 150–162. [CrossRef]
26. Cheng, S.B.; Liu, Z.J.; Wang, Q.F.; Feng, B.; Wei, X.L.; Liu, B.Z.; Qin, L.Y.; Zhao, B.J.; Shui, P.; Xu, L.; et al. SHRIMP zircon U–Pb dating and Hf isotope analyses of the Muniushan Monzogranite, Guocheng, Jiaobei Terrane, China: Implications for the tectonic evolution of the Jiao–Liao–Ji Belt, North China Craton. *Precambrian Res.* **2017**, *301*, 36–48. [CrossRef]
27. Xie, S.W.; Wu, Y.B.; Zhang, Z.M.; Qin, Y.C.; Liu, X.C.; Wang, H.; Qin, Z.W.; Liu, Q.; Yang, S.H. U–Pb ages and trace elements of detrital zircons from Early Cretaceous sedimentary rocks in the Jiaolai Basin, north margin of the Sulu UHP terrane: Provenances and tectonic implications. *Lithos* **2012**, *154*, 346–360. [CrossRef]
28. Goss, S.C.; Wilde, S.A.; Wu, F.Y.; Yang, J.H. The age, isotopic signature and significance of the youngest Mesozoic granitoids in the Jiaodong Terrane, Shandong Province, North China Craton. *Lithos* **2010**, *120*, 309–326. [CrossRef]
29. Ma, L.; Jiang, S.Y.; Dai, B.Z.; Jiang, Y.H.; Hou, M.L.; Pu, W.; Xu, B. Multiple sources for the origin of Late Jurassic Linglong adakitic granite in the Shandong Peninsula, eastern China: Zircon U–Pb geochronological, geochemical and Sr–Nd–Hf isotopic evidence. *Lithos* **2013**, *162–163*, 251–263. [CrossRef]
30. Li, X.H.; Fan, H.R.; Santosh, M.; Yang, K.F.; Peng, H.W.; Hollings, P.; Hu, F.F. Mesozoic felsic dikes in the Jiaobei Terrane, southeastern North China Craton: Constraints from zircon geochronology and geochemistry, and implications for gold metallogeny. *J. Geochem. Explor.* **2019**, *201*, 40–55. [CrossRef]
31. Ballard, J.R.; Palin, J.M.; Williams, I.S.; Campbell, I.H.; Faunes, A. Two ages of porphyry intrusion resolved for the super-giant Chuquicamata copper deposit of northern Chile by ELA-ICP-MS and SHRIMP. *Geology* **2001**, *29*, 383–386. [CrossRef]
32. Nasdala, L.; Hofmeister, W.; Norberg, N.; Martinson, J.M.; Corfu, F.; Dörr, W.; Kamo, S.L.; Kennedy, A.K.; Kronz, A.; Reiners, P.W. Zircon M257—A Homogeneous Natural Reference Material for the Ion Microprobe U-Pb Analysis of Zircon. *Geostand. Geoanalytical Res.* **2008**, *32*, 247–265. [CrossRef]
33. Black, L.P.; Kamo, S.L.; Allen, C.M.; Aleinikoff, J.N.; Davis, D.W.; Korsch, R.J.; Foudoulis, C. TEMORA 1: A new zircon standard for Phanerozoic U–Pb geochronology. *Chem. Geol.* **2003**, *200*, 155–170. [CrossRef]
34. Sang, H.Q.; Wang, F.; He, H.Y.; Wang, Y.L.; Yang, L.K.; Zhu, R.X. Intercalibration of ZBH-25 biotite reference material untilized for K-Ar and Ar-40-Ar-39 age determination. *Yanshi Xuebao/Acta Petrol. Sin.* **2006**, *22*, 3059–3078.
35. Stuart, F.M.; Burnard, P.G.; Taylor, R.P.; Turner, G. Resolving mantle and crustal contributions to ancient hydrothermal fluids: He Ar isotopes in fluid inclusions from Dae Hwa W Mo mineralisation, South Korea. *Geochim. Cosmochim. Acta* **1995**, *59*, 4663–4673. [CrossRef]
36. Song, Y.X.; Song, M.C.; Ding, Z.J.; Wei, X.F.; Xu, S.H.; Li, J.; Tan, X.M.; Hu, H.; Cao, J. Major Advances on Deep Prospecting in Jiaodong Gold Ore Cluster and Its Metallogenic Characteristics. *Gold Sci. Technol.* **2017**, *25*, 4–18.
37. Hou, M.L.; Jiang, S.Y.; Jiang, Y.H.; Ling, H.F. S-Pb isotope geochemistry and Rb-Sr geochronology of the Penglai gold field in the eastern Shangdong province. *Acta Petrol. Sin.* **2006**, *22*, 2525–2533.
38. Yang, K.F.; Jiang, P.; Fan, H.R.; Zuo, Y.B.; Yang, Y.H. Tectonic transition from a compressional to extensional metallogenic environment at ~120Ma revealed in the Hushan gold deposit, Jiaodong, North China Craton. *J. Asian Earth Sci.* **2018**, *160*, 408–425. [CrossRef]
39. Feng, L.Q.; Gu, X.X.; Zhang, Y.M.; Wang, J.L.; Ge, Z.L.; He, Y.; Zhang, Y.S. Geology and geochronology of the Shijia gold deposit, Jiaodong Peninsula, China. *Ore Geol. Rev.* **2020**, *120*, 103432. [CrossRef]
40. Yang, L.Q.; Deng, J.; Wang, Z.L.; Guo, L.N.; Li, R.H.; Groves, D.I.; Danyushevsky, L.V.; Zhang, C.; Zheng, X.L.; Zhao, H. Relationships Between Gold and Pyrite at the Xincheng Gold Deposit, Jiaodong Peninsula, China: Implications for Gold Source and Deposition in a Brittle Epizonal Environment. *Econ. Geol.* **2016**, *111*, 105–126. [CrossRef]
41. Xiao, J.D.; Xie, Z.J.; Xia, Y.; Lan, T.G.; Zhang, L.; Pan, L.C.; Hu, H.L.; Wang, H.; Tan, Q.P.; Xu, Y. Predominant fracture gold in the Jiaodong gold province, China: Constraints on gold mineralization processes. *Ore Geol. Rev.* **2024**, *165*, 105915. [CrossRef]
42. Cao, G.S.; Zhang, Y.; Chen, H.Y. Trace elements in pyrite from orogenic gold deposits: Implications for metallogenic mechanism. *Acta Petrol. Sin.* **2023**, *39*, 2330–2346. [CrossRef]
43. Yu, B.; Zeng, Q.D.; Frimmel, H.E.; Fan, H.R.; Xue, J.L.; Yang, J.H.; Wu, J.J.; Bao, Z.A. Spatio-temporal fluid evolution of gold deposit in the Jiaodong Peninsula, China: A case study of the giant Xiling deposit. *J. Geochem. Explor.* **2024**, *260*, 107455. [CrossRef]
44. Cai, W.Y.; Song, M.C.; Santosh, M.; Li, J. The gold-telluride connection: Evidence for multiple fluid pulses in the Jinqingding telluride-rich gold deposit of Jiaodong Peninsula, Eastern China. *Geosci. Front.* **2024**, *15*, 101795. [CrossRef]
45. Wei, Y.J.; Yang, L.Q.; Qiu, K.F.; Wang, S.R.; Ren, F.; Dai, Z.H.; Li, D.P.; Shan, W.; Li, Z.S.; Wang, J.H.; et al. Geology, mineralogy and pyrite trace elements constraints on gold mineralization mechanism at the giant Dayingezhuang gold deposit, Jiaodong Peninsula, China. *Ore Geol. Rev.* **2022**, *148*, 104992. [CrossRef]
46. Yang, Q.Y.; Shen, J.F.; Li, S.R.; Santosh, M.; Luo, Z.H.; Liu, Y. Oxygen, boron, chromium and niobium enrichment in native Au and Ag grains: A case study from the Linglong gold deposit, Jiaodong, eastern China. *J. Asian Earth Sci.* **2013**, *62*, 537–546. [CrossRef]

47. Goldfarb, R.J.; Hofstra, A.H.; Simmons, S.F. Critical Elements in Carlin, Epithermal, and Orogenic Gold Deposits. In *Rare Earth and Critical Elements in Ore Deposits*; Verplanck, P.L., Hitzman, M.W., Eds.; Society of Economic Geologists: Littleton, CO, USA, 2016; Volume 18, pp. 217–244.
48. Wei, W.F.; Hu, R.Z.; Bi, X.W.; Jiang, G.H.; Yan, B.; Yin, R.S.; Yang, J.H. Mantle-derived and crustal He and Ar in the ore-forming fluids of the Xihuashan granite-associated tungsten ore deposit, South China. *Ore Geol. Rev.* **2019**, *105*, 605–615. [CrossRef]
49. Hu, R.Z.; Burnard, P.G.; Bi, X.W.; Zhou, M.F.; Pen, J.T.; Su, W.C.; Wu, K.X. Helium and argon isotope geochemistry of alkaline intrusion-associated gold and copper deposits along the Red River–Jinshajiang fault belt, SW China. *Chem. Geol.* **2004**, *203*, 305–317. [CrossRef]
50. Burnard, P.G.; Hu, R.; Turner, G.; Bi, X.W. Mantle, crustal and atmospheric noble gases in ailaoshan gold deposits, Yunnan Province, China. *Geochim. Cosmochim. Acta* **1999**, *63*, 1595–1604. [CrossRef]
51. Zhang, L.C.; Shen, Y.C.; Li, H.M.; Zeng, Q.D.; Li, G.M.; Liu, T.B. Helium and argon isotopic compositions of fluid inclusions and tracing to the source of ore-forming fluids for Jiaodong gold deposits. *Acta Petrol. Sin.* **2002**, *18*, 559–565.
52. Zhu, R.X.; Xu, Y.G.; Zhu, G.; Zhang, H.F.; Xia, Q.K.; Zheng, T.Y. Destruction of the North China Craton. *Sci. China Earth Sci.* **2012**, *55*, 1565–1587. [CrossRef]
53. Ying, J.F.; Zhang, H.F.; Kita, N.; Morishita, Y.; Shimoda, G. Nature and evolution of Late Cretaceous lithospheric mantle beneath the eastern North China Craton: Constraints from petrology and geochemistry of peridotitic xenoliths from Jünan, Shandong Province, China. *Earth Planet. Sci. Lett.* **2006**, *244*, 622–638. [CrossRef]
54. Wu, F.Y.; Yang, J.H.; Xu, Y.G.; Wilde, S.A.; Walker, R.J. Destruction of the North China Craton in the Mesozoic. *Annu. Rev. Earth Planet. Sci.* **2019**, *47*, 173–195. [CrossRef]
55. Zheng, Y.F.; Xu, Z.; Zhao, Z.F.; Dai, L.Q. Mesozoic mafic magmatism in North China: Implications for thinning and destruction of cratonic lithosphere. *Sci. China Earth Sci.* **2018**, *61*, 353–385. [CrossRef]
56. Yang, J.H.; Xu, L.; Sun, J.F.; Zeng, Q.D.; Zhao, Y.N.; Wang, H.; Zhu, Y.S. Geodynamics of decratonization and related magmatism and mineralization in the North China Craton. *Sci. China Earth Sci.* **2021**, *64*, 1409–1427. [CrossRef]
57. Cheng, S.B.; Liu, Z.J.; Wang, Q.F.; Wang, F.J.; Xue, Y.S.; Xu, L.; Wang, J.P.; Zhu, B.L. Mineralization age and geodynamic background for the Shangjiazhuang Mo deposit in the Jiaodong gold province, China. *Ore Geol. Rev.* **2017**, *80*, 876–890. [CrossRef]
58. Zhu, G.; Lu, Y.C.; Su, N.; Wu, X.D.; Yin, H.; Zhang, S.; Xie, C.L.; Niu, M.L. Crustal deformation and dynamics of Early Cretaceous in the North China Craton. *Sci. China Earth Sci.* **2021**, *64*, 1428–1450. [CrossRef]
59. Wang, Q.F.; Liu, X.F.; Yin, R.S.; Weng, W.J.; Zhao, H.S.; Yang, L.; Zhai, D.G.; Li, D.P.; Ma, Y.; Groves, D.I.; et al. Metasomatized mantle sources for orogenic gold deposits hosted in high-grade metamorphic rocks: Evidence from Hg isotopes. *Geology* **2023**, *52*, 115–119. [CrossRef]

MDPI AG
Grosspeteranlage 5
4052 Basel
Switzerland
Tel.: +41 61 683 77 34

Minerals Editorial Office
E-mail: minerals@mdpi.com
www.mdpi.com/journal/minerals

www.ingramcontent.com/pod-product-compliance
Lightning Source LLC
Chambersburg PA
CBHW041359210926
51850CB00013B/3131